# Urban Planning During Socialism

*Urban Planning During Socialism* delves into the evolution of cities during the period of state socialism of the 20th century, summarising the urban and architectural studies that trace their transformations.

The book focuses primarily on the periphery of the socialist world, both spatially and in terms of scholarly thinking. The case study cities presented in this book draw on cultural and material studies to demonstrate diverse and novel concepts of "periphery" through transformations of socialist cityscapes rather than homogenous views on cities during the period of state socialism of the 20th century. In doing so the book explores the transversalities of political, economic and social phenomena; the places for everyday life in socialist cities; and the role of professional communities on production and reproduction of space and ecological thinking.

This book is aimed at scholarly readership, in particular scholars in architecture, urban planning and human geography, as well as undergraduate, graduate and post-graduate students in these disciplines studying the urban transformation of cities after WWII in socialist countries. It will also be of interest for planning officials, architects, policymakers and activists in former socialist countries.

**Jasna Mariotti** is a senior lecturer in architecture at Queen's University Belfast, UK. Her research focuses on the relationship between urban history, planning and architecture in the 20th and 21st centuries, linking two main themes. The first one focuses on architecture and urban planning under the influence of political organisations and mechanisms of production of space in socialist and post-socialist countries. The second one relates to the architecture of mass housing in particular to the spontaneous and planned practices of transformation of housing estates and changing notions of habitation. Her current research is funded by the EPSRC and The Great Britain Sasakawa Foundation. Previously she was an architect and urban designer at WEST 8 Urban Design and Landscape Architecture in Rotterdam.

**Kadri Leetmaa** holds a PhD in human geography from the University of Tartu, Estonia. Currently, she works as the Head of the Department of Geography and the Associate Professor of Human Geography at the Centre for Migration and Urban Studies at the University of Tartu. She is a member of the Scientific Board of the Leibniz Institute for Regional Geography, Leipzig, Germany. Her research topics include urban geography under and after socialism, inequalities in urban and rural space, urban planning and housing policies affecting inequalities, migration, residential preferences, neighbourhood change and inter-ethnic contacts in society and space.

# Routledge Research in Historical Geography

This series offers a forum for original and innovative research, exploring a wide range of topics encompassed by the sub-discipline of historical geography and cognate fields in the humanities and social sciences. Titles within the series adopt a global geographical scope and historical studies of geographical issues that are grounded in detailed inquiries of primary source materials. The series also supports historiographical and theoretical overviews, and edited collections of essays on historical-geographical themes. This series is aimed at upper-level undergraduates, research students and academics.

**Earth, Cosmos and Culture**
Geographies of Outer Space in Britain, 1900–2020
*Oliver Tristan Dunnett*

**Recalibrating the Quantitative Revolution in Geography**
Travels, Networks, Translations
*Edited by Ferenc Gyuris, Boris Michel and Katharina Paulus*

**American Colonial Spaces in the Philippines**
Insular Empire
*Scott Kirsch*

**Empire, Gender and Bio-Geography**
Charlotte Wheeler-Cuffe and Colonial Burma
*Nuala C Johnson*

**Dissertating Geography**
An inquiry into the making of student geographical knowledge, 1950–2020
*Mette Bruinsma*

**Urban Planning During Socialism**
Views from the Periphery
*Edited by Jasna Mariotti and Kadri Leetmaa*

For more information about this series, please visit: https://www.routledge.com/Routledge-Research-in-Historical-Geography/book-series/RRHGS

# Urban Planning During Socialism

Views from the Periphery

Edited by Jasna Mariotti and
Kadri Leetmaa

LONDON AND NEW YORK

First published 2024
by Routledge
4 Park Square, Milton Park, Abingdon, Oxon OX14 4RN

and by Routledge
605 Third Avenue, New York, NY 10158

*Routledge is an imprint of the Taylor & Francis Group, an informa business*

© 2024 selection and editorial matter, Jasna Mariotti and Kadri Leetmaa; individual chapters, the contributors

The right of Jasna Mariotti and Kadri Leetmaa to be identified as the authors of the editorial material, and of the authors for their individual chapters, has been asserted in accordance with sections 77 and 78 of the Copyright, Designs and Patents Act 1988.

The Open Access version of this book, available at www.taylorfrancis.com, has been made available under a Creative Commons Attribution-NonCommercial-ShareAlike (CC-BY-NC-SA) 4.0 license.

The Open Access version of this book was funded by City University of Hong Kong, Estonian Academy of Arts, Queen's University Belfast, Faculty of Architecture and Design, Slovak University of Technology, Bratislava, XM Architekten GmbH, and University of Tartu.

*Trademark notice:* Product or corporate names may be trademarks or registered trademarks, and are used only for identification and explanation without intent to infringe.

*British Library Cataloguing-in-Publication Data*
A catalogue record for this book is available from the British Library

*Library of Congress Cataloguing-in-Publication Data*
Names: Mariotti, Jasna, editor. | Leetmaa, Kadri, editor.
Title: Urban planning during socialism / edited by Jasna Mariotti and Kadri Leetmaa.
Description: Abingdon, Oxon ; New York, NY : Routledge, [2023] | Series: Routledge research in historical geography series | Includes bibliographical references and index.
Identifiers: LCCN 2023028254 (print) | LCCN 2023028255 (ebook) | ISBN 9781032355979 (hardback) | ISBN 9781032355986 (paperback) | ISBN 9781003327592 (ebook)
Subjects: LCSH: Cities and towns--Communist countries. | City planning--Environmental aspects--Communist countries. | Economic development--Communist countries--History--20th century. | Socialism--History--20th century.
Classification: LCC HT119 .U6953 2023 (print) | LCC HT119 (ebook) | DDC 909/.09717--dc23/eng/20230825
LC record available at https://lccn.loc.gov/2023028254
LC ebook record available at https://lccn.loc.gov/2023028255

ISBN: 978-1-032-35597-9 (hbk)
ISBN: 978-1-032-35598-6 (pbk)
ISBN: 978-1-003-32759-2 (ebk)

DOI: 10.4324/9781003327592

Typeset in Times New Roman
by MPS Limited, Dehradun

**To our families**

# Contents

*List of Contributors*   *x*
*List of Figures*   *xvi*
*List of Tables*   *xviii*
*Acknowledgements*   *xix*

Revisiting urban planning during socialism:
Views from the periphery. An introduction   1
JASNA MARIOTTI AND KADRI LEETMAA

**PART I**
**Urban planning, politics and power: Relations in the periphery**   13

1 Urbanising the Virgin Lands: At the frontier
of Soviet socialist planning   15
GIANNI TALAMINI

2 From Breslau to Wrocław: Urban development
of the largest city of the Polish "Regained Lands"
under socialism   31
AGNIESZKA TOMASZEWICZ AND JOANNA MAJCZYK

3 Dreaming the capital: Architecture and urbanism
as tools for planning the socialist Bratislava   50
HENRIETA MORAVČÍKOVÁ, PETER SZALAY, AND
LAURA KRIŠTEKOVÁ

4 The Yugoslav Skopje: Building the brutalist
city, 1970–1990   66
MAJA BABIĆ

5 From reverse colonial trade to antiurbanism:
  Frustrated urban renewal in Budapest, 1950–1990  79
  DANIEL KISS

**PART II**
**Architects and urban planners in the socialist city:**
**Roles and positions in the periphery**  99

6 Passive agents or genuine facilitators of citizen
  participation? The role of urban planners under
  the Yugoslav self-management socialism  101
  ANA PERIĆ AND MINA BLAGOJEVIĆ

7 The influence of nuclear deterrence during the
  Cold War on the growth and decline of the
  peripheral town of Valga/Valka  119
  KADRI LEETMAA, JIŘÍ TINTĚRA, TAAVI PAE, AND
  DANIEL BALDWIN HESS

8 The role of architects in fighting the monotony
  of the Lithuanian mass housing estates  134
  MARIJA DRĖMAITĖ

**PART III**
**The non-politics of everyday life in spatial peripheries**
**during socialism**  151

9 Courtyards, parks and squares of power in
  Ukrainian cities: Planning and reality of
  everyday life under socialism  153
  KOSTYANTYN MEZENTSEV, NATALIIA PROVOTAR,
  AND OLEKSIY GNATIUK

10 Planning urban peripheries for leisure:
   The plan for Greater Tallinn, 1960–1962  176
   EPP LANKOTS

11 Gldani: From ambitious experimental project
   to half-realised Soviet mass-housing district
   in Tbilisi, Georgia  191
   DAVID GOGISHVILI

## PART IV
## Ecology and environment in the socialist periphery    207

12  New ecological planning and spatial assessment
    of production sites in socialist industrial
    Yekaterinburg (formerly Sverdlovsk) in the
    1960s–80s                                            209
    NADEZDA GOBOVA

13  Peripheral landscapes: Ecology, ideology
    and form in Soviet non-official architecture         227
    MASHA PANTELEYEVA

14  Conceptions of 'nature' and 'the environment'
    during socialism in Albania: An ecofeminist
    perspective                                          244
    DORINA POJANI AND ELONA POJANI

*Index*                                                  *268*

# Contributors

**Maja Babić** is an architectural and urban historian. She is an assistant professor of history and theory of architecture and urbanism at the University of Groningen in the Netherlands. Dr. Babić's scholarship focuses on the intersection of architecture and ideology during the state-socialist and post-socialist periods in East-Central and Southeastern Europe, particularly focusing on former Yugoslavia. Maja studies the intertwined nature of architectural production, urban planning and socio-political events in the state-socialist countries of the 20th century and the socio-political and architectural developments in contemporary post-communist Europe. Her current work explores the production and negotiations of urban identities in post-socialist cities.

**Mina Blagojević** is an architect with a master's degree focused on urban planning and design from the University of Belgrade. With several years of professional engagement in scientific institutes, public authorities and the private sector in Serbia and Switzerland, she is experienced in historical urban research, urban planning and urban design practice and consulting related to international collaboration in urban development processes. Her professional interest revolves around topics related to public space, urban heritage and building cultures, as well as participatory planning procedures. She is fluent in Serbian, English and German.

**Marija Drėmaitė** holds a PhD in History of Architecture (2006) and is a professor at Vilnius University, Department of History. Her interest is focused on 20th-century architecture, housing and industrial heritage. In 2017 she published the book *Baltic Modernism. Architecture and Housing in Soviet Lithuania* and in 2020 co-edited *Lithuanian Architects Assess the Soviet Era: 1992 Oral History Tapes* with John V. Maciuika.

**Oleksiy Gnatiuk** is an assistant professor at the Department of Economic and Social Geography, Taras Shevchenko National University of Kyiv. His PhD thesis, defended in 2015, addresses territorial identity of the population of Podolia, Ukraine. After that (2015–2017) he worked as a research fellow at the lab "Regional problems of economics and politics" and participated in several domestic and international research projects and collaborations. His principal interests are urban geography with specific foci on suburban development and post-socialist urban transformations, urban and regional identity, memory politics, critical place name studies and perceptual and cultural geography.

**Nadezda Gobova** is an architectural historian and practicing architect. She completed her PhD at the Bartlett School of Architecture, UCL in 2020, where she studied design, planning and construction of Soviet industrial cities in the Ural region with a focus on Yekaterinburg (formerly Sverdlovsk). Previously Nadezda studied architecture at the Academy of Art University in San Francisco, as a Fulbright Scholar, and at the Ural State Academy of Architecture and Arts in Yekaterinburg. Her research interests include histories and theories of industrial urbanism, socialist and post-socialist city planning and architectural endeavour in the epoch of environmental concerns.

**David Gogishvili** is a senior researcher at the Department of Geography at the University of Lausanne. His research focuses on the role large-scale urban development projects, such as mega-events and cultural flagship institutions, play in cities across the globe. He has research expertise in the cities of Central Asia and the South Caucasus and is currently developing in the Gulf Region. While primarily relying on qualitative research tools, he has also been involved in using spatial analysis methods in research and has developed several large databases for academic research.

**Daniel Baldwin Hess** earned a PhD in urban planning from the University of California, Los Angeles. He is a professor in the Department of Urban and Regional Planning at the University at Buffalo, State University of New York. He was formerly a Marie Skłodowska-Curie fellow at the University of Tartu, Estonia. Hess's research addresses interactions between housing, transportation, land use and other public concerns. He develops new pathways for understanding the complex socio-economic and ethnic landscape of cities and spatial inequalities. He is the winner of the Teaching Excellence Award from the Chancellor of the State University of New York and serves as co-editor of the journal *Town Planning Review*.

**Daniel Kiss** holds a PhD in planning history from ETH Zurich, where he is a lecturer of urban design and spatial planning. His research interests comprise theories of urban form, strategic planning and the history and theory of urbanisation. His book Modeling Post-Socialist Urbanization: The Case of Budapest (2018) presents an explanatory model of post-socialist urbanisation, abstracted from an empirically deep and richly detailed single case study of Budapest's recent urban development, planning and governance. He is a co-author of Relational Theories of Urban Form. An Anthology (2021), a volume outlining the concept of urban form within the relational field of space and agency.

**Laura Krišteková** is an architect and researcher at the Department of Architecture in the Institute of History SAS and a member of the Slovak DOCOMOMO working group. In 2017, she received her doctorate at the Faculty of Architecture at the Slovak University of Technology in Bratislava. Her research field is 20th- and 21st-century architectural history and theory, focusing on military architecture and urban history. She is a co-author of the publication *Bratislava (Un)planned City* (2020). In 2022, she received the Young Scientist Award, organised by the Slovak Academy of Sciences.

**Epp Lankots** is an architectural historian and senior researcher at the Estonian Academy of Arts. Her research interests include architecture and material culture under socialism and the historiography of modern architecture. She has written on the social differentiation in the Soviet domestic sphere and the mingling of avant-garde and historiographic practices in Soviet architectural historiography. Her present research centres around the ideas of the socialist "leisure society" and how they were mediated into the everyday environment during the 1960s and 1970s. She is the co-editor and co-curator of the award-winning book and the exhibition *Leisure Spaces: Holidays and Architecture in 20th Century Estonia* (2020).

**Joanna Majczyk** is an architect and assistant professor at the Faculty of Architecture/Wrocław University of Science and Technology and former head of the Department of Contemporary Architecture at the Museum of Architecture in Wrocław (Poland). She does research on architecture and urban planning of the 20th century, socialist realism and European modernism. A curator of architectural exhibitions, she promotes post-war architecture and urbanism in cooperation with the trade press and non-governmental institutions.

**Kostyantyn Mezentsev** is a professor and Head of the Department of Economic and Social Geography at Taras Shevchenko National University of Kyiv. His recent research examines the transformation of post-Soviet urban regions and cities, public spaces and new-build gentrification, suburban development, urban geopolitics and IDP issues. He is a co-editor of the book *Urban Ukraine: In the Epicentre of the Spatial Changes (2017)* and has recently contributed to T. Kuzio's, P. D'Anieri's and S. Zhuk's *Ukraine's Outpost: Dnipropetrovsk and the Russian-Ukrainian War (E-International Relations, 2021)*. He is involved in the international multi-disciplinary research project *Ukrainian Geopolitical Fault-line Cities: Urban Identities, Geopolitics and Urban Policy* (2018–2023).

**Henrieta Moravčíková** is a professor of Architecture History at the Faculty of Architecture Slovak University of Technology, Head of the Department of Architecture at the Slovak Academy of Sciences, and Chair of the Slovak DOCOMOMO chapter. Her field of interest is 20th- and 21st-century architecture and modern architecture heritage. Her book on Jewish architect, *Architect Friedrich Weinwurm* (2014), was awarded the International DAM Book Award. Together with her team, she prepared the first complex analytical monograph on modern town planning of the capital of Slovakia *Bratislava (un)planned city* (2020). The International Creative Media Award recently awarded the monograph the Gold Award.

**Taavi Pae** holds a PhD in human geography from the University of Tartu, Estonia. He is an associate professor of Estonian geography at the Geography Department of the University of Tartu. His major research is related to the history of Estonian geography, cartography and cultural geography. As the compiler of the Estonian National Atlas and other key publications on cultural geography and Estonian cartography, he has played a significant role in shaping the understanding of the country's geography and cultural heritage. Taavi Pae is an active member in the academic organisations such as Estonian Learned Society and Estonian Geographic Society.

**Masha Panteleyeva** received her PhD in architectural history from Princeton University and is currently teaching at the Architecture Art and Planning Department at Cornell University and at the Pratt School of Architecture. Her work has been supported by the Graham Foundation and the Canadian Centre for Architecture and has been featured at the Venice Biennale, the Museum of Modern Art in Warsaw and the State Moscow Architecture Museum. Her recent book *New Element of Settlement: City of the Future* examines the evolution of Soviet formal architectural language as a discursive tool used to outline fundamental differences between the socialist urban fabric and its Western counterpart.

**Ana Perić** holds a PhD in urban planning from the University of Belgrade and works as an assistant professor in the School of Architecture, Planning and Environmental Policy, University College Dublin. Revolving the research around the domain of comparative studies, she explores various topics across scales of institutional landscapes and planning cultures with a specific focus on collaborative and participatory planning instruments, methods and theory. Dedicated to combining research with practical assignments, she has been active in several policy and professional working groups, including the Western Balkan Network on Territorial Governance, ISOCARP (International Society of City and Regional Planners), serving as a Board Member (2016–2022), and UN-Habitat/ISOCARP Community of Practice on Urban Innovation.

**Dorina Pojani**, originally from Albania, is an associate professor in urban planning at The University of Queensland, Australia. Her research interests encompass built environment topics (urban design, transport and housing) in both the Global North and South. Her latest books are *Trophy Cities: A Feminist Perspective on New Capitals* (2021) and *Alternative Planning History and Theory* (Routledge, 2023).

**Elona Pojani** is an associate professor in environmental economics at the University of Tirana, Albania. Her research focuses on climate change and disaster risk management in the Western Balkan region. In 2011, she was a visiting fellow at the University of Denver, USA, funded through the Junior Faculty Development Program of the American Councils. Since 2010, she has served as a consultant for the UNDP-GEF Climate Change Programme.

**Nataliia Provotar** is an associate professor at the Department of Economic and Social Geography, Taras Shevchenko National University of Kyiv. Her recent research interests include post-socialist urban transformations, urban public spaces, suburban development, gender issues, socio-spatial inequality and migration in Ukraine. She is a co-editor of *Urban Ukraine: In the Epicenter of the Spatial Changes* (2017). She is involved in research projects on Kyiv metropolitan region (2019–2023), comparative study of changing everyday practices in suburban spaces in Austria and Ukraine (2019–2020) and urban spatial transformation in Ukraine (2016–2017).

**Peter Szalay** is an architectural historian based in Bratislava. He is a researcher at the Department of Architecture, Institute of History, Slovak Academy of Sciences in Bratislava. His research focuses on modern movement architecture and town planning, specifically on its social and political context, as well as on the heritage protection of the 20th-century architecture. He is a managing editor of the scientific journal *Architektúra & Urbanizmus* and is a member of DOCOMOMO International. He is the author and co-author of more scientific studies and monographs, among others (Bratislava Atlas of Mass Housing, 2012, or Bratislava (Un) planned city 2020).

**Gianni Talamini, PhD**, is an award-winning architect, urbanist and scholar. His PhD focused on Central Asia urbanism. He is an associate professor at the City University of Hong Kong, where he leads the Master of Urban Design and Regional Planning. Gianni researches the notion of symbiotic urbanism and investigates the relationship between society and space. He works for an environmentally innocuous, culturally leavened and spatially just society.

**Jiri Tintera** holds a PhD in architecture from the Tallinn University of Technology, Estonia. He is a senior lecturer in Tallinn University of Technology (Tartu College) and the town architect in Valga municipality. He graduated from Czech Technical University in Prague and defended his doctoral thesis "Urban Regeneration Strategies for Shrinking Post-Soviet European Communities: A Case Study of Valga, Estonia" at Tallinn University of Technology in 2019. His research focuses on outcomes of population shrinkage on urban space.

**Agnieszka Tomaszewicz** is an architect, professor at the Faculty of Architecture/ Wrocław University of Science and Technology (Poland), vice Editor-in-Chief of the scientific journal "Architectus", member of the City Urban Planning and Architectural Commission and an advisory body to the Mayor of Wrocław in the field of city planning and development (2015–2018). She authored many publications, studies and projects in the revaluation and conservation of historical buildings, including those built during the modernism period. Her research interests include architecture and urban planning of the 19th and 20th centuries. She is a member of the Wrocław Branch of the Polish Academy of Sciences and EU expert in the New European Bauhaus project.

# Figures

| | | |
|---|---|---|
| 1.1 | The Virgin Lands Campaign Area and Astana | 17 |
| 1.2 | Palace of Youth in 1975 | 24 |
| 2.1 | The first General Development Plan of the City of Wrocław, designed by team at Wrocław Design Office led by Tadeusz Ptaszycki, 1949 | 35 |
| 2.2 | Wrocław-Południe district, competition project inspired by modernist urban planning, designed by Kazimierz Bieńkowski, Tadeusz Izbicki, Wacław Kamocki and Julian Łowiński, 1962 | 41 |
| 3.1 | Proposal for a new commercial and social centre on Obchodná ulica, Ivan Matušík, 1970 | 60 |
| 3.2 | Study of the gradual development of the linear centre of the Petržalka City Sector, Jozef Chovanec and Stanislav Talaš, 1972 | 63 |
| 4.1 | Skopje City Wall, 1970s | 68 |
| 4.2 | Ss. Cyril and Methodius University, 1970s | 74 |
| 5.1 | Streetscape from the 1965 integrative renewal scheme of Budapest's eighth district (János Brenner et al., BUVÁTI) | 87 |
| 5.2 | The eighth district's tabula rasa plan from 1971 (Árpád Mester et al., BUVÁTI) | 88 |
| 5.3 | Façade sketches and masterplan from the 1987 gradualist, block-scale renewal scheme of Budapest's eighth district (Anna Perczel et al., VÁTI) | 92 |
| 5.4 | Frustration spanning ages: scarce remnants of socialism's ambitious but largely thwarted urban renewal programme hovering above a dilapidated 19th-century tenement in Budapest's eighth district. Photograph by the author | 94 |
| 6.1 | Centre of the first local commune in New Belgrade | 110 |
| 6.2 | Main phases and participants in making the Belgrade Master Plan of 1972 | 113 |
| 7.1 | Effects of military installations on Valga/Valka and surrounding landscapes | 123 |
| 7.2 | Valga/Valka combined general plan, 1970 | 128 |

Figures xvii

| | | |
|---|---|---|
| 8.1 | Architect Birutė Kasperavičienė at her drawing desk at the State Urban Construction Design Institute in Vilnius | 141 |
| 8.2 | Architects Genovaitė Balėnienė and Aida Lėckienė with colleagues working on the detailed plans of Lazdynai mass housing area. Photo: T. Žebrauskas, 1973 | 143 |
| 9.1 | Main square in the large ordinary city | 159 |
| 9.2 | Central city park in the large ordinary city | 163 |
| 10.1 | A caricature by Edgard Valter in the nature journal *Eesti Loodus* | 177 |
| 10.2 | The scheme by Asta Palm depicting the Tallinn recreational area in 1963 | 182 |
| 11.1 | The plan view of Gldani mass-housing district | 192 |
| 11.2 | The aerial photo of Gldani shot in 1981 | 196 |
| 12.1 | Fragment of the Plan of Mikrorayon Shartash showing the interweaving condition of industrial (shaded) and residential infrastructure | 216 |
| 12.2 | Drawing of a model of Mikrorayon Sinie Kamni in Yekaterinburg surrounded by industrial railways | 223 |
| 13.1 | A. Baburov et al., *The Ideal Communist City*, book cover | 232 |
| 13.2 | The NER Group, urban settlement integrated into the surrounding landscape. Plasticine model for Milan Triennale, 1968 | 237 |
| 14.1 | Untitled, Emi Skënderi, 2022. This drawing illustrates the contrast between environmental ideals and realities under socialism | 249 |
| 14.2 | "Planting trees", by Edi Hila (1972). Source: National Art Gallery of Albania. Artwork in the public domain. One can perceive this *aksion* scene as bursting with euphoria and youth energy, or as an agonising cry for help. Nature is as tormented as the people and seeks to escape regimentation | 261 |
| 14.3 | Summary of findings | 264 |

# Tables

6.1  Timeline of key federal (Yugoslav) and national (Serbian) legislative documents and their main substantive and procedural features     104
6.2  The overview of main substantive and procedural aspects of the planning process in leading international and national events and policies     107
14.1 Analysed works (in chronological order)     250

# Acknowledgements

Ideas about this book were first formed in preparation for a session on "Urban planning during socialism", co-chaired by Jasna Mariotti and Kadri Leetmaa at the European Architectural History Network (EAHN) conference in Edinburgh 2020. As the pandemic interrupted the world, the conference was moved from the summer of 2020 to June 2021. Despite the pause of one year, regular meetings in preparation for our session at the EAHN gave us an invaluable opportunity to frequently discuss and frame the outline for this book.

As editors we want to express our deep gratitude to Richard Williams and Richard Anderson from the University of Edinburgh, who gave us an opportunity to chair a session on "Urban planning during socialism" at the EAHN in Edinburgh, that initiated the idea about this book. Some of the papers presented in this book are from participants to our session at the EAHN in Edinburgh. Several other scholars were invited to contribute to the book as well, giving an opportunity to extend the context and geographical cover of this book, providing distinct views on different notions of periphery during socialism. Here we wish to thank all authors who contributed chapters in this book for the discussions that we had in the past couple of years and the thought-provoking exchanges that led to its publication. The meetings with authors that we held in the past few years provided invaluable opportunities to share ideas amongst colleagues and we look forward to continuing these discussions as we set up new research agendas for the socialist city. We hope the book will help readers from both sides of the former Iron Curtain and will further challenge socialist ideological assumptions which triggered production and reproduction of space in socialist cities.

We would also like to sincerely thank Routledge for the opportunity to publish this book and for their interest and ongoing support. In particular, we would like to thank Faye Leerink and Prachi Priyanka from the Routledge Research in Historical Geography. The positive feedback that we got from them was an important push for this book.

Generous funding from the following institutions supported the Open Access publication of this book (in alphabetical order): City University

Hong Kong, Estonian Academy of Arts, School of Natural and Built Environment at Queen's University Belfast, Slovak University of Technology in Bratislava (VEGA – project no.1/0286/21), The Estonian Research Council Grant PRG1919 and XM Architekten Basel. Thank you.

Finally, here we would also like to thank our families. The past couple of years were challenging for everyone, but giving us a space to work on this book, while dealing with multiple lockdowns and irregular schedules, were invaluable. This book would not have been here without their support, encouragement, joy, inspiration and everything else they give.

# Revisiting urban planning during socialism
Views from the periphery. An introduction

*Jasna Mariotti and Kadri Leetmaa*

> Although some of the substance of Communism is still active in all Communist countries – chiefly the one-party system and the party bureaucracy monopoly over the economy – the international relations and international position of each country differ so radically today from any other that to treat them all as the same would be the gravest conceivable mistake.
>
> (Milovan Djilas, *The Unperfect Society*, 1969)

During the state socialism of the 20th century, the urban transformations of cities "have been quite different from those of neighbouring non-socialist countries" (Szelenyi 1983:1). The transformation of cities under socialism was often integrated with five-year plans and involved domination by the state in all activities (French and Hamilton 1979; Bater 1980; Szelenyi 1983; Andrusz 1984; Smith 1996; Musil 2005). Fisher writes that "city planning in socialist countries is integrated with the overall economic planning of the state" and highlights four principles for socialist city planning: standardisation; town sizing and its dependence on its productive function and the number of people employed; the city centre and its distinctive character as a political, cultural and administrative core; and the neighbourhood unit concept as a tool for construction of cities (Fisher 1962:251-3). Spatially, socialist cities were initially transformed through construction of large industrial complexes and housing estates that were amongst the first projects that were planned and built in cities during this period (Musil 1980; Szelenyi 1983; Meuser and Zadorin 2016; Zarecor 2018). In such setting, "[t]he task of the planner was clear. He had to prepare the physical environment for the fulfilment of specific objectives within a set time frame" (Bater 1980:26), the specific objectives mostly referring to the grand goals of building capable industrial cities.

Sotsgorod, a key concept for the development of the future socialist city, had a great impact on the transformation of cities across wider geographies during this period. In his seminal book "*Sotsgorod: The Problem of Building Socialist Cities*", Miliutin (1974) proposes a new spatial vocabulary for the socialist city, a linear city, but also a reorganisation of the way of life in it throughout the settlement system.

> The reconstruction of our way of life on new socialist principles is the next problem facing the Soviet Union. Along with this, we are confronted with the overall problem of sanitary and health improvements in settlements throughout the USSR; nor can we allow the kind of criminal anarchy in construction procedures that characterizes the capitalist world. The Soviet village must be built in such a way as not to perpetuate the very conditions we are struggling against, but rather to create the basis for organization of a new socialist, collective way of life.
>
> (Miliutin 1930/1974:50)

These changes followed the transformation of political, economic and social life based upon the Marxist-Leninist thinking and resulted with an unprecedented expansion of cities during the period of state socialism, fostering record levels of urbanisation in socialist countries but also making smaller settlements more urban.

Literature, to date, focuses on urban planning during socialism predominately from the perspective of the capital city, being the locality of political authorities and their centres, and adequately covering topics including planning, housing, industrialisation and population growth (Musil 1987; Gentile & Sjöberg 2013; Steinberg 2021; van der Straeten and Petrova 2022). Many authors have conducted work on socialist cities including book projects (Molnár 2013; Crowley and Reid 2002; DeHaan 2013; Hess and Tammaru, 2019) and journal special issues which adequately cover the topics of architectural and planning history of cities during socialism (Bocharnikova and Kurg 2019).

Yet, at the same time, clear research enquiries remain – most importantly into studies that explore socialist cities beyond the limits of the centre, challenging the generalised and often preconceived assumptions of cities during the period of state socialism and their social and spatial equality from the perspective of "periphery". Throughout this book, we address this research gap using new case studies that enable further detailed exploration and comparative analysis of cities whose transformation is challenging the preconceived notions of the socialist city and its centre, as a considerably homogenous space, dominated by normative planning processes.

The term periphery assumes that wherever or whatever the periphery is, there is always a centre. In urban studies, periphery is often a spatial concept and therefore has spatial implications. The core-periphery model is often used to delineate uneven developments, spatial polarisation and inequality (Wellhofer 1989; McLoughlin 1994; Krugman 1998). The concept of peripheralisation in a spatial scale is also closely linked to the notion of marginalisation (Herrschel 2011; Danson and De Souza 2012), highlighting the link between peripheralisation and exclusion.

Wallerstein (1974) continues to be a critical source for scholarship on "modern world systems", the importance of centre-periphery concepts and the tensions between them. He distinguishes three zones of the world economy, namely, semiperiphery, the core and the periphery, where "periphery" means to be subordinated, while its resources tend to become redistributed to the core. Flint and Taylor (2018) have developed it further by integrating the global core-periphery politics across geographical scales and relating global processes to daily experiences. In this book, we make the argument that in the Eastern Bloc, the centre and the periphery together made state socialism and formed the concept of the socialist city, yet urban planning under state socialism is rarely analysed from the perspective of the system's periphery.

In defining the periphery during state socialism in this book, we took as a starting point the notion of periphery as "situated on the fringe" (Kühn 2015) and the broad definition of periphery as "the distance in relation to the core, in terms of geographic, economic, political or social factors" (Bourne 2010). Here, we argue that the concept of periphery is about dependences and transversalities too, as well as developing an understanding of what the periphery is distant from. In addressing this, we take a closer look at how different notions of periphery impacted the development and transformation of cities during state socialism of the 20th century. We further argue that interrelating different notions of periphery or constructing relations amongst different concepts are necessary, thereby extending the focus and understanding of the notion of periphery beyond the solely spatial core-periphery configurations that are currently dominant in urban studies. These relationships may be complex and contested, yet they are critical to understanding the nature of the socialist city and urban planning under socialism.

It is in this context that we propose to address the concept of periphery during the period of state socialism – of relational approaches and connectivity of different conditions – political, economic, social and spatial, that highlight particularities beyond the sole hierarchical considerations of the core and the periphery as a spatial system. By using this notion of periphery, contributors in this book explore complex processes of production of space in socialist cities, and the chapters presented here also complement this notion of the concept of "periphery" in the socialist world in the 20th century.

At the core of our book, there are two arguments. First, we argue that the largely unfinished project of the socialist city, neither homogenous nor anticipated, contributes to defining its periphery: economic, political, social and spatial, sometimes changing the centre-periphery interrelations. Second, we argue that the periphery of the socialist city is highly diverse and heterogeneous and cities in any of the peripheries during state socialism were often places for visionary urban experimentations at different scales.

Chapters in this book address the asymmetries and preconceptions of socialist cities, providing detailed contextual evidence from the perspective of the periphery. Furthermore, the chapters in this book advance the state of the art of socialist cities in two areas: first, through the diversity of case studies and experiences from a wide area of socialist countries and second, through the construction of interrelations and juxtapositions of the notions of periphery that extend discussions on cities during the period of state socialism of the 20th century. The studies presented in this book reveal spatial aspirations, experimentation and exchanges of architectural and planning ideas in the periphery during socialism. The studies also exhibit that in the complex mosaic of the socialist city, the periphery has functional properties too and articulate diverse urban experiences that contribute towards a renewed understanding of the socialist city that exceed regional geopolitical conceptualisations. The chapters furthermore encourage us to reflect on the significance of core-periphery, East-West and North-South polarities. It is in this context that the chapters in this book address the notion of periphery, embracing scholarship that inspires future urban studies with new concepts and theoretical considerations.

Recognising the complexities of the transformation of cities during the period of state socialism, we propose four takeaway messages on the notions of periphery in the socialist city as a future reference in urban studies. These takeaway messages are not exhaustive but provide critical thinking on the understanding of the concept of periphery during the period of state socialism in the 20th century, indicating interactions at different spatial scales and networks of power, beyond state institutions and beyond the centre as a place where "top decision makers are situated" (Langholm 1971:273).

1 **The concept of periphery in the socialist city is created through multiple interrelated processes, and its diverse spatial configurations depend on the transversalities of political, economic and social phenomena.**

The political, economic and social transformations during the period of state socialism had an impact on urban planning of cities and were intrinsic to every facet of urban space and life in socialist cities. These phenomena were intertwined, producing distinct forms and spaces in the socialist city that were not only spatial but also abstract accounts of the concept, resulting from the transversalities of political, economic and social relations.

In Kazakh Soviet Socialist Republic, a country at the spatial periphery of the USSR, a Virgin Lands campaign promoted by Nikita Khrushchev in the 1950s displayed the complex relations between urban planning and politics in the periphery through the transformation of Tselinograd. Tselinograd, established as the capital of the Virgin Land Territory in 1961, shortly after became a showcase for new town planning concepts and standardisation of buildings in wider parts of Central Asia, contextualising the evolution of the city into a new socio-economic centrality (Talamini 2024).

Economic variables during socialism had an impact on the development of the socialist city. Wrocław, formerly Breslau, a city in the "Regained Lands" in Polish People's Republic, during the period of state socialism, was transformed under a limited budget, in contrast to other cities in the country after WWII (Tomaszewicz and Majczyk 2024). Directly dependent on the finances from the centre – Warsaw, the projected multiple centres for socialist Wrocław were never realised. In the early 1960s priority in the city was assigned to the construction of residential estates and the reconstruction of the central district and not to any of the service centres located in the housing estates (cf. Leetmaa and Hess 2019), despite them being indicated in the post-WWII city development plans. This was seen also as a measure of marginalisation of the city on the national arena.

For the city of Bratislava, a city with a longstanding peripheral status within Czechoslovakia, the administrative confirmation of the city as the capital city of the Federative Republic of Slovakia in 1968 marked a period of the most ambitious development of the city. During this period the city grew nearly twice its size and new districts, infrastructures, as well as a renovation of the entire city centre were planned. Yet, most of these ambitious plans were realised only in fragments or remained only in the form of ambitious intentions as a result of financial constraints (Moravčíková, Szalay and Krišteková 2024).

In the city of Skopje, a capital city in the most southern republic of SFR Yugoslavia, the reconstruction following the 1963 earthquake was a delicate international game of Cold War politics (Babić 2024). The reconstruction was led by local and Yugoslav architects that introduced the city as an international blueprint of modernist planning, facilitated through transfer of knowledge that was sponsored by the UN and global collaborations established by the Yugoslavs. The technological advancement of the Yugoslav and Macedonian construction industry and the know-how of the architects were on full display in the city, where throughout the 1970s and 1980s brutalist constructs exemplified its urban identity.

Complex social relations and structures, in relation to the division of labour, produced specific spatial relations in Budapest during the period of state socialism. The construction of housing and the urban renewal of the city can be asserted to the decay of the historical housing during socialism in Budapest's second urban belt, contributing to the segregation of working-class communities in the city as new mass housing estates were predominantly inhabited by middle-class families (Kiss 2024). As a result, the working-class in the case of Budapest, although being central to state propaganda during socialism, remained on the social peripheries in Hungary during this period.

2 **The role of the architect and the planner during socialism is often seen as one that fulfils the requirements set by the state. Yet, architects and urban planners that found themselves at any of the peripheries were often "freed" from meeting specific requirements set by the state, professionally challenging the spatial East-West, North-South, core-periphery divides and that of the periphery viewed as a disadvantaged space.**

Urban planners that found themselves at any of the peripheries of the socialist city often applied globally trendy theories for the future socialist city and were in close communication with their Western colleagues (Ferenčuhová 2021), for example, visiting them on various exchange trips even during the tensest periods of international relations. The location of housing estates and their construction was one of the main tasks for the architects and urban planners during the period of state socialism, turning political propaganda into spatial reality.

In socialist Yugoslavia, the interrelationship between planning and citizen participation served as a tool for pursuing self-management socialism in the country which was considered peripheral to both the West and the East (Perić and Blagojević 2024). In order to foster local community needs, public participation in the country was introduced through national planning acts, which also established the roles of different actors in the planning processes. Within it, the role of the planner was a neutral professional service, contributing significantly to the societal emancipation, modernisation and welfare.

The ability of countries to adapt space to military needs is always prioritised over conventional spatial planning; this mostly happens when there is no direct war and the countries are in deterrence mode. As the period of state socialism of the 20th century coincided with the accumulation of international tensions, allocating land and reserving the locations for military use diverged from regular spatial planning – in many cities, "white areas", so to speak, were on the map for the planner. Leetmaa et al. (2024) present the case of a small peripheral Estonian-Latvian border town of Valga/Valka, which, due to the influence of Cold War priorities – the need to place medium-range missiles within firing range – shaped an insignificant "small place" into an important location on the world map due to the influence of a global "large issue". Gobova's chapter in Part IV of this book also refers to the need to adapt socialist urban planning to the country's military needs that always serve as priorities (Gobova 2024).

Individualised design approaches and regional differences to Soviet mass housing and standardised architecture of large housing estates also existed during the period of state socialism (cf. Drėmaitė 2019; Leetmaa and Hess 2019; Drėmaitė 2024). In the Lithuanian Soviet Socialist Republic, a Western periphery of the Soviet Union, the aspirations of local professional architects were taken into consideration and architects enjoyed greater freedom compared with other creative professions during this period. Architects were regarded as experts and specialists, and despite standardisation, there were numerous attempts to improve the design of housing estates and neighbourhood planning, often with strong regionalist approaches in architecture in residential neighbourhoods.

3 **Production and reproduction of places for everyday life in the socialist city were often dependent on social equity, therefore defining the complex notions of periphery through the non-politics of socio-spatial relations.**

The transformation of everyday life in the socialist city was one of the targets in the organisation of the socialist state, ideologically positioning equal distribution of resources and national socio-spatial ambitions. Places for everyday life, processes and activities enabled citizens to experience the socialist city, and through the design of places for everyday life activities, architects and town planners directly impacted the organisation of the socialist state and its commitment to equity. Symbolically, the places for everyday life in the socialist city were situated between ideological meaning and social equity.

Planning for spaces for everyday life in Ukrainian large ordinary cities, considered as a political periphery in the Soviet Union's urban network, was conditioned by standardisation (Mezentsev, Provotar and Gnatiuk 2024). In the large ordinary cities of Vinnytsia and Cherkasy, the desire of the Soviet authorities to establish certain patterns of planning organisation on the use of public spaces was met with reaction from the local residents. They accepted the city squares as places of power, while their collective interests were displayed in cities central parks and courtyards of large residential estates which remained oases for allowed freedom during the period of state socialism of the 20th century.

During the period of state socialism, the outer parts of cities and natural areas outside urban territories remained peripheral in post-war planning inquiries. In the spatial periphery of Tallinn in the 1960s, a peri-urban zone envisioned also in "The Project for Greater Tallinn" became a place for a new lifestyle for the city's residents and a site for family life (Lankots 2024). In this regard, the spatial periphery of the city had a strategic role in the planning of the socialist city where summer house settlements operated as an extension to the everyday urban environment, while not only providing places for relaxation but also becoming sites for family life, domestic duties as well as freedom and self-realisation in the socialist city (cf. Nuga et al. 2016).

Mass housing estates dominated the urban landscapes of socialist cities. Gldani, a mass housing neighbourhood at the northern edge of Tbilisi, was developed by the Georgian Soviet Socialist Republic to meet the city's growing population (Gogishvili 2024). The neighbourhood, designed by Bochorishvili, although developed according to strict standards of Soviet urban planning, had original features too. Yet, the experimental elements in the plan, including social spaces, public halls, recreational areas and the vertical axis, were never fully realised due to lack of funding but also due to the prioritisation of Soviet building standards and economic principles over the more idealistic architectural visions.

## 4 Environmental and ecological consciousness under state socialism highlights moral disengagement with political reasoning at different scales.

In order to maximise economic growth, natural resources during state socialism were often exploited while industrial growth contributed to the growth of cities during this period (Whitehead, 2005). Bater (1986) writes that the environmental context of the Soviet city was tantalisingly elusive. Yet, at the same time, environmental concerns and incorporation of nature in the city during this period were challenging the political structures and the industrial assets of the socialist city.

For cities in the spatial periphery in the USSR, growth was supported by the growth of their industrial enterprises, causing also ecological problems (Gobova 2024). In the Urals, contrary to the notions of centrally planned cities during state socialism, urban planning in Yekaterinburg (formerly Sverdlovsk) in the 1960s also took into consideration ecological and environmental influences, revealing also different narratives of power relations in the city planning process. In this spatial periphery of the USSR, architects and planners in the 1960s–80s took into consideration the growing industrial enterprises in the city, ensuring to reduce their negative ecological and environmental impact and setting up standards for future ecological and functional planning of industrial and residential districts.

Departing from Marxist ecological critique of capitalism as well as Soviet ecological thought, studies on Soviet unofficial architecture provide valuable insights on the environmental movements that emerged during late socialism (Panteleyeva 2024). This is linked with the emerging understanding of architecture as environment as well as its formal "identification" with the concept of nature which triggered the conceptualisation of "national" landscape and "nature" itself as agencies of political change. Architecture collectives, such as *NER* group's visions for the city, embraced new Soviet material models and realities, suggesting also reconciliation of urbanism with the natural domain through formal experimentations.

During the period of socialism in Albania, a country that was extremely isolated even within the Eastern Bloc, the conceptions of "environment" and "nature" served the government propaganda. Analysis of literature, film, music and painting of this period reveals some of the official conceptions of "nature" and "the environment" in the country (Pojani and Pojani 2024). The analysis of these symbolic products reveals emerging environmental consciousness and ideas under socialism through the theoretical lens of ecofeminism, incorporating notions of environmental exploitation and women's oppression.

**Further research**

Contributions in this book highlight different notions of periphery and peripheral in cities during the period of state socialism and their interrelations and transversalities – political, economic, social and spatial, contributing therefore indirectly to the understanding of the notion of the centre during this period too. These novel contributions present multiple experiences from socialist cities across wider geographies and social realities of the socialist world. The book aims to inspire a renewed research agenda on the socialist city, focusing on understanding its periphery as a considerably different urban scene compared to the centre (and its conventional case studies in urban research on socialist city). The different trajectories that are presented in this book outline some ideas for such future studies and enhance our appreciation of the socialist city through architectural and urban planning considerations. These perspectives also call for revisiting of the existing centre-periphery definitions in socialist cities and beyond. The case study cities presented in this book also aim in contributing to a wider scholarship in other contexts and provide a foundation for further research on cities whose transformation is pervasive and whose peripheries are in flux.

Finally, we would like to emphasise the timing of the completion of this book. When launching the collection in 2021, our initial goal was to highlight a specific and also an understudied chapter of the history of urban planning, namely, views from the periphery on urban planning under socialism. During the evolution of the book, however, global uncertainties have come about with the war in Ukraine – the future, nature and duration of which none of us unfortunately has still a precise idea about. Therefore, the case studies presented in this book are of special value. Very likely for some time, fieldwork (like the work with archival materials) will be limited for researchers in some of the cities represented in this collection. However, the material presented here uniquely reflects what the specific features of cities were in one urban system dictated by the ideological, political, social and economic motivations. We assume that the views on the history of urban planning presented on these pages also favour an understanding of cities that we may have the opportunity to explore more closely in the future.

## References

Andrusz, G. 1984. *Housing and Urban Development in the USSR*. Albany, NY: SUNY Press.
Babić, M. 2024. The Yugoslav Skopje: building the brutalist city, 1970–1990. In: Mariotti, J., and Leetmaa, K. (Eds.), *Urban Planning During Socialism: Views From the Periphery*. London: Routledge.
Bater, J.H. 1980. *The Soviet City: Ideal and Reality*. Edward Arnold: London.
Bater, J.H. 1986. Some recent perspectives on the Soviet city. *Urban Geography* 7(1):93–102. 10.2747/0272-3638.7.1.93
Bocharnikova, D., and Kurg, A. 2019. Introduction: urban planning and architecture of late socialism. *The Journal of Architecture* 24(5):593–603. 10.1080/13602365.2019.1671658
Bourne, L.S. 2010. Living on the edge: conditions of marginality in the Canadian urban system. In: Lithwick, H., and Gradus, Y. (Eds.), *Developing Frontier Cities: Global Perspectives – Regional Contexts*, pp. 77–97. Netherlands: Kluwer Academic Publishers.
Crowley, D., and Reid, S.E. (Eds.). 2002. *Socialist Spaces: Sites of Everyday Life in the Eastern Bloc* (Vol. 1). Oxford: Berg.

DeHaan, H.D. 2013. *Stalinist City Planning: Professionals, Performance, and Power.* Toronto: University of Toronto Press.

Djilas, M. 1969. *The Unperfect Society: Beyond the New Class.* London: Methuen.

Drėmaitė, M. 2019. The exceptional design of large housing estates in the Baltic countries. In: Hess, D.B., and Tammaru, T. (Eds.), *Housing Estates in the Baltics: The Legacy of Central Planning in Estonia, Latvia, and Lithuania*, pp. 71–93. Cham: Springer.

Drėmaitė, M. 2024. The role of architects in fighting the monotony of the Lithuanian mass housing estates. In: Mariotti, J., and Leetmaa, K. (Eds.), *Urban Planning During Socialism: Views from the Periphery.* London: Routledge.

Danson, M., and De Souza, P. (Eds.). 2012. *Regional Development in Northern Europe. Peripherality, Marginality and Border Issues*, pp. 49–64. London: Routledge.

Ferenčuhová, S. 2021. Thinking relationally about socialist cities: cross–border connections in Czechoslovak post-war urban planning and housing construction. *Planning Perspectives* 36(4):667–687. 10.1080/02665433.2020.1844042

Fisher, J.C. 1962. Planning the city of socialist man. *Journal of the American Institute of Planners* 28(4):251–65. 10.1080/01944366208979451

Flint, C., and Taylor, P.J. 2018. *Political Geography: World-Economy, Nation-State and Locality.* 7th edition. London: Routledge.

French, R.A., and Hamilton, F.E.I. (Eds.). 1979. *The Socialist City: Spatial Structure and Urban Policy.* Chichester: Wiley.

Gentile, M., and Sjöberg, Ö. 2013. Housing allocation under socialism: the Soviet case revisited. *Post-Soviet Affairs* 29(2):173–95. 10.1080/1060586X.2013.782685

Gobova, N. 2024. New ecological planning and spatial assessment of production sites in socialist industrial Yekaterinburg (formerly Sverdlovsk) in the 1960s–80s. In: Mariotti, J., and Leetmaa, K. (Eds.), *Urban Planning During Socialism: Views from the Periphery.* London: Routledge.

Gogishvili, D. 2024. Gldani: from ambitious experimental project to half-realised Soviet mass-housing district in Tbilisi, Georgia. In: Mariotti, J., and Leetmaa, K. (Eds.), *Urban Planning During Socialism: Views from the Periphery.* London: Routledge.

Herrschel, T. 2011. Regional development, peripheralisation and marginalisation—and the role of governance. In: Herrschel, T., and Tallberg, P. (Eds.), *The Role of Regions? Networks, Scale, Territory*, pp. 85–102. Kristianstad: Kristianstad Boktrycker.

Hess, D.B., and Tammaru, T. (Eds.). 2019. *Housing Estates in the Baltic Countries: The Legacy of Central Planning in Estonia, Latvia and Lithuania.* Springer Open.

Kiss, D. 2024. From reverse colonial trade to antiurbanism. Budapest's frustrated urban renewal between 1950 and 1990 in the face of the Soviet world order's anomalous centre-periphery relations. In: Mariotti, J., and Leetmaa, K. (Eds.), *Urban Planning During Socialism: Views from the Periphery.* London: Routledge.

Krugman, P.R. 1998. Space: The final frontier. *Journal of Economic Perspectives* 12(2):161–74. 10.1257/jep.12.2.161

Kühn, M. 2015. Peripheralization: theoretical concepts explaining socio-spatial inequalities. *European Planning Studies* 23(2):367–378. 10.1080/09654313.2013.862518

Langholm, S. 1971. On the concepts of center and periphery. *Journal of Peace Research* 8(3/4):273–8.

Lankots, E. 2024. Planning urban peripheries for leisure: the plan for Greater Tallinn, 1960–1962. In: Mariotti, J., and Leetmaa, K. (Eds.), *Urban Planning During Socialism: Views from the Periphery*. London: Routledge.

Leetmaa, K., and Hess, D.B. 2019. Incomplete service networks in enduring socialist housing estates: Retrospective evidence from local centres in Estonia. In: Hess, D.B., and Tammaru, T. (Eds.), *Housing Estates in the Baltic Countries: The Legacy of Central Planning in Estonia, Latvia, and Lithuania*, pp. 273–99. Dordrecht: Springer Open.

Leetmaa, K., Tintěra, J., Pae, T., and Hess, D.B. 2023. The influence of nuclear deterrence on the growth and decline of the peripheral town of Valga/Valka during the Cold War. In: Mariotti, J., and Leetmaa, K. (Eds.), *Urban Planning During Socialism: Views from the Periphery*. London: Routledge.

McLoughlin, J.B. 1994. Centre or periphery? Town planning and spatial political economy. *Environment and Planning A*, 26(7):1111–22. 10.1068/a261111

Meuser, P., and Zadorin, D. 2016. *Towards a Typology of Soviet Mass Housing: Prefabrication in the USSR 1955-1991*. Berlin: DOM Publishers.

Mezentsev, K., Provotar, N., and Gnatiuk, O. 2024. Courtyards, parks and squares of power in Ukrainian cities: planning and reality of everyday life under socialism. In: Mariotti, J., and Leetmaa, K. (Eds.), *Urban Planning During Socialism: Views from the Periphery*. London: Routledge.

Miliutin, N.A. 1974. *Sotsgorod: The Problem of Building Socialist Cities* (trans. Arthur Sprague). Cambridge: The MIT Press.

Molnár, V. 2013. *Building the State: Architecture, Politics, and State Formation in Postwar Central Europe*. London: Routledge.

Moravčíková, H., Szalay, P., and Krišteková, L. 2024. Dreaming the capital: architecture and urbanism as tools for planning the socialist Bratislava. In: Mariotti, J., and Leetmaa, K. (Eds.), *Urban Planning During Socialism: Views from the Periphery*. London: Routledge.

Musil, J. 1980. *Urbanization in Socialist Countries*. White Plains, New York: Sharpe.

Musil, J. 1987. Housing policy and the socio-spatial structure of cities in a socialist country: the example of Prague. *International Journal of Urban and Regional Research* 11(1):27–36. 10.1111/j.1468-2427.1987.tb00033.x

Musil, J. 2005. City development in Central and Eastern Europe before 1990: historical context and socialist legacies. In: Hamilton, F.E.I., Dimitrovska Andrews, K., and Pichler-Milanovic, N. (Eds.), *Transformation of Cities in Central and Eastern Europe: Towards Globalization*, pp. 22–43. Tokyo: United Nations University Press.

Nuga, M., Leetmaa, K., and Tammaru, T. 2016. Durable domestic dreams: exploring homes in Estonian socialist-era summer-house settlements. *International Journal of Urban and Regional Research* 40:866–83. 10.1111/1468-2427.12403

Panteleyeva, M. 2024. Peripheral landscapes: ecology, ideology and form in Soviet non-official architecture. In: Mariotti, J., and Leetmaa, K. (Eds.), *Urban Planning During Socialism: Views from the Periphery*. London: Routledge.

Perić, A., and Blagojević, M. 2024. Passive gents or genuine facilitators of citizen participation? The role of planners under the Yugoslav self-management socialism. In: Mariotti, J., and Leetmaa, K. (Eds.), *Urban Planning During Socialism: Views from the Periphery*. London: Routledge.

Pojani, D., and Pojani, E. 2024. Conceptions of 'nature' and 'the environment' during socialism in Albania: an ecofeminist perspective. In: Mariotti, J., and Leetmaa, K. (Eds.), *Urban Planning During Socialism: Views from the Periphery*. London: Routledge.

Smith, D.M. 1996. The socialist city. In: Andrusz, G., Harloe, M., and Szelenyi, I. (Eds)., *Cities after Socialism*, pp. 70–99. Oxford: Blackwell.

Steinberg, M.D. 2021. The new socialist city: building Utopia in the USSR, 1917–1934. *International Critical Thought* 11(3):427–49. 10.1080/21598282.2021.1966819

Szelenyi, I. 1983. *Urban Inequalities Under State Socialism*. Oxford: Oxford University Press.

Talamini, G. 2024. Urbanising the Virgin Lands: at the frontier of Soviet socialist planning. In: Mariotti, J., and Leetmaa, K. (Eds.), *Urban Planning During Socialism: Views from the Periphery*. London: Routledge.

Tomaszewicz, A., and Majczyk, J. 2024. From Breslau to Wrocław. Urban development of the largest city of the Polish "Regained Lands" under socialism. In: Mariotti, J., and Leetmaa, K. (Eds.), *Urban Planning During Socialism: Views from the Periphery*. London: Routledge.

van der Straeten, J., and Petrova, M. 2022. The Soviet city as a landscape in the making: planning, building and appropriating Samarkand, c. 1960s–80s. *Central Asian Survey* 41(2):297–321. 10.1080/02634937.2022.2060937

Wallerstein, I. 1974. *The Modern World-System*. New York: Academic Press.

Wellhofer, E.S. 1989. Core and periphery: territorial dimensions in politics. *Urban Studies* 26(3):340–55. https://doi.org/10.1080/00420988920080341

Whitehead, M. 2005. Between the marvellous and the mundane: everyday life in the socialist city and the politics of the environment. *Environment and Planning D: Society and Space* 23(2):273–94. 10.1068/d372t

Zarecor, K.E. 2018. What was so socialist about the socialist city? Second World urbanity in Europe. *Journal of Urban History* 44(1):95–117. 10.1177/0096144217710229

# Part I
# Urban planning, politics and power
Relations in the periphery

# 1 Urbanising the Virgin Lands
## At the frontier of Soviet socialist planning

*Gianni Talamini*

**Introduction**

This chapter focuses on the Kazakh Soviet Socialist Republic to investigate the Soviet project to urbanise the Central Asian steppes, an enormous, pioneering project that remains largely underinvestigated. Starting in the 1930s, the Soviets began to build an extensive network of canals, railways and roads to exploit the region's natural resources. The Virgin Lands Campaign, promoted by Nikita Khrushchev in the 1950s, further contributed to the human colonisation of a vast barren area by providing a territorial infrastructure for further spatial development. Tselinograd was established as the capital of the USSR Virgin Lands in 1961. The planned city ultimately showcased a new conception of town planning and the standardisation of buildings under the leadership of Nikita Khrushchev. Over a large span of time and space, new typologies were synthesised and reproduced. These typologies eventually became fundamental components of the cultural history of Central Asia. Moreover, the Soviet territorial infrastructure is still shaping the region's socio-economic development trajectory today. This chapter contextualises the association between systems of signs and modes of spatial production, discussing the socialist phase of territorial development from a *longue durée* perspective, that of the evolution of a peripheral settlement into a new centrality.

\* \* \*

The production of space as an apparatus of power has been the focus of extraordinary research, partially inspired by and building on the works of French intellectuals such as Henri Lefebvre and Michel Foucault. Although a small portion of the urban space is intentionally conceived as an apparatus, urban and architectural spaces often unintentionally express, shape and operationalise the hegemonic values of a society. A crucial theoretical attempt to investigate the association between modes of production and the production of space was carried out by Henri Lefebvre (1991; 1996). As David Harvey (2010) put forward, both the basis and superstructure – as well as their inherent social, economic, political and ideological attributes, including the relationships between them – can be rendered through the study of the production of space. Notably, no necessary link of reciprocity exists between spatial forms and modes of production. Yet a reasonable assumption is "that a dominant mode of production will be characterised by a dominant urbanistic form and, perhaps, by a certain homogeneity in the built form of the city" (Harvey 2010:204).

As a premise for this research, the etymology of the word *periphery* (from *peri* – "around" + *pherein* "to bear") could help clarify how this notion depends on that of a *centre*. The periphery is also commonly intended in geometry as the outer surface, the outside boundary of a closed figure, thus indicating the limit of a finite system. How this boundary is interpreted and shaped can manifest a society's ideological superstructure and the spatial relations of production. In the USSR, since the 1920s, the need to intensify and extend the territorial control and exploitation of natural resources in non-capitalist ways drove an intensive debate on the periphery as the place to experiment with new spatial forms for building communism. The periphery at this historical juncture coincided with the notion of the frontier, intended in both physical and intellectual connotations.

Notably, Andrei Platonov set some of his most famous and controversial works at the periphery of the USSR in the depth of Central Asia. The village of *Chevengur*, the imagined last reserve of a group of people searching to realise communism in the aftermath of the October Revolution, is located in an undefined area of steppes; both *The Sea of Youth* and *Dzhan* are set in areas that could be identified in the steppes of modern Turkmenistan. These works not only reflected the vibrant intellectual debate about giving a coherent spatial form to the newly established communist society but also questioned the pioneers' role in the utopian construction of real socialism.

In parallel to writers – *Engineers of the Soul* (Westerman 2010), such as Platonov, Gorky and Olesha – urbanists such as Milyutin, Leonidov, May and Forbat were confronted with the need to put ideas into practice and give a physical form to communism. Soon after the 1930s, the urban planning discourse initially clustered around two polarised positions: the so-called *urbanist* and *disurbanist*. The utopianism of such early proposals left space for more pragmatic approaches, which served for the development of industrial and agricultural production under the leadership of Stalin and Khrushchev, respectively.

This chapter focuses on the case of Astana as a peripheral Soviet space in which it is possible to read the stratified layering of the spatial crystallisation of power but also as a place semanticised through territorialisation. Astana is the current name of a location in the heart of Central Asia, previously known as, in order of time, Akmolinsk, Tselinograd, Akmola, Astana and Nur-Sultan. It has developed from a tsarist Russian outpost on the north bank of the Ishim River to the current capital of Kazakhstan (Figure 1.1). Crucial evolutions of the site followed the establishment of Tselinograd and Astana, respectively.

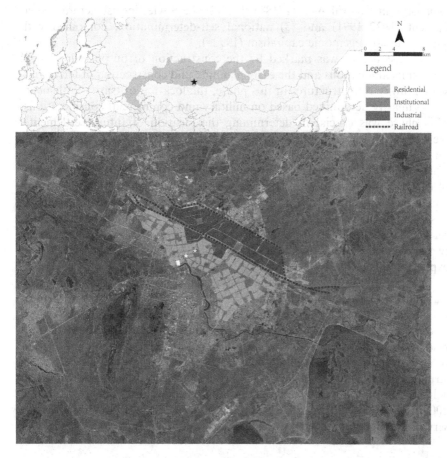

*Figure 1.1* Top: In grey, the Virgin Lands Campaign area and Astana; bottom: Astana's urban area, with the superimposition, in the centre, of the 1963 plan for Tselinograd; white square: Akmolisk fort; white star, left: Palace of Tselinograd Virgin Lands Developers (today, Concert Hall "Astana"); and white star, right: Palace of Youth (today, Zhastar Sarayy).

*Source:* Author; adapted from the 1963 plan for Tselinograd.

## Central Asia urbanism from a *longue durée* perspective

A paucity of research exists on the Central Asian spatial palimpsest, both as a space in which to read the stratified layering of the spatial crystallisation of power and as a place signified by urban planning and architecture. Drawing on the theoretical framework proposed by Giovanni Arrighi in his world-system analysis (Arrighi 1994), the production of space in Central Asia can be read as part of a *longue durée* process fostering the material expansion of the world system. This transformation unfolded in three centuries-long stratifications of the territorial palimpsest: (1) the Russian Empire territorialist expansion in Central Asia (1718–1917); (2) the Soviet infrastructural development (1922–1991) and (3) national self-determination, conjoined with supranational hegemonic capitalism (1992–).

The first period was marked by the construction of outposts in critical geographical locations and the early planning and construction of territorial-scale railway infrastructure. In this phase, the location of tsarist settlements was primarily determined based on military and commercial considerations. This phase was crucial in determining the location of future settlements. Nevertheless, despite being relatively long, this phase left only minor traces on the territorial palimpsest.

During the second period, the Soviets began to build an extensive network of canals, railways and roads to rationally exploit the natural (underground and agricultural) resources of the barren deserts and steppes surrounding the oases and fertile valleys, such as the Chorasmia and Fergana valleys, that had served as cradles of civilisations for centuries. This period was marked by two crucial phases: industrialisation in the 1930s and 1940s and agricultural expansion in the 1950s and 1960s.

The third phase started with the dissolution of the USSR. Since then, new state entities have overtaken former Soviet republics and strived to project a coherent image of cultural and political independence. The new states also concurrently opened to foreign investments, exploiting both natural resources and locational advantages. As a consequence, a double production of space occurred. On one end was the expression of an ideological superstructure through the synthesis of a new architectural idiom that aimed to achieve internal stability and international recognition (Fauve & Gintrac 2009; Fauve 2015; Köppen 2013); on the other end, the production of space served the functional needs of the external resource acquisition.

**Phase 1: A point for control**

The first phase coincided with the territorialist expansion of the Russian Empire and configured a diffusion of strategic points in an untamed environment. Initially military outposts and then trading hubs, these points functioned as the early footholds in the following infrastructural development of the territory. Astana was established, although with a different name, two centuries ago as one such early point of territorialist expansion.

*A tsarist outpost on the Ishim*

The Kazakh capital stands in a place whose urban history is short but particularly eventful. A crossing point of the Ishim river, this place was traversed by the Kazakh nomads who left the first faint traces of human life in the kurgans, the burial mounds used by these people. The site was the northernmost place in the vast area over which the routes of the cyclical transhumance of the Kazakh nomads extended.[1] The name of the first tsarist-era camp that arose here, Akmolinsk, was reminiscent of the original necropolis that stood next to the camp: Akmola (meaning "white tomb"). Established in 1824 and developed from 1830 onwards in the southern expansion within the larger framework of the *Great Game*, the tsarist camp arose as the site of a small military garrison resembling the many others established in the same period. The garrison was permanently inhabited by Cossack officers and guards; it served as a shelter, a sort of caravanserai, for the merchant caravans passing through. As Andrey Fyodorovich Dubitsky (1986) mentions, the place was crossed by an ancient caravan route known as the "Blood Road" due to the bloody attacks of marauders – ambushes that usually occurred in the thick bushes on the south bank of the Ishim. The need to control the ford is why Colonel Fyodor Kuzmich Shubin II established the garrison near the Qara-Ötkel (meaning "black ford" in Kazakh). The military garrison initially consisted of only stone pavement and adobe barracks. However, constant attacks by nomadic populations quickly forced the tsarist government to take defensive countermeasures and thus erect a fortification to shelter the barracks: the complex had a square plan and five bastions. In 1839, a low defensive rampart and a moat were built to defend three sides of the fort, and in 1840, the central bastion on the northern side (the most exposed to attack) was crowned by a squat tower.

The small agglomeration, little more than a village, overgrew for two main reasons: its location and exceptional benefits, including the cancellation of customs duties. These same reasons also attracted merchants, particularly Tatar merchants who arrived in large numbers. As a result, despite having a military origin, the settlement's dominant nature soon became commercial: the place began to act as a distributor of Russian goods transported there to meet increasing local demand. The town also began to develop administrative functions by enhancing the commercial vocation and embryonic manufacturing industry. By the end of the 19th century, the small town had a settled population of a few thousand people outside the fortress in spatially bounded social groups: the Cossack, the Tatar and the Kazakh villages. The cattle trade drove the economy of the town, whose appearance was marked by the many surrounding windmills. Regarding housing, buildings of usually one or two storeys were built of wood, brick and wood or entirely of brick. Adobe brick buildings were relatively rare. The small town along the Ishim then continued developing until the time of the revolution. In the following period, the civil war paralysed the development of the town: on 1 January 1912, Akmolinsk had 14,756 inhabitants but only 10,686 on the same date in 1923. A sudden growth due to an abnormal and forced wave of migration occurred during WWII: Germans, Chechens, Koreans, Poles and other ethnic groups present in the territory of the Soviet Union were deported to the less reachable regions of the USSR. However, this considerable population displacement only partly touched Akmolinsk since most of this population was assigned to major industrial centres, such as Karaganda.

## Phase 2: A line of a network

Since the 1920s, the urbanists in the Soviet Union have been confronted with an essential question: which urban form has a socialist city? This question pushed the frontiers of the intellectual debate into the Virgin Lands, literally and figuratively. The initial debate, which involved European intellectuals, has been widely reported. The following phase – one ruled out by Stalin coming to power and the consequential departure of Europeans such as André Lurçat, Ernst May, Hans Schmidt and Mart Stam – has been less avant-garde and remains largely underinvestigated. What emerges from the case of the Kazakh Soviet Socialist Republic is the pragmatic application of the linear city model that was coherently proposed for Magnitogorsk and typically developed along an extended railway network. This model was first implemented in the early process of industrialisation of Central Asia, for example, in Fred Forbat's plan for Karaganda of 1932. Under the leadership of Nikita Khrushchev, the model was adopted in an attempt to eliminate the "'contradictions between town and country' in Bolshevik parlance. Khrushchev wished to turn the peasant into skilled agricultural labourers, a rural proletariat whose mindset and way of life would differ little from that of urban industrial workers" (Tompson 2016:96). Driven by the industrialisation of agriculture, a new impetus towards the heavy anthropisation of Central Asia occurred. The momentum resulted in human history's largest policy-induced cropland expansion – with the anthropogenic effects on climate recently identified through science (Rolinski et al. 2021). Concurrently, the standardisation of construction, although lowering spatial and material qualities of the built environment, provided a large population with new residential spaces. Thus, the Virgin Lands of Central Asia provided Khrushchev with the ideal space to test the industrialisation of agriculture and construction at an unprecedented scale. In such historical circumstances, the intellectual effort focused on speeding up production via simple construction techniques and standard design to quickly implement projects across the Union.

*The network – The railway as a territorialist infrastructure*

The railway development played a crucial role in the territorialist expansion of tsarist Russia and, later under the USSR, in the construction of socialism. The first project for a railway line connecting Tyumen to Tashkent, passing through Akmolinsk, dates back to 1878: the Russian Ministry of Railways was considering the possible realisation of the work at the time, but the line stopped in Omsk in 1895 due to lack of funds. By the end of the first decade of the 20th century, Akmola Oblast had only about 40 kilometres of narrow-gauge track. Outside the administrative boundaries, British concessionaires built a section between Karaganda and the Spassky copper factory from 1906 to 1908. Shortly before the WWI, Russian and foreign capitalists formed a company to build the South Siberian Railway. The route would pass through Orsk, Akmolisnsk and Semipalatinsk. Construction occurred during the WWI and spurred the mass employment of prisoners of war. However, the October Revolution, followed by the Russian Civil War, halted the expansion of the rail infrastructure until Lenin – who recognised the urgency of completing the work to be able to transport food and supplies to those starving from the great famine that hit the southern Union provinces in the early 1920s – officially started the urgent construction of the infrastructure connecting the cities of the Kazakh steppe on 5 August 1920. Workers were sent from other cities, and the area's army and population were mobilised. In 1922, the railway reached Kokshetau, one of the northernmost regional capitals of present-day Kazakhstan, but the work was later interrupted due to the enormous economic difficulties the country was experiencing; work was not resumed until the late 1920s. Accordingly, the first train arrived in Akmolinsk on 8 November 1929, and from then on, the city quickly became an important railway hub due to the strategic location. Karaganda was joined by the railway two years later in 1931; the Akmolinsk-Kartaly line's construction began in 1939 and was completed in 1945; and the Akmolinsk-Pavlodar link was completed in 1952. However, Akmolinsk had to wait until 1962 for a new, modern station.

Meanwhile, Kazakh SSR played an essential role in the early 1940s due to the war's events in the geographical European side of the Soviet Union. Far from the seas and possible enemy invasions, Kazakh's cities were industrialised to compensate for losses on the Western Front and support war needs. These reasons also fuelled the crucial mining discoveries in the Kazakh underground. Akmolisk quickly found itself equipped with many state-of-the-art production facilities, laboratories and equipment.

*The linear city – A line in the Virgin Lands: the plan for Tselinograd*

From the beginning of the 1950s, the infrastructural expansion of Central Asia gained renewed impetus; the Virgin Lands Campaign, promoted by Khrushchev, further contributed to the human colonisation of a vast barren area. This expansion provided a territorial infrastructure for further spatial development. Stalin's successor was a son of peasants convinced that the development of the Soviet Union had to pass through that of its agriculture; Nikita Sergeyevich Khrushchev launched the Virgin Lands Campaign in 1954. The aim was to expand agriculture to those areas of the Soviet Union that had never been cultivated to permanently remedy the Union's food shortage. To host a large migrant population employed within the expanded agricultural sector, the conversion of a vast territory of the Eurasian steppes to arable land entailed the expansion of urban areas in the region. The urbanisation of the Virgin Lands was accomplished via the movement of people and ideas from the Western and – at that time – more advanced parts of the Union. Eventually, the campaign allowed Khrushchev to conceptualise ideas he developed as head of the Communist Party of the Ukrainian Soviet Socialist Republic, such as collectivising agriculture and eliminating the difference between town and countryside. To achieve the latter, Khrushchev conceived the concepts of "rural proletariat" and "agro-town" – larger than the typical rural village, such a settlement would have offered public facilities and services typical of a town (Tompson 2016). To rapidly achieve this goal, a new residential typology was proposed in conjunction with the Virgin Lands Campaign. Popularly known as *khrushchyovka*, the typology was adopted throughout the USSR territory and exported to other socialist countries. In the early 1950s, within two decades, this new typology was used to supply dwellings for around 60 million people in the USSR. The *khrushchyovka* contributed to shaping the socialist society into an essential part of USSR history, popular culture and collective memory.

The Virgin Lands Campaign is a heroic period in the region's history (Dubitsky 1986). The campaign yielded extraordinary results in the short term but was later run aground for various reasons, starting with the sudden impoverishment of the soil. Trainloads of volunteers, following the call of the supreme leader, enlisted from all over the USSR and arrived at Akmolisnk. The city was renamed Tselinograd (literally "city of the Virgin Lands") on 20 March 1961 and elevated as the *capital* (leading centre) of the Virgin Lands. Agricultural policies marked the history of these places for a long time, starting with the development of a town plan to regulate the considerable population growth.

The Lengorstrojproekt, Leningrad's planning institute – with the assistance of the Urban Development Institute, the Promstrojproekt and other planning institutions in the Union – was therefore commissioned to draw up the plan for the new city. Vyacheslav Alekseevich Shkvarikov directed the work; the architects Knyazev, Varlamov, Yargina, Zhukov, Lukyanov and Zarudko collaborated on drafting the plan. The plan was finally drafted at the end of 1962 and approved in February 1963. Only a few paper fragments of the graphic design remain, but the city's urban structure results are still legible today (Khairullina 2015; Figure 1.1). The plan envisaged a linear development in three functional zones: an industrial zone on the northern side of the city (extending from the railway line), a median residential zone (containing five districts of 50–100 thousand inhabitants) and a final strip to the south, near the river, dedicated to institutions and recreational activities. The linear scheme then has a significant advantage in that it can potentially be extended indefinitely. The plan was the occasion for Shkvarikov, a leading academician and director of the Central Scientific Research and Design Institute for Town Planning of the Soviet Union, to experiment with innovative approaches in the organisation of the *microraion* (microdistric) and the dynamic growth of the urban area. According to Elvira Khairullina (2015), the plan makes clear reference to the 1934 Van Eesteren's plan for Amsterdam in both the relationship with the historic settlements and the green areas as structural articulation elements of the plan. Each *microraion* was planned to be served by community facilities in urban green spaces, and it was provided with direct access to the green network and public transportation system (Khairullina 2015). The size and layout of such superblocks adapted to the existing condition; in historical central areas, the size of the block was reduced to conform it to the pre-existing urban materials.

The new plan showed a significant discontinuity with the old tsarist settlement: the balkanisation of Akmolinsk's population – its spatial division into ethnically homogeneous settlements of Cossacks, Tatars and Kazakhs, with toponyms such as Mechetnaya Street (Mosque Street) and Cerkovnaya Street (Church Street) – contrasted with the homogeneity of the new neighbourhoods, and the old religious buildings were replaced, per socialist dictates, with the headquarters of the new secular institutions.

Immediately after approval, the plan became operational, and construction of the city began to provide an immediate response to the pressing demand for housing. The construction was sponsored by experienced builders who had travelled to Tselinograd from Moscow and Leningrad to construct the first demonstration buildings. In just six months, the city measured an expansion of 650 dwellings and comprised four, five-storey school buildings. The first street in the plan to be realised was the Mira, on which five-storey buildings were constructed. At the same time, the main square had also begun to take shape: the seven-storey Soviet Palace, as was the housing, was erected with prefabricated concrete modules. The blocks of flats were mostly five storeys, and after learning the techniques for assembling the modules, the local teams quickly learned how to erect the buildings themselves. By the end of 1962, the city's housing had increased by some 115,000 square metres, and the network of educational institutions boasted an impressive 2,270 seats and a dining hall seating 300. In addition, a network of paved roads and pavements stretched a total of about 36 kilometres.

The city's construction proceeded apace, and to meet the demand for building materials, on-site production of expanded clay, bricks, cement, asphalt, precast concrete blocks and other concrete products was started. However, these efforts were not sufficient to meet the needs. Therefore, during June 1964 alone, 600 loads of materials, technicians, plumbers, painters, plasterers and electricians were sent from Moscow. The settlement expansion also brought further material demands for water and electricity. The first was solved by building the vast Vyacheslavkoe reservoir 60 kilometres east of the urban centre. The electricity network, on the other hand, was implemented thanks to a line running alongside the new road infrastructure: in 1964, the electrical connection between Tselinograd and Karaganda was built in record time, covering a distance of over 250 kilometres. Aviation then developed as well: on the 46th anniversary of the Great October Revolution, the first scheduled turboprop plane (an IL-18) landed at the city's new airport on 4 November 1963, inaugurating the direct Tselinograd–Moscow route. The non-stop flight to Moscow took four hours and twenty minutes, and the flight to Alma–Ata only one hour and forty minutes: the large distances between the city and the two capitals could thus be bridged in no time. Finally, in 1964, television signal transmission became possible thanks to electrification.

During this period of fast urban development, while anonymous residential blocks were constructed according to standardised models, the most iconic buildings were designed by prominent architects: emblematic is the case of the Palace of Youth, designed by a team of architects led by the Russian Anatoly Polyansky (Figure 1.2). Polyansky gained fame for the pavilion of the USSR at the International World Fair in Brussels in 1958 and was later appointed as the chairman of the Union of Architects of the USSR. The Palace of Youth was completed on 20 March 1975 and designed by architect Kirill Mironov, engineer Tsilya Nakhutina, and artists Dmitry Merpert and Nelli Mironova; the palace housed 1,200 seats, a convertible stage, a 400-seat sports hall, a swimming pool with diving boards, exhibition halls, a library, group work rooms, a 150-seat bar, a banquet hall for 50 people and several other rooms. The construction of the building was achieved using imported materials and local Taskol marble, but construction was put on hold for a decade after the dismissal of Khrushchev (Iskakov 2020). A copy of the Palace of Youth of Tselinograd was constructed in the city of Donetsk in 1975; evidently, design references were circulating across and within the boundaries of the Union, despite the international isolation.

*Figure 1.2* Palace of Youth in 1975.
Source: Photo courtesy of Vasily Toskin.

Another notable case is the Palace of Tselinograd Virgin Lands Developers, whose design was picked up by Khrushchev at the National Exhibition of Economic Achievements in Moscow: "We need such a building in Tselinograd", dictated Nikita Khrushchev (Gudro and Krastiņš 2019). The building was originally designed in 1960 by the Latvian architects Kraulis, Danneberga and Fogels as a panoramic cinema for Riga, a cinema which was never built. The palace's auditorium had 2,355 seats, about 50 speakers, a dozen projectors and a 34-metre by 13-metre monitor. Architect Daina Danneberga, a key member of the design team, was indirectly influenced by the work of Finnish masters Armas Lindgren and Eliel Saarinen, with whom Danneberga's professor Andrei Olj had worked. The Palace of Tselinograd Virgin Lands Developers was developed by five institutes based in Moscow, Leningrad and Riga, together with the Latvian State Urban Design Institute. Furniture for the palace was also produced in Riga and transported on-site (Gudro and Krastiņš 2019).

In the same year the Palace of the Youth was completed, on the occasion of the 58th anniversary of the October Revolution on 5 November 1975, the new monument to Lenin was also inaugurated, crowning the large city square. The 16-metre-high bronze statue of Lenin dominated the square, his back to the tallest building (the headquarters of the Giprosel'hoz). The buildings overlooking the ample open space were characterised by the regular rhythm of the openings, the absence of decoration and achromia. The only note of colour was the faces of Lenin and Marx, painted large on the massive volume of the House of Soviets. As Takashi Tsubokura (2010:16) noted, "the main point of the central square of the Tselinograd days was a combination of representational figures of socialist heroes and anonymous architecture on the background. It was nothing else but a visual representation of 'an orderly society led by socialism'".

**Phase 3: A radial city**

The last of the three phases was inaugurated by the collapse of USSR and marked by the mutation of the Union's internal administrative borders into external national boundaries. The balkanisation of Central Asia resulted in the elevation of regional centres into new pivotal centralities. Such mutation, accompanied by the restructuring of the socio-economic basis, produced new urban forms and new architectural idioms.

*When the periphery became the centre: The construction of a nation in Astana*

In 1992, the government of the newly formed Republic of Kazakhstan changed toponyms to mark a distance from the Soviet past. Alma-Ata became Almaty, and most of the pre-Soviet cities regained the name they had before the 70-year communist era. Such was the case with Dzhambul, now Taraz; the same fate befell Tselinograd, which took back its original toponym to become Akmola. However, apart from such a formal mutation and concurrently with the removal of Soviet vestments and effigies, no transformation of a structural nature was recorded. On the contrary, in the precarious economic and political situation of the period following the dissolution of the USSR, a general shrinkage of means, people and knowledge led to social atrophy and impoverishment. Along with losing centralised power, Moscow's managerial and organisational functions also disappeared: internal restructuring began but was long and difficult to implement given the shortage of means. The lowest point in this historical transition occurred in 1996 when the country's economic situation reached its lowest point.

To exit the economic impasse – and in all likelihood to also free himself from the increasingly suffocating grip of the elites of the old capital – the then president of the Republic of Kazakhstan made official in 1997 what seemed to be a somewhat risky and doomed move: the relocation of the capital from Almaty to Akmola. The official reasons were many: they ranged from the need for a strategic relocation, which would avert potential dangers due to the peripheral location, to the strong seismicity of the territory where the old capital stood. Whatever the reason, the move was clearly to avoid the danger of an internal uprising and the consequent detachment of the northern part of the territory, where the Russians were still the majority. Subsequently, the city name was changed from Akmola to Astana ("Capital" in Kazakh). The official reasons ranged from the possible misfortune that a name as funereal as Akmola (meaning "white tomb") could bring, to the easy pronunciation in many languages of Astana. However, many saw in Astana a temporary name and, in Nazarbayev's decision, the wish to emulate Peter the Great in naming the new capital after himself. Indeed, the project was hatching both Peter the Great's and Atatürk's footsteps. In 1996, a national competition for the design of the new capital was called; the winner was a design studio from Almaty with a somewhat emblematic name: Ak Orda ("White Horde"). The plan completely overturned the inspirational principles of the 1963 plan, undermining the linear development itself: the scheme envisaged an expansion southwards, beyond the river, where the administrative and commercial functions would be primarily housed. The basic principles of this proposal, headed by Kazakh architect Kaldybaj Montahaev – who designed Almaty's Republic Square in 1980 – were preserved by subsequent plans. The relocation of the capital and its government institutions was remarkably rapid, and what had been Lenin Square in Tselinograd was equipped as a temporary seat of government.

Dissatisfied with the results of the above competition and to give the operation international prominence and prestige, Nazarbayev decided to expand the scope of the call beyond national borders. The Kazakh government therefore launched an international competition that was announced and published in foreign media in April 1998. The call was sent to 40 participants from 19 different countries; 27 participants submitted a project proposal. On 6 October 1998, the president of the Republic of

Kazakhstan officially awarded the first prize to the Japanese architectural firm Kisho Kurokawa & Associates. The concept behind the plan is structured around the critical notions of Kurukawa's theory: *symbiosis*, *metabolism* and *abstract symbolism*. The first of these three concepts, foundational in the Japanese architect's intellectual production, was expressed as a dialogic relationship between the different groups of buildings. The plan was to leave the existing structure, buildings and trees untouched. In contrast, the new addition was to be built across the river as stipulated by government dictates but located along the watercourse – the symbiotic relationship was intended to be between the natural environment and the artificial additions – in continuity with the original idea of linear development. According to Tsubokura (2010), who worked with Kurokawa on the plan, the *metabolic city* was an enlarged reproduction of the linear zoning advanced by the 1963 land use plan, and the three zones were to be articulated into seven: a green buffer zone to protect the north side of the city from north sandstorms; a regenerated industrial zone; an intermediate green zone for environmental protection; a retained urban area; a new residential zone; a new urban centre; and an ecological park in the south. As stated by Tsubokura, Kurokawa held the 1963 plan in high regard, witnessing in it the capacity to guide urban development in a balanced order.

Not long afterwards, an unexpected event upset the outcome of the competition: in December 1999, a master plan for the city of Astana, drafted by the Saudi Binladin Group, was delivered to the municipality of the new capital. The Saudi master plan appeared to be developed in continuity with the plan submitted in 1996 by Ak Orda and was to be implemented with Arab funds. On 10 February the following year, the new plan was approved. Kurokawa was confronted with the difficult decision to either abandon the table or take the paternity of the Saudi plan; he chose the latter by combining the features of the three plans into one. In August 2001, the government of the Republic of Kazakhstan approved the new Japanese plan and shelved the Saudi one.

*A mutated perspective: From a linear to a radial plan*

Since 1997, the physical transformation of the built environment in the city was initially carried out primarily as façadism; buildings were covered with new effigies and decorations taken from the pre-soviet Kazakh tradition. "Simple solids of the Tselinograd period were thoroughly covered with this kind of superficial graphics, which changed these buildings out of all recognition [...] It was a change of city image, from a provincial utilitarian town reflecting the well-being and status of a basically industrial society into a high-status urban capital city" (Tsubokura 2010:16). The following phase of urban development, still ongoing, has been marked by the expansion on the south bank of the Ishim River, with the addition of the Millennium Axis clustering a collection of buildings designed by renowned international firms.

The construction of the new capital proceeded quickly, initially per the indications given by the Japanese firm. However, the prescriptions of the plan were not always adhered to, and the plan had to quickly accommodate Nazarbayev's aspirations and pressures from local and foreign investors. The planning body in charge of controlling urban development is the GenPlan: the bureau where Nazarbayev's requests were promptly answered, in front of an enormous three-dimensional model of the future urban layout. The continuous modifications to Kurokawa's plan eventually distorted it, and GenPlan was commissioned to develop a further elaboration that was finally drafted at the end of 2010 and approved in January of the following year. With the new plan, the innovative charge of the 1963 plan came to an end. Whilst the Kurokawa's second plan already proposed a peripheral green belt protecting the agglomeration, superseding the linear articulation of Tselinograd with a concentric development, the new plan completed that morphological mutation, accentuating a radial form. From a formal point of view, the new plan seems to be inspired by the one developed for Copenhagen under Peter Bredsdorff in 1947. Like Fingerplan and unlike Kurokawa's 2001 plan, the new plan for Astana also has a regional vocation, paying particular attention to the development of connections between the capital and the surrounding territory: a vision that was aimed at directing the city's growth until 2030.

The new capital had a little over 250,000 inhabitants when it was elected to house the institutions of the newly founded republic. A quarter of a century later, the official population is now about 1.3 million. The city seems to be increasingly oriented towards self-sufficiency, having promoted and begun to develop a construction industry that encompasses the entire supply chain. On the other hand, the recent urban expansion, like every new urban area, lacks the charm of the patina of history (Keeton 2011) and the variety that comes from stratification. Concurrently, many city inhabitants still reside in a *khrushchyovka*; the old Tselinograd is still the city's core. Today, Astana is a rich palimpsest, comprising what Françoise Choay (1986) would define as *hypersignificant built-up systems*, where extraordinary architectural artefacts illustrate a history of powers within the layering of diverse planning models (Khairullina 2015).

## Conclusion

This chapter investigated the spatial production in the former Soviet republics of Central Asia, the Kazakh Soviet Socialist Republic, building on the Schmittian reciprocity between spatial forms and forms of power (Schmitt 2003). Thus, the chapter aimed to contribute to the Foucaultian project to write a whole history "of spaces – which would at the same time be the history of powers" (Foucault 1980:149). The intuition of this investigation was that, on the periphery – of the world system – the correspondence between spatial forms and forms of power is more readable than in the centre due to the absence (or irrelevant presence) of spatially fixed social structures. Such a peculiar condition makes the periphery the ideal testing ground for urbanism and the space where the planning experiments often occur on a tabula rasa, manifesting themselves purely in stratified spatial structures. As emerged from the investigation, the three phases of development of the city went along with the transformation of the city's socio-economic base. The first was marked by the necessity of strategic control of the territory, with an outpost on a ford. The second made the city a linear development along the railway infrastructure to showcase Khrushchev's aspiration to industrialise agriculture and construction while dissolving the difference between town and countryside into the "agro-town". Finally, the third phase elevated the city to a new capital from which power and control irradiate concentrically. In the first phase, the economy of the small settlement was based on trade, while in the following two phases, the economy was sustained by organised exploitation of natural resources – through agriculture and materials extraction, respectively. Concurrently, architecture expressed the hegemonic values of the Russian territorialist expansion, the Soviet infrastructural development and the Kazakh national self-determination.

**Acknowledgement**

This research was supported by a grant from the City University of Hong Kong (project title: Superimpositions of Spatial Orders, project no. 7005771). The author thanks the editors for the insightful comments and enthusiastic commitment and Elvira Khairullina for sharing her extraordinary knowledge about the 1963 plan; he acknowledges Weike Li and Chenxi Huang for their assistance and Maria Babak for her crucial help in getting permission for Figure 1.2. Finally, the author is thankful for Vasily Filippovich Toskin's generosity.

**Note**

1 Those who were called Kyrgyz by the Russians: the term Kazakh was only introduced in Soviet times to differentiate the Kyrgyz from those who were called "black Kyrgyz". Since then, the former, nomads of the steppes, started being referred to as "Kazakhs" and the latter as "Kyrgyz".

**References**

Arrighi, G. 1994. *The Long Twentieth Century: Money, Power, and the Origins of Our Times*. Verso.
Choay, F. 1986. Urbanism and semiology. In: Gottdiener, M., and Lagopoulos, A. Ph. (Eds.), *The City and the Sign: An Introduction to Urban Semiotics*, pp. 160–175. Columbia University Press.
Dubitsky, A.F. 1986. *City on Ishim [original in Russian]*. Kazakhstan.
Fauve, A. 2015. Global Astana: nation branding as a legitimisation tool for authoritarian regimes. *Central Asian Survey* 34(1):110–124. 10.1080/02634937.2015.1016799

Fauve, A., and Gintrac, C. 2009. Production de l'espace urbain et mise en scène du pouvoir dans deux capitales «présidentielles» d'Asie Centrale. *L'Espace Politique. Revue en ligne de géographie politique et de géopolitique*, (8). 10.4000/espacepolitique.1376

Foucault, M. 1980. *Power/Knowledge: Selected Interviews and Other Writings, 1972–1977*. Vintage.

Gudro, I., and Krastiņš, J. 2019. Contribution of architect Daina Danneberga to the architecture in the second half of the 20th century. *History of Engineering Sciences and Institutions of Higher Education* 3:11–34. 10.7250/HESIHE.2019.002

Harvey, D. 2010. *Social Justice and the City*. University of Georgia Press.

Iskakov, T. 2020. Lost palaces of the capital. The Zhastar Palace [original in Russian], *Vlast*. https://vlast.kz/gorod/40062-utracennye-dvorcy-stolicy-dvorec-zastar.html

Keeton, R. 2011. *Rising in the East: Contemporary New Towns in Asia*. International New Town Institute, SUN.

Khairullina, E. 2015. Tres pasos en una distancia: los planes y manzanas en Astaná durante el periodo soviético (1957–1987). In: *VII Seminario Internacional de Investigación en Urbanismo, Barcelona-Montevideo, junio 2015*. Departament d'Urbanisme i Ordenació del Territori. Universitat Politècnica de Catalunya. 10.5821/siiu.6115

Köppen, B. 2013. The production of a new Eurasian capital on the Kazakh steppe: architecture, urban design, and identity in Astana. *Nationalities Papers* 41(4):590–605. 10.1080/00905992.2013.767791

Lefebvre, H. 1991. *The Production of Space*. Oxford: Blackwell.

Lefebvre, H. 1996. *Writings on Cities*. Trans. H. Kofman, and E. Lebas. Oxford: Blackwell.

Rolinski, S., Prishchepov, A.V., Guggenberger, G., Bischoff, N., Kurganova, I., Schierhorn, F., Müller, D., and Müller, C. 2021. Dynamics of soil organic carbon in the steppes of Russia and Kazakhstan under past and future climate and land use. *Regional Environmental Change* 21:1–16. 10.1007/s10113-021-01799-7

Tsubokura T. 2010. 'The next metropolis of Central Asia: Astana, a geopolitical node in the steppes'. In: *Atlas: Asia and the Pacific: Architectures of the 21st Century*, 16. Fundación BBVA, Bilbao.

Tompson, W. 2016. *Khrushchev: A Political Life*. Springer.

Schmitt, C. 2003. *The Nomos of the Earth*. New York: Telos Press.

Westerman, F. 2010. *Engineers of the Soul: The Grandiose Propaganda of Stalin's Russia*. Abrams.

# 2 From Breslau to Wrocław

## Urban development of the largest city of the Polish "Regained Lands" under socialism

*Agnieszka Tomaszewicz and Joanna Majczyk*

### City of Wrocław: Pre-WWII spatial structure

Wrocław, historical capital of the Silesia region, situated by the Oder River in the south-western part of today's Poland, is currently the city inhabited by nearly a million citizens and has an exceptionally complicated history. The city used to constitute the centre of the Piast Duchy of Wrocław. Since 1335, it belonged to Czech kings and later was the seat of the Prussian regency of Breslau (Regierungsbezirk Breslau), annexed to the Silesian province (Provinz Schlesien) after the Congress of Vienna. Before WWII, Wrocław was the German city of Breslau and consisted of the Old Town, a densely built-up downtown with neighbouring garden housing estates. In the outskirts of the urban arrangement, former suburban villages and two small towns of Leśnica (Deutsch Lissa) and Psie Pole (Hundsfeld) were located, incorporated into the city in 1928. The oldest part of Old Town centre of Wrocław is constituted by Ostrów Tumski – former island on the Oder River, on which the gord of the first representatives of the Piast dynasty was situated in the 10th century, together with the 13th century, regularly arranged charter-based town with a quadrilateral Market Square (172×207 m) and an auxiliary square connected with it by its corners (Solny Square, 80×120 m). Charter-based town was located on the left bank of the Oder River, which eventually became its northern border, while the other borders – southern, eastern and western – were formed by the city walls built and successively rebuilt from the mid-13th century to the end of the 18th century. Fortifications were demolished after the Napoleonic Wars, but the moat was partially preserved, and it became part of the Wrocław "ring" – a green strip arranged within former post-fortification areas. The promenade connected the Old Town with the downtown, initially evolving along main exit roads and, from the mid-19th century, within the areas between them and divided mainly by checkered street network. The scale of the downtown was determined by four-, five- and six-storey tenement houses, densely located along the established side building line, sometimes separated by front gardens from street bordering line. Spatial arrangement of the downtown was complemented by the integrated squares as well as a limited number of small green areas. The silhouette of the Old Town would distinguish itself on this background, dominated by the towers - of the City Hall and churches.

In the 19th century, Breslau was an important urban centre in eastern Prussia, a university seat and the capital of the Silesian Province. In 1900, the city's population was approaching 425,000, making Breslau the third-largest city in Germany after Berlin and Hamburg. During WWI, Breslau avoided damage, but due to the provisions of the Treaty of Versailles and the related territorial changes, the city found itself on the far eastern border of the country. After Poland's statehood was restored and a part of Upper Silesia was annexed to it, Breslau retained its position as the capital of the new, but much smaller province of Lower Silesia (Provinz Niederschlesien), while gaining the status of a "front" city as it was only several dozen kilometres from the Polish and Czecho-Slovak borders.

The peripheral location of the city in relation to the centre of the state had a negative impact on its internal economic and political situation. Production plants lost markets in the east, and poor railway connections and unfavourable transport tariffs hindered domestic exports. The situation was aggravated by high unemployment and significant overpopulation (one of the highest in the country) (Kulak 2006:287). Despite a large increase in population (about 630,000 in 1939), Breslau fell to the ninth position in the ranking of the largest German cities. In the 1920s, the area of Breslau was doubled, and initiatives were taken to loosen the spatial structure of the city. New garden estates were located on the outskirts of downtown, and until the outbreak of WWII, the city developed concentrically around the centre of the Old Town.

### City of Wrocław: New beginning – People and politics

A new chapter in the history of Wrocław came in 1945. On 6 May, German troops defending the city surrendered, and during the Potsdam Conference, eastern borderlands of Germany were annexed to Poland. The German City of Breslau thus became the biggest city within the so-called "Regained Lands" – a strip of land situated on the eastern side of the Oder and Lusatian Neisse rivers, in the northern and western parts of the country. Thus, the former, peripherally located Breslau gained a completely new political status – a significant urban and industrial centre of Poland and a symbol of victory over Nazi Germany. The "Regained Lands" were presented in the propaganda message as "the future" or "the wealth of Poland" and "ancient Slavic lands that were illegally [...] seized by the Germans and then deceitfully Germanized" (Batowski 1946:3–4). The key to building a new identity of the region was its mythical connections with the Piast dynasty, a dynasty considered to be the progenitors of Polish statehood. The clarity of the message was at the same time to cover up the loss of Eastern Borderlands (a strip of land in the east of the country, currently within the borders of Ukraine, Belarus and Lithuania) to the Soviet Union.

Totally new political, economic and social factors conditioned further development of the city. Within the first few years after WWII, the population of the city was completely "exchanged" – former inhabitants of Wrocław were resettled deep into Germany and replaced with migrants, mainly from the central and eastern, peripheral parts of Poland. Efforts were made to remove German inscriptions and symbols (coats of arms, monuments) from public space, and new street names were introduced, in which historical links with Poland were particularly emphasised. A transitional period lasting several years, characterised by socio-political instability, ended in the late 1940s with the establishment of the Polish United Workers' Party and the sealing of political changes in the country. The introduction of the socialist system was crucial for the development of all Polish cities, which determined the entry of the country into the periphery of the Western World as well as the directions of construction and investment policy. Against this background, the situation of Wrocław was extremely unfavourable – the city had a poor road and rail connection with central Poland, and investments were suspended due to the lack of regulated legal relations with Germany. It was not until 1970 that an agreement was signed with the then Federal Republic of Germany on the inviolability of the border drawn along the Oder and Lusatian Neisse rivers. Until then, the affiliation of the "Regained Lands" to Poland was considered uncertain. As a result, despite extensive propaganda affirming the "Regained Lands", funds from the state budget for the reconstruction and development of Wrocław were disproportionately smaller than in the case of other cities. Wrocław also became a victim of the "reclaiming bricks" campaign to rebuild the capital of the country (Tyszkiewicz 2020:78). It is estimated that at the turn of the 1940s and 1950s, several hundred million bricks from the rubble and demolition of buildings were transported to Warsaw. Moreover, the vast majority of the inhabitants were migrants/resettlers from villages and small towns in the east of the country. The low cultural capital, combined with the "uncertainty of borders" and a negative attitude towards Germany (including German material heritage), resulted in the degradation of the urban tissue and plundering of the abandoned estates in the first years after the end of WWII.

Paradoxically, the huge wartime and immediate post-WWII destruction of Wrocław opened up a chance for the socialist authorities to introduce new values to the city plan. The reconstruction of Wrocław according to the theses of socialism was included in the electoral programme of the ruling party. The "new" city was to be built in contrast to pre-WWII Breslau.

The following part of the chapter presents key stages of the reconstruction and extension of the City of Wrocław in the years 1945–1989, i.e., from the moment of the city finding itself under Polish jurisdiction until the collapse of Polish People's Republic (PRL) in 1989.

## Stage 1: Post-WWII damage of Wrocław and the first development plan (1945–1949)

During WWII, Wrocław was damaged by nearly 70%. Its southern and western parts were nearly totally annihilated (Ptaszycka 1956:204). Destruction of the Old Town centre was around 50%, but the buildings in its north-eastern section were nearly completely swept away, in particular in the area of the Nowy Targ Square. In light of the magnitude of damage, it was being considered to reduce the scale of Wrocław and transform it from a pre-WWII metropolis of over 600,000 inhabitants into a city with three times smaller population (Bukowski 1980:18–9).

Works connected with the reconstruction of the City began with rubble removal, unblocking main communication routes and securing the buildings intended for future use. At the same time, nearly immediately after the liberation of the city, studies of its functional and spatial structure were undertaken. They formed the basis for systemic planning solutions written down in the first General Development Plan of the City of Wrocław (1949). In the design, prepared within the Wrocław Design Office established especially for this purpose, general rules and directions for the development of the city were defined, corresponding to the guidelines of the first post-WWII economic plan. The document focused in particular on the functional division of territories, with special attention paid to locating the complexes of public utility buildings, residential and industrial areas (see Figure 2.1). The plan suggested to reorganise communication arrangement of the city, liberating the historical centre from vehicle traffic as well as to erect the "new centre", situated in the southern part of the Old Town, within the area limited by the course of former Lower Silesian railway. According to the concept outlined by Wrocław Design Office, the Market Square was to become the city's main cultural centre, while today's General Tadeusz Kościuszko's Square became the initiating element of the "banking district" (Grotowski 1949:1). The plan assumed the establishment of two academic centres, located in the vicinity of pre-WWII main edifices of the University (1 Uniwersytecki Square) and University of Technology (27 Stanisław Wyspiański Shore), as well as the construction of the new administrative centre in the eastern section of the Old Town, in the vicinity of today's Powstańców Warszawy Square. An innovative solution was provided for greenery system, within which the territories of urban parks and forests were connected with one another, and from the windward side, green wedges were designed, facilitating the ventilation of the Old Town[1]. At the same time, works on the first detailed plan for the development of the Old Town Centre were in progress in Wrocław Design Office, assuming the restoration of spatial arrangement within mediaeval defensive walls, preserving historical construction lines and at the same time liberating quarter interiors. It was assumed to transform the Old Town into a residential area for 21,000 citizens, while, as it was written,

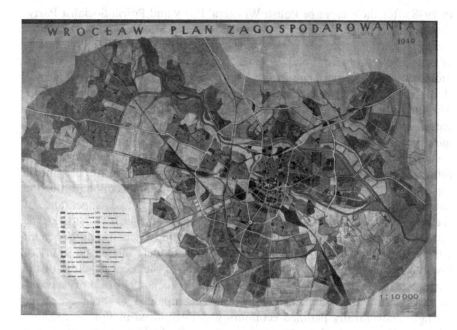

*Figure 2.1* The first General Development Plan of the City of Wrocław, designed by team at Wrocław Design Office led by Tadeusz Ptaszycki, 1949.

Source: Museum of Architecture in Wrocław, signature MAt-V-744F.

"historical bourgeois tenement houses were also supposed to serve residential purposes, preserving their former external architectural structure, adapted to modern residential needs. The outbuildings will generally be removed and green areas introduced into block interiors under the form of vegetable gardens, orchards or flowerbeds [...]" (Kaliski 1949:518). The plan was developed by a group of Polish architects led by Tadeusz Ptaszycki. Most of them graduated from the Warsaw University of Technology in the interwar period and had little experience in spatial planning. Ptaszycki and his wife Anna Ptaszycka (both were 1936 graduates) ran a studio in Warsaw until the outbreak of WWII, where they successfully designed modernist architecture. The devastating losses among the Polish intelligentsia, including the deaths of about one-third of Polish architects (Mrówczyński 1979:1), meant that almost all survivors were involved in the reconstruction of towns and villages.

### Stage 2: The episode of Socialist Realism; reconstruction of the Old Town of Wrocław (1949–1956)

The first plans of the development of Wrocław were completed in 1949, but before they were forwarded for implementation, the political situation in Poland changed – Polish United Workers' Party (PZPR), established in 1948 after the merger of Polish Workers' Party and Polish Socialist Party, took over the power. Already during the first unification congress, references were made to the necessity of "establishing the premises for the development of socialist culture in Poland" (Deklaracja 1949:140). The basis of art, architecture and urban planning was supposed to be constituted by Socialist Realism, relying on the models introduced in the 1930s in the Soviet Union. The doctrine of Socialist Realism was considered binding for Polish architecture and urban planning in June 1949 during the session of architects – members of Polish United Workers' Party. The resolution issued at the time specified main threats in the area of the "construction of cities", being "the influence of Anglo-Saxon deurbanisation trends, promoting the pessimistic mistrust in the city as a centre of social, political and economic life, the necessity to escape from the city, as a result leading to the disappearance of the notion of city composition as a whole" (Rezolucja 1949:162). At the same time, designers were instructed not to follow "the schematic approach of Corbusier's super-urban doctrine", quoting as models the assumptions of Soviet urban planning, in particular the concept of the reconstruction of the City of Moscow dating back to the 1930s (Rezolucja 1949:162). The transformation of Polish political system was accompanied by deep economic changes, as centrally planned economy was introduced in Poland together with the command-and-quota system, as a result of which the private sector was eliminated and the state became the only ordering party, contractor and recipient of all investment projects. The architects could no longer conduct their activities in their own private studios, as huge state-owned design offices were established, in which the recommendations of state administration were implemented.

The first concept of a "socialist district" was established in Wrocław in late 1949, and it referred to the development of the area on the right bank of the Oder River, in the eastern part of the city centre (Tomaszewicz and Majczyk 2019). The design assumed the creation of a new administrative and residential centre with its composition based on the pre-WWII wide avenue (Kaiserstrasse, today's Grunwaldzki Square), connecting in a straight line two bridges and leading the traffic out of Wrocław Old Town to the east. The existing avenue was in the design connected with giant "defilade" squares, rectangular and oval communication nodes with smaller squares, axis arrangement of streets defined by the rows of monumental buildings with stepped outline. Urban structures within the new development corresponded to Soviet projects, and the frame of main communication axis was for sure supposed to resemble Gorki's Street in Moscow, socialist model *via trumphalis*. The concept of erecting the district in the area of ruined buildings forming the city centre constituted to some extent the response to the provisions of the new state economic plan, the so-called six-year plan (1950–1955), which "forces the establishment of the new centre according to socialist theses, ensuring new opportunities for the development of the city"[2]. The idea of locating the centre outside Old Town borders was justified by too small dimensions of the historical centre "to store new socialist content". The head of the studio in which the plan of the Grunwaldzka Axis was created was Marian Spychalski, an architect, general and communist who was removed from power for political reasons and sent "to the provinces" – to Wrocław. Shortly after finishing work on the design, Spychalski was arrested and imprisoned, where he was held until 1956. The design of the new district was not forwarded for implementation, and the only bigger urban project completed during the period of Socialist Realism was Kościuszko's Residential District, designed as a new "package" for the 19th-century city square.

Simultaneously with the works on the creation of a new "socialist" centre, activities to protect the monuments of Wrocław commenced. In the first place, public buildings were secured, including churches and the town hall, erected in the Gothic style (14th and 15th centuries), a process associated with the "Piast", and thus implicitly – Polish history of the city. The project of rebuilding the Old Town referred to the theory of Jan Zachwatowicz, which postulated the reconstruction of residential buildings in the forms from the late 18th century, taking care to restore the silhouette of the city typical of the Middle Ages. However, while the proposed method could be used to rebuild the Old Town in Warsaw, it turned out to be difficult in Wrocław. Wrocław belonged to the Kingdom of Prussia in the 18th century, so the architecture of that period was characterised by "Germanness" and could not be a model for a socialist state. In practice, the partially rebuilt Old Town has become a historicising creation by local architects.

### Stage 3: The Thaw, prefabrication and modernisation of Wrocław's architecture (late 1950s and early 1960s)

In 1956, Socialist Realism was rejected in Poland, and the objective undertaken in the five-year economic plan that followed (1956–1960) referred to making the construction sector mass-scale, typified and industrialised. Already from the end of the 1940s, architectural and urban designing activity was subjected to various regulations – normative standards were introduced first, determining among others the density and intensity of building developments, biologically active surfaces and the number and arrangement of service facilities. In the late 1950s, in turn, regulations concerning standardisation in the construction industry were adopted, aiming at ensuring the progress and at the same time reducing building costs and accelerating the "production" of apartments. It was assumed that "residential districts [...] should be performed with the use of industrial methods from multi-dimensional prefabricated components" with the use of "typical building designs" (Uchwała 1959). The introduction of normative standards as well as standardisation and prefabrication of construction components importantly influenced urban composition, to a large extent adjusted to technological possibilities of construction management. On the other hand, numerous restrictions somehow forced the creation of residential districts following the model of neighbourhood units, i.e., complexes of residential buildings equipped with service facilities and green areas, excluded from pass-through vehicle traffic.

In mid-1950s, the situation in the Wrocław housing sector was dramatic – the offer of new apartments would not in any way follow the demand, constantly increasing in connection with migrations resulting from the development of industrial facilities as well as the post-WWII baby boom. It was calculated that ongoing housing needs in Wrocław could be satisfied by commissioning ca. 15,000 residential rooms per year, while in reality, there were only 4,000 of them made available. As a result, in the late 1950s, the city began to depopulate (Kusy budżet 1958). City authorities would unsuccessfully apply for increasing the expenditure on residential construction at central administration bodies, while the then chief architect of Wrocław would even claim that in connection with insufficient funding, amendments were introduced into five-year development plan resulting in reducing by nearly 70% the number of new investment projects in the housing sector (W ankiecie 1958).

In mid-1950s, works on the General Development Plan of the City of Wrocław for the period until 1975 and detailed plans were simultaneously established for selected areas in the city together with the Stage Plan for the Development of the City of Wrocław in the years 1955–1960. The documents outlined among other functional division of the central part of the city, were preserved with minimal amendments in the decades that followed. The Old Town, together with its immediate neighbourhood within the city centre, was intended for residential purposes – the area was divided into housing estates equipped with service buildings subjected to normative standards (Przyłęcka 2012:100). Within the zones including preserved pre-WWII buildings, "punctual" filling of the gaps in frontages with section blocks was planned, together with the erection of new educational facilities. Big centres providing supra-local services were situated by the most important communication nodes, and the plans included the creation of a university campus within the areas adjacent to the Grunwaldzka Axis.

According to the adopted plans, at the turn of the 1950s and 1960s, the emphasis was put on the reconstruction of ruined areas within downtown and the eastern part of the Old Town, where two bigger residential districts were erected in the vicinity of the Polish Committee of National Liberation (PKWN) Square and Nowy Targ Square.

Both projects preserved, due to the necessity of saving the funds, the pre-WWII street network in the city. Only a few corrections were introduced, aimed at improving the transport in the entire city. The basis for developing new residential districts was constituted by repetitive, four- and five-storey blocks, situated along the streets and squares in the city. Clearances were left between the buildings in order to facilitate the ventilation and illumination of quarter interiors. The arrangement was complemented with "experimental" tower blocks and those with balcony access, situated in locations with insufficient access to sunlight. Quarter interiors were equipped with educational pavilions as well as green areas accessible for all residents. The ground floors of houses erected by main communication routes included shops and restaurant facilities. Single tall buildings of 10 up to 16 floors were designed at the city squares, emphasising main public spaces, and thus the silhouette of the downtown was complemented. The majority of buildings in both residential districts were performed with the use of traditional technologies, with prefabricated components introduced in some of them, the so-called Plattenbau. In the "PKWN Square District" that took up the surface of 13 hectares, 4,100 "residential rooms" were commissioned intended for 6,000 citizens, together with two kindergartens, a nursery and a primary school. Nowy Targ district became the home of 3,000 citizens for whom a dozen sectional and tower blocks were constructed, together with a nursery, a kindergarten and a primary school (Majczyk and Tomaszewicz 2017). The "PKWN Square District" was designed by a team of architects led by Kazimierz Bieńkowski and implemented in the years 1956–1962. The design of the Nowy Targ estate was developed by Włodzimierz Czerechowski, Ryszard Natusiewicz and Anna and Jerzy Tarnawski, and its implementation was carried out in the years 1959–1965. The authors of both concepts were young architects, graduates of the local Faculty of Architecture in the years 1951–1952, who, after completing their studies, got jobs in state design offices. The involvement of "young people" in the reconstruction of the city had a symbolic and practical dimension. On one hand, they represented the change that took place in Polish architecture after 1956, and on the other hand, they constituted a "particularly valuable element" for the authorities, because "they had already completed their studies – as it was written in the press – in the conditions of People's Poland. This fact undoubtedly determined their social attitude" (Wrocławskie Biura 1950:4). In practice, the generation of architects who graduated in the 1950s played the greatest role in shaping the city under socialism.

### Stage 4: The Athens Charter, large-panel technology and large housing estates of Wrocaw (the 1960s)

In view of the lack of available investment grounds in the very centre of Wrocław, the decision was made in the early 1960s to lead the construction activity "outside" the downtown district. It was planned to transform the southern and western part of the city – damaged almost entirely during the war and with the rubble gradually removed after its end – into new residential districts: Wrocław-Południe (South) and Wrocław-Zachód (West).

The southern part of the city, bordered from the north with the railway line connecting the oldest train stations in Wrocław, was gradually extended in the 19th and early 20th centuries. Before WWII it was inhabited by ca. 120,000 people, while after the war, the number decreased almost ten times. After the demolition of ruined tenement houses, the area of nearly 200 hectares was obtained, where the construction of the new residential area was planned, intended for 60,000 citizens and, as it was written in the press, "the area of the southern districts constitutes a natural expansion zone for satisfying burning needs of the housing sector thanks to its convenient location, very favourable health conditions as well as exceptional values represented by land development plan (street paving and the network of underground installations)" (Duchowicz and Majerski 1957:373).

In 1962, competition for the design of the Wrocław-Południe district was announced. The work by a team of young architects: Kazimierz Bieńkowski, Tadeusz Izbicki, Wacław Kamocki and Julian Łowiński – was awarded with the first prize. The winning concept initiated the "new era" in Wrocław urban planning, in which housing estates were composed of repetitive, large-dimensional residential blocks erected following industrialised technology. Competition jury appreciated the "simplicity of the arrangement, the possibility of its fast performance, perfect communication solutions, greenery (...)" (Główny architekt 1962), the repetitiveness of structures and economical approach. The design of Wrocław-Południe district included regular arrangement of buildings along the north-south and east-west axes, according to the principles ensuring the appropriate access of sunlight (see Figure 2.2). Obvious inspiration for the plan for the architects constituted the Athens Charter, popularised in Poland after 1956. Witold Molicki, an architect and witness of those times, recalled that architects were also inspired by the designs of housing estates implemented in Western Europe, such as, for example, Alton Estate in London, Roehampton, and "The Grand Ensemble" of Sarcelles in Paris (Molicki 1996:38). Due to cost-saving reasons, pre-WWII street network of Wrocław was preserved, together with single still usable public utility and residential buildings. However, it was decided to resign from erecting border buildings, instead concentrating on independent "blocks immersed in greenery". The district was divided into four smaller estates, with each of them consisting of two, three or four neighbourhood units equipped with basic services and each intended for 5,000 residents. "No housewife will have to – as it written in the press – walk more than 120 m to buy milk and bread, and each child will walk to school by green alleys, without crossing any road with vehicle traffic" (Chwieduk 1962). Units forming part of each residential area were connected with one another by a common "neighbours' garden". It was planned that in the Wrocław-Południe district, mainly 11-storey blocks would be erected, 5-storey buildings (40%) and a few 22-storey skyscrapers. A supra-local service centre was also intended to be built, located along the main axis heading towards the Old Town.

Already a year after announcing the competition for the design of the

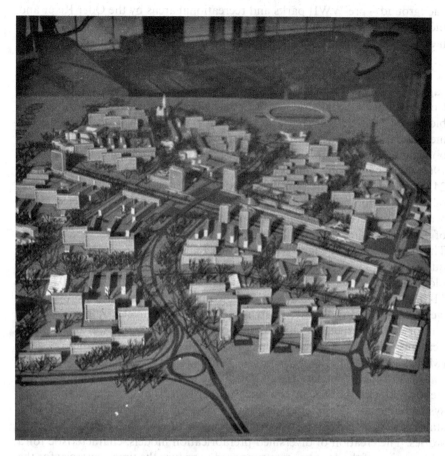

*Figure 2.2* Wrocław-Południe district, competition project inspired by modernist urban planning, designed by Kazimierz Bieńkowski, Tadeusz Izbicki, Wacław Kamocki and Julian Łowiński, 1962.

Source: Museum of Architecture in Wrocław, signature MAt-AB-558F.

Wrocław-Południe district, the concept for the development of the city changed completely. The concentric model, promoted immediately after the war, was rejected and replaced by strip arrangement shaped along the east-west axis. The plan to build huge residential districts in the southern part of the city was abandoned in connection with the intention to construct a new civilian airport by the south-western border of Wrocław as well as the necessity to protect arable lands (!) at the time situated by the southern border of the city (Przyłęcka 1996:10). The northern part of the city became temporarily excluded from investment undertakings and intended for special, mainly military, purposes. At the same time, along the east-west strip, the most important industrial plants were located, together with the city's green background – pre-WWII parks and recreational areas by the Oder River and the Ślęza River. New public utility investment projects were also planned in the west: central wastewater treatment plant and a CHP plant.

Construction activity within the already mentioned western part of the city began with building the "Zachód I" housing district, situated at the edge of the Old Town along main historical transport axis (Gabiś 2018). Composition of the district would come down to the even distribution of blocks of flats designed for the Wrocław-Południe district following the meridional arrangement. More interesting shapes were given to the neighbouring estate – "Zachód II", today called Szczepin, designed in the years 1966–1968 by Witold Molicki. The composition of this area, intended for 24,000 inhabitants, relied on the construction of residential units consisting of repetitive components: doubled 11-storey blocks of flats connected at the ground floor level by a single-level pavilion as well as lower 5-storey blocks of flats consisting of several staircases and stepped segment arrangement. The combination of high-rise and lower buildings formed partially closed quasi-quarters with small service pavilions arranged in the inner parts of the blocks, positioned also in vast green areas (Sobolewski and Czajka 2022). All 11-storey buildings were placed along the north-south axis, five-storey blocks were constructed crosswise, referring to pre-WWII street arrangement. The author of the design saw in this composition "reminiscences of tower fortifications of mediaevalcities, surrounding the castles from the outside" (Molicki 1996:38).

The so-called Wrocław Large Panel was used for the construction of the stepped blocks of flats in the Szczepin district. This local prefabrication system made it possible to perform blocks of flats consisting of several staircases with transverse structural arrangement and the height of up to 12 stories. The creation of diversified prefabrication models, characteristic for a given region of the country, was supposed to reduce the time necessary for the construction of flats – Wrocław Large Panel was manufactured in Wrocław in order to facilitate the transport of components to the construction site and reduce costs. The different shapes of large-size building elements also influenced the differentiation of the architectural forms of the blocks of flats and gave them a "regional" character. Local design offices worked on unusual solutions for façade panels, loggias and balconies, and the Wrocław Large Panel was used only in Lower Silesia.

**Stage 5: Wrocław Large Panel in search of individuality (the 1970s)**

The first bigger residential complex erected fully using Wrocław Large Panel was the Popowice district, designed in 1970 by the already mentioned architect Witold Molicki in the western part of Wrocław. The estate was intended for nearly 20,000 inhabitants and its spatial concept completely rejected the recommendations of the Athens Charter by replacing rectangular arrangement with diagonal one, in which residential buildings situated in lines along the streets were slightly tilted in the eastern or western direction. Five-storey blocks were erected from the south, while from the north – 11-storey buildings. Access roads with parking complexes were designed in their shade. Areas included between the buildings were intended for playgrounds and recreational green areas. Schools and kindergartens were situated in the centre of the district, while service facilities found themselves in its outskirts. The designer also managed to "liven up" the architecture of the blocks of flats thanks to individually designed trapezoid balconies harmonised with stepped arrangement of buildings.

In December 1970, the year when the design of the Popowice district was performed, Edward Gierek became First Secretary of Polish United Workers' Party. The new leader initiated the ten-year period of economic "modernisation" of the country, relying to a large extent on loans from foreign banks. Thanks to the influx of money and a certain "opening" towards the technologies applied in Western Europe, sudden development of the construction industry took place in the 1970s. New W-70 open prefabrication system was introduced, based on modular mesh (dimensions 60×60) enabling the gradation of the spacing of load-bearing walls and facilitated the differentiation of flat typologies. In 1974, new architectural normative standards were introduced, enabling the increase of flat surfaces, which brought the improvement of living conditions of citizens. In the 1970s, ca. 250,000 flats were commissioned annually, and in the record-breaking year 1978 their number reached 283,600 (Rocznik 1979). However, at the same time the method for erecting housing estates began to be criticised for its excessive unification, with over-scaling, low performance quality and marginalisation of the role of architects in the design process. As mentioned by Jacek Nowicki, one of the architects involved in the discussion on the future of residential construction, published in 1970 in *Architektura*, the most important Polish trade magazine: "regression began when the architect's role was reduced to the function not compliant with the character of this job. In short, the architect is nowadays required to provide the construction industry with assistance. However, the architect's basic role is to set tasks to be completed by the construction industry. [...] We operate under blackmail conditions – that the plant [manufacturing prefabricated components – author's note] will not do it because they can't. In face of such threat, we accept everything in order for anything to be done" (Paszyński 1970:373). There were also calls for the shift in urban design paradigms postulating: clear establishment of district borders, giving them their own characteristic features, providing comprehensive residential solutions together with landscaping and greenery, division of big housing estates into smaller neighbourhood units and finally – the participation of future residents in the design process (Siemiński 1974:434–5).

In the early 1970s, Wrocław authorities undertook the objective of "establishing" within the historical urban tissue "new structural components". Among them, there were modern city centre for all citizens, residential estates and districts based on modern settlement units, hierarchical network of lower- and higher-level service centres, separated recreational and sports as well as industrial and warehousing areas and transport network enriching pre-WWII road system (Bieńkowski 1970:36). At the same time, it was finally decided to abandon the designing of "Corbusier's" urban arrangements, with priority being granted to nest-like compositions, the construction of which was made possible thanks to the introduction of trapezoidal connectors into the Wrocław Large Panel system, as well as with the popularisation of the W-70 technology. One of the earliest examples of a housing estate designed following the nest-like arrangement was Gądów-Lotnisko intended for 33,000 inhabitants. Its concept, developed by architects, Kazimierz Bieńkowski, Andrzej Chachaj, Zbigniew Malinowski and Daniela Przyłęcka won the first prize in the competition organised by the Association of Polish Architects (SARP) in 1983. The residential area situated in the western part of the city was supposed to consist of nine structural units in the form of irregular nests of multi-family buildings, elongated in the north-south direction and having a diversified number of stories (Nasterski 1974:450–3). Building complexes were separated from one another with two latitudinal composition axes – a pedestrian route for commercial purposes as well as linear park in the southern section of the area. It was planned to close both axes with a supra-local service centre (from the east) and a sports complex (from the west). Basic services, such as schools, kindergartens and nurseries, were designed inside each of the units and connected with recreational grounds by the buildings. The concept of this estate assumed complete separation of pedestrian and vehicle traffic.

Numerous residential districts designed in Wrocław with the use of prefabricated technology referred to the same arrangement scheme (among others, Kozanów, Nowy Dwór and Różanka). They were erected not only within the previously undeveloped areas, but also in the vicinity of pre-WWII garden residential estates, and even within the complexes of single-family houses. In the immediate vicinity of the colony of two-storey detached houses in the Krzyki district in Wrocław, "Osiedle Przyjaźni", with its original name constituting the expression of appreciation for Polish-Soviet friendship, was erected according to the design by Witold Molicki. The new residential complex consisted of 12-storey tower blocks and rising cascade polyline buildings of 4 up to 11 stories surrounding octagonal courtyards. New buildings forming the estate, strikingly opposing the dimensions of the existing houses, dominated the entire district with their height. It is, however, worth to notice that in the early 1970s, design rules for multi-family residential buildings assumed their isolation from inconvenient vehicle and railway transport, shaping the structure of residential estates in a way to ensure appropriate connection of their function as well as social integration, erecting educational, cultural and sports facilities, introducing vast green areas inside the estates as well as eliminating the collision between road and pedestrian traffic.

Administrative reform was introduced in Poland in 1974, as a result of which the status of Wrocław changed. The city was previously the capital of the region that after 1974 was divided into four small voivodeships. Wrocław became the capital of only one of them, which resulted in important reduction of central subsidies intended for investment projects. Similar degradation affected other large cities in the country, while medium-sized cities benefited from the reform, often advancing to the position of regional capitals. The growing economic crisis in the second half of the 1970s also caused significant limitations in the implementation of housing estate projects. In the decade that followed, attempts were made to finish the construction of already initiated investment projects, but supra-local service centres were not erected either in the district of Wrocław-Południe or within any of the housing estates in the Wrocław-Zachód district. Building activities were suspended following the construction of facilities serving basic educational and service purposes.

**Stage 6: Return to Downtown (the 1980s)**

Deep economic crisis, together with the collapse of the implementation of central industrialised construction programme, led to the emergence of a new initiative – cooperative building activity. In 1981, Central Association for Cooperative Residential Construction adopted the resolution legalising the establishment of small, society-driven housing cooperatives that were allowed to get involved in the development of limited areas or even single plots specified by local authorities (Przyłęcka 2012:246). Several dozen cooperatives of this kind were at the time established in Wrocław and, in connection with the absence of big, undeveloped areas having access to municipal infrastructure, they got involved in arranging the areas within the downtown district. In the early 1980s, city authorities prepared the list of 280 locations in the city centre intended for the erection of residential and service buildings, with the majority of them being constructed within this decade. The houses were inserted between existing pre-WWII buildings or entire housing complexes were built, corresponding to the outlines and dimensions of historical quarters. Following the wave of post-modernism emerging in Poland, the value and type of urban heritage began to be appreciated. It is claimed that the constructed infill buildings constituted a characteristic feature and specialty of Wrocław architects (Sawa-Borysławski 2011:140).

In 1984, city authorities adopted the document entitled "Revitalisation of the Old Town Historical Complex" (designed by a team led by Andrzej Gretschel), which became the basis for large-scale renovation activities within Wrocław downtown, referring not only to buildings but also to public spaces and green areas. Even though the document noticed the potential of the Oder River in shaping the landscape of the Old Town, it was necessary to wait until the first decade of the 21st century for the postulated "reconstruction and renovation of boulevards". It is worth adding that the initiative was not accompanied by the previously present propaganda regarding the reconstruction of the "Piast stronghold" and "Gothic" architecture. The legal status of Wrocław and the "Regained Lands" has been regulated, and the city has been inhabited by Polish citizens for two generations, and the works undertaken were related to the improvement of the general technical and aesthetic condition of the Old Town. The 1980s, however, were particularly difficult for the construction industry due to the already mentioned economic crisis but also political turmoil – martial law between 1981 and 1983 and a wave of strikes in industrial plants in 1988, which a year later led to the fall of the regime of the Polish Republic People's.

**"Regained" Wrocław**

The transformation of German Breslau into Polish Wrocław was based on the tragedy of WWII – destruction, expulsions and misfortune of entire nations and individual people. Poland got the "Regained Lands" back, regardless of its opinion on the subject, and it soon turned out that the liberation by the Red Army meant the country's complete subordination to the Soviet Union. Socialism was imposed on Poland with its planned economy, central management and apparent egalitarianism. For this reason, the spatial policy pursued in Wrocław in the years 1945–1989 could not be fundamentally different from urban planning activities undertaken in other Polish cities of historical origin. People and their attitude towards the city, which evolved from complete negation through indifference to acceptance, initially mainly of socialist achievements and then also of the surviving heritage of predecessors, constituted its distinguishing feature.

Negation was associated with an unwanted change that the new inhabitants of the city were forced to make. Wrocław was not their dream place to live; they got there often by accident, transferred from the Eastern Borderlands to the western outskirts of post-war Poland. Wrocław overwhelmed them with its size, the scale of destruction, often – ruined modernity, but above all – the German past. That is why the signs and symbols of the old Breslauers were systematically destroyed, that is why mainly those monuments that could be considered "Piast" were renovated, and that is why there was no protest against the robbery of the resources of Wrocław and other cities of the "Regained Lands". After a temporary, largely propaganda-related, co-financing of the reorganisation of Wrocław, there was a long period of stagnation, during which various plans for the reorganisation of the Old Town and downtown were created, but there was no money to implement them. It was only in 1952 that the then Deputy Prime Minister, Stefan Jędrychowski, announced financial support for "comprehensive development of the central districts" of Wrocław, while shaping "their architectural and urban appearance, taking into account valuable local traditions, following the example of Warsaw and other modern socialist cities" (Umacniajmy jedność 1952). In Wrocław, however, it was impossible to define "local traditions" for emotional as well as ideological reasons. Gothic, which was associated with the mythical Piast past of the city, was obviously unsuitable for use in post-WWII design practice, while the architecture of the 19th century, which was the closest to modernity, was associated not only with "Germanness", but also with capitalism.

The preserved structure of the city was shown indifference at best, pre-WWII housing complexes were treated in an utilitarian manner, monumental buildings were neglected, some of them were demolished, as if they had no purpose in the socialist society. After a short episode of Socialist Realism, urban planning practices changed, and economics was assigned a superior role. The city spaces destroyed during WWII began to be filled with prefabricated housing estates, the forms of which were subordinated to the large panel technology, as well as architectural and urban planning norms. Industrialised residential estates were of course characteristic not only for the city of Wrocław or for Poland. Similar solutions were also applied in other European countries, but the difference consisted here in the fact that while in Western Europe, housing estates with blocks constructed following prefabrication-based technology constituted one of the applicable solutions, while the countries forming part of the so-called Eastern Block there was no alternative.

In the early 1960s, the concept for the development of Wrocław was completely remodelled, from the previous concentric arrangement to latitudinal strip solution. Due to constant scarcity of flats, priority was assigned to the construction of vast residential estates at a growing distance from the city centre. At the turn of the 1960s and 1970s, the search began for more individualised urban housing solutions, enabled by the manufacturing of the Wrocław Large Panel and then the W-70 prefabrication system, accepted for use in Wrocław in the mid-1970s. Priority assigned to the construction of housing estates led to inequalities in the development of the city with its big residential areas situated far away from the city centre, which in turn remained neglected. In socialist Wrocław, not a single city-wide service centre was built, despite the fact that their location was indicated in all post-WWII city development plans. The lack of funds for the implementation of large public investments was a measure of the marginalisation of Wrocław on the national arena.

As early as in the mid-1960s, Wrocław was proclaimed a city of young people, as it could boast the largest percentage of inhabitants under 20 in Poland. Most of the youngest inhabitants of Wrocław were born in former Breslau and naturally considered this city as "their own". The slow process of accepting the history of Wrocław began in the period of socialism, and today the multiculturalism of the city, still visible in urban planning, is an element of the identity of its inhabitants.

### Notes

1 On the works by Anna Ptaszycka, the author of the concept of Wrocław greenery cf.: (Majczyk and Tomaszewicz 2019).
2 Spychalski's speech during the conference concerning the revision of the original spatial arrangement plan for the City of Wrocław, 27 April 1950 (Protokół 1950:7).

## References

Batowski, H. 1946. Odzyskane ziemie słowiańszczyzny. *Życie Słowiańskie*, December 8:3–6.
Bieńkowski, K. 1970. Wrocław dziś i jutro. Rozwój Wrocławia w 25-leciu i w perspektywie. *Miasto* 7/8:33–8.
Bukowski, M. 1980. [Tadeusz Ptaszycki] Pożegnanie w imieniu Oddziału Wrocławskiego SARP oraz Prezydium RN m. Wrocławia i województwa wrocławskiego. *Komunikat SARP* 5:18–19.
Chwieduk, T. 1962. Jaka będziesz nowa dzielnico?. *Słowo Polskie*, December 12.
Deklaracja. 1949. Deklaracja ideowa PZPR. In: *Podstawy ideologiczne PZPR*, pp. 129–43. Warszawa: Książka i Wiedza.
Duchowicz, J., and Majerski, Z. 1957. Wrocław-Południe. *Architektura* 10:373–7.
Gabiś, A. 2018. *Całe morze budowania. Wrocławska architektura 1956–1970*. Wrocław: Muzeum Architektury we Wrocławiu.
Główny architekt 1962. Główny architekt Wrocławia o nagrodzonym projekcie. *Słowo Polskie*, December 12.
Grotowski, Z. 1949. Wrocław – punkt obserwacyjny przeszłości i przyszłości Polski. *Zwierciadło – dodatek niedzielny do Słowa Polskiego*, July 24:1.
Kaliski, E. 1949. Znaczenie wrocławskiego ośrodka historycznego w planie zagospodarowania. *Przegląd Zachodni* 11:518.
Kulak, T. 2006. Historia Wrocławia. In: Harasimowicz, J. (Ed.), *Encyklopedia Wrocławia*, pp. 282–290. Wrocław: Wydawnictwo Dolnośląskie.
Kusy budżet. 1958. Kusy budżet Wrocławia ogranicza niezbędne inwestycje i to w szerokim zakresie. *Słowo Polskie*, April 4.
Majczyk, J., and Tomaszewicz, A. 2019. Anna Ptaszycka. Harcerka, architektka, urbanistka. In: *Pionierki*, pp. 13–45. Kraków: Wydawnictwo EMG
Majczyk, J., and Tomaszewicz, A. 2017. Just after Architecture and urban planning in Wrocław in the late 1950s and early 1960s. *Wiadomosci Konserwatorskie – Journal of Heritage Conservation* 49:181–91. DOI:10.17425/WK49SOCREALISM.
Molicki, W. 1996. Polskie domy we Wrocławiu. In: Zasada, J. and Zwierzchowski, A. (Eds.), *Architekci Wrocławia 1945–1995*, pp. 33–45. Wrocław: Stowarzyszenie Architektów Polskich.
Mrówczyński, T. 1979. Lista strat polskich architektów w wojnie światowej 1939–1945. *Komunikat SARP* 8–9.
Nasterski, Z. 1974. Spółdzielcze budownictwo mieszkaniowe we Wrocławiu. *Architektura* 11–12:438–57.
Paszyński, A. 1970. Trasą polskiej architektury (I). *Architektura* 10:373–75.
Protokół. 1950. Protokół konferencji w sprawie rewizji wstępnego planu zagospodarowania przestrzennego m. Wrocławia [manuscript], at National Archives in Wrocław, signature 213/2/18
Przyłęcka, D. 1996. O planowaniu przestrzennym Wrocławia w okresie 1945–1995. In: Zasada, J. and Zwierzchowski, A. (Eds.), *Architekci Wrocławia 1945–1995*, pp. 7–26. Wrocław: Stowarzyszenie Architektów Polskich.
Przyłęcka, D. 2012. *Nie od razu Wrocław odbudowano. Plany zagospodarowania przestrzennego, koncepcje oraz projekty urbanistyczne i architektoniczne, a ich realizacja w latach 1945–1989*. Wrocław: Oficyna wydawnicza ATUT.
Ptaszycka, A. 1956. Zagospodarowanie Wrocławia w latach 1945–55. In: *Wrocław: rozwój urbanistyczny*, pp. 187–325. Warszawa: Budownictwo i Architektura.

Rezolucja. 1949. Rezolucja Krajowej Partyjnej Narady Architektów w dniu 20–21 czerwca 1949 r. w Warszawie. *Architektura* 6-7-8:162.
Rocznik. 1979. *Rocznik statystyczny 1979*. Warszawa: Główny Urząd Statystyczny.
Sawa-Borysławski, T. 2011. Rewaloryzacja miasta historycznego i zabudowa plombowa. In: Eysymontt, R., Ilkosz, J., Tomaszewicz, A., and Urbanik, J. (Eds.), *Leksykon architektury Wrocławia*, pp. 139–146. Wrocław: Via Nova.
Siemiński, W. 1974. Nowi mieszkańcy w nowych osiedlach. *Architektura* 11–12:434–7.
Sobolewski, A., and Czajka, R. 2022. Zachód II housing estate in Szczepin in Wrocław – a place built anew. *Architectus* 1(69):119–28. 10.37190/arc220110
Tomaszewicz, A., and Majczyk, J. 2019. Town planning and Socialist Realism: new in Wroclaw (Poland) – unfinished projects from the 1950s. *Planning Perspectives* 34. 10.1080/02665433.2018.1437556
Tyszkiewicz, J. 2020. Wokół specyfiki dolnośląskiej w okresie powojennym. In: Fica, M. (Ed.), *Powrót do Macierzy? Ziemie Zachodnie i Północne w Polsce Ludowej*, pp. 75–85. Katowice: Wydawnictwo Uniwersytetu Śląskiego.
Uchwała. 1959. Uchwała nr 285 Rady Ministrów z dnia 2 lipca 1959 r. w sprawie przyjęcia tez dotyczących typizacji w budownictwie. *Monitor Polski* vol 70/365.
Umacniajmy jedność. 1952. Umacniajmy jedność i zwartość narodu wokół sprawy przyszłości i rozwoju Ziem Odzyskanych. Przemówienie wicepremiera Stefana Jędrychowskiego wygłoszone na Kongresie Ziem Odzyskanych we Wrocławiu, *Słowo Polskie*, September.
W ankiecie. 1958. W ankiecie "Słowa" kandydaci na radnych RN m. Wrocławia mówią o pilnych "sprawach do załatwienia". *Słowo Polskie*, January 15.
Wrocławskie Biura. 1950. Wrocławskie Biura Projektów czekają na nowych inżynierów i architektów. *Gazeta Robotnicza*, July 1:4.

# 3 Dreaming the capital
## Architecture and urbanism as tools for planning the socialist Bratislava

*Henrieta Moravčíková, Peter Szalay, and Laura Krišteková*

At midnight on 31 December 1968, a sizeable crowd gathered in the courtyard of Bratislava Castle – not merely to welcome in the new year but also to gather in commemoration of the new federal organisation of the Czechoslovak Republic, which assumed legal force on 1 January 1969. This spontaneous celebration of the national emancipation efforts of the Slovak people contrasted starkly with the tragic situation facing the whole of Czechoslovak society after the invasion of Warsaw Pact troops in August of the very same year. Clearly, one should not assume that Slovakia's inhabitants were incapable of recognising the devastating effects of the occupation: quite the reverse. However, they compensated the nationwide frustration by a sense of satisfaction from the finally acquired independence. Bratislava and its inhabitants could finally dream of assuming the status of a capital city. And paradoxically, despite the ongoing process of political repression that affected the whole Czechoslovakia after 1968, this dream soon began to come to reality.

The city of Bratislava, previously known trilingually as Pressburg, Pozsony or Prešporok, is determined by its geographical location on the edge of historical state borders and its secondary position within them. It only reached the status of the capital city of a major state once before in its history, in the period between 1536 and 1783, when the central offices of the Hungarian Kingdom left Budapest in the wake of Turkish expansion for Bratislava. After these years, the city was consigned to a mere secondary role within the wider kingdom. And its position as secondary to another capital remained unchanged with the creation of Czechoslovakia in 1918. Bratislava found itself literally on the state border of the new republic and was even connected to the former centre of the Habsburg Monarchy by an ordinary tram. Bratislava was, of course, the natural centre of the Slovak part of the state, yet the true capital with all associated privileges was Prague. The first serious chance to surmount this historical handicap arose in 1939 with the creation of the nationalist Slovak Republic. Ambitious plans drawn up in those years for transforming Bratislava into a modern capital city, though, remained consigned to paper because of the wartime economic situation. After 1945,

DOI: 10.4324/9781003327592-5
This chapter has been made available under a CC-BY-NC-SA 4.0 license.

Bratislava found itself once more in the restored state of Czechoslovakia, in the same secondary status, with all the state governing structures still situated in Prague. Only in 1968 did another opportunity to escape peripheral status arise, with the passing of the law transforming Czechoslovakia into a federation. It was from this legislation that Bratislava was confirmed as the capital city of the Slovak Socialist Republic. The following two decades represented the period when the city's most ambitious development was planned and partially realised: in these years, the number of Bratislava's residents grew by over one-third, while its area practically doubled. Plans were made for new urban districts, new transport infrastructure, new buildings for state institutions or urban renewal of the central core. Yet most of these plans were realised only in fragmentary form or remained as mere intentions. One of the reasons was undoubtedly their excessive ambitions, related precisely to this heartfelt need to confirm Bratislava as the region's genuine capital.

The chapter focuses on the period after the federalisation of Czechoslovakia in 1968. It studies the era's urbanistic and architectural designs created towards radically changing the city's character. Furthermore, the chapter discusses why these planning instruments became possible, who were the main actors that influenced it and what were the circumstances stimulating such ambitious urban planning for Bratislava. From this, the chapter turns to analysis of concrete examples illustrating these processes after 1968 in the city. Bratislava, a city in the periphery or centre on the edge, accompanies the investigation of this chapter, often considered a topic unworthy of interest due to its fragmentary and polemic nature.

**A new capital city at the edge of the Eastern Bloc**

Of key importance for the changes in Bratislava's status and its development were the social and political changes through which Czechoslovakia underwent in the 1960s. The efforts towards social renewal characterising the era of the "Prague Spring" were directly reflected even in the question of reorganising the previously centralistic, unified state. The question of a symmetrical organisation of the Czechoslovak Socialist Republic assumed relevance at the start of 1968, and on 14 March 1968, the Slovak National Council (*Slovenská národná rada*) approved Act no. 43/1968 establishing Bratislava as the capital of the Slovak Socialist Republic. The law also specified the territorial divisions of the city and the organs of state power and administration, these being the National Committee of the city and the district national committees, and their authority. Paragraph 7 of the law stated: "The foremost task of the National Committee of the capital of Slovakia, Bratislava, and the district national committees is to ensure the multidimensional development of Bratislava as the capital city of Slovakia, specifically its planned construction, development of services, protection of cultural heritage, creation and protection of a healthy living environment, and the securing of public order" (Act no. 43/1968). The culmination of this legislative process was the constitutional act on the Czech-Slovak federation, approved by the National Assembly of the Czechoslovak Socialist Republic on 27 October 1968. Signed in Bratislava Castle on 30 October, the federation act took effect as of 1 January 1969 (Act no. 143/1968).

The common term invoked to describe the changes occurring in Czechoslovakia after the collapse of the reformist forces in the wake of the Warsaw Pact military invasion, the pressures imposed by the USSR on the Czechoslovak authorities and the removal from power of key proponents of reform is the word "normalisation". In reality, it formed a revisionist and repressive process of reinforcing the state-socialist order. Paradoxically, though, precisely this period of social decline brought Bratislava a wide range of significant investments in the construction of new buildings and plans for the thorough rebuilding of its material essence. No small influence on this process was applied by the fact that the Czechoslovak Communist Party (KSČ) was headed from April 1969 up until 1987 by Bratislava native Gustáv Husák – who also served from 1975 until 1989 as Czechoslovakia's president. For this reason, "during the entire state-socialist period, Bratislava was one of the most rapidly growing Czechoslovak towns" (Spurný 2020:33). Equally interesting was the city's social standing. The more the authoritative power of the state and the ideology of the KSČ were demonstrated towards the social atmosphere in the traditional national centre of Prague, the less were these forces concentrated on the atmosphere in the new capital of the Slovak part of the federation. Bratislava thus not only drew upon its new political standing but also enjoyed certain advantages in its cultural and geographical peripherality. Lying right at the southern border, Bratislava lay only 200 km from the Hungarian capital Budapest, where a more open form of state socialism was in force. And the Iron Curtain itself, dividing Europe into a sharply defined East and West, literally ran along the Bratislava city limits. Vienna, the capital of Austria and the former Habsburg monarchy, lay on the other side of the Iron Curtain, yet only 55 km as the crow flies from the centre of Bratislava. Thanks to Austrian television and radio broadcasts, which Bratislava's residents could follow without trouble, Vienna formed a still-relevant source of influence. Attesting to the high following of these media was the interest displayed in purchasing the only Austrian daily paper then accessible, Die Volksstimme – especially its Friday edition with the television programme.

This position at the edges of the two sections of a bipolar world further underscored the strange ambivalent character of Bratislava as a capital city on the periphery. Bratislava was on the geographical edge of the socialist bloc and at the same time still in the centre of Central Europe, which was its historical home region. Plans were made for investments in constructing buildings for new institutions of governance, education or culture that would ensure the operation of the Slovak Socialist Republic within the federation, along with new residential districts to house the rapidly swelling population. At the same time, these aims clashed with the geographic limits of the city located right on the national border. The growth of Bratislava, namely, still reflected the directions of development established in the City Regulation and Development Plan from 1917. One characteristic form of this ambivalence was the expansion of the growing city to the right bank of the Danube, right up immediately against the Austrian border.

## The new generation of architects and urbanists and their ambitious plans for socialist Bratislava

Despite its longstanding peripheral status within Czechoslovakia, Bratislava was well prepared for the new challenges connected with construction. From the start of the 1960s, the city had at its disposal an institutionalised form of urban planning with the Office of the Chief City Architect. Founded in 1962, this office was headed by architect and planner Milan Hladký, author of the first post-WWII city master plan; though dating from 1956, it too had assumed exceptional growth. This plan took a perspective of 15 years, during which Bratislava was expected to expand into a metropolis of 300,000 residents. In proportion, the plan assumed a growth in the housing fund of up to 60%. Based on this plan, new residential districts should focally spread outward to the northeast and east, though partially extending as well to the northwest to the foothills of the Lesser Carpathians or even the hilly terrain to the west. Hladký's plan similarly envisioned the city expanding towards the right bank of the Danube; towards this aim, it included two further bridges across the river. Likewise included in the plan was extensive rebuilding in the city centre, matched by an equally great extent of planned demolition.

Though this directive plan by the team headed by Milan Hladký was never approved by the government, its influence on the city's future form was significant. The radically modernist ideas it proposed were well matched in the 1960s to the exceptional dynamism of investment construction. It was on the basis of this plan, as well as a series of urban-architectural competitions in the 1960s and 1970s, that the greatest quantity of earlier construction was demolished within the wider city centre and several ambitious construction projects were launched. The most significant changes in the physical substance of Bratislava had their intellectual grounding precisely in the directive plan from 1956. Doubtless, the fulfilment of these ideas was furthered by the status of Hladký, as the head of the team that drew up the plan, who from 1962 to 1964 held the position of chief architect and then (1964–1969) the post of chairman of the City National Committee of Bratislava (Mestský národný výbor – MsNV), the highest-level governing body of the city. The Czech historian Matěj Spurný even voiced a hypothesis in this connection that it was "precisely the experts – technocrats who, in the era of post-war modernity, held onto the real power over the organisation of space, aesthetics of buildings, or forms of housing, in other words, over matters that significantly structured, indeed determined, the everyday life of the urban dweller" and thus met the essence of technocratic socialism (2020:33). In the case of

Milan Hladký, there can be no doubt that he decided, from the power of his position, to bring his vision of the city to fruition. Immediately upon his assuming the post of MsNV Bratislava head, he transmitted information about all key intents for rebuilding the city to the chair of the Slovak National Council.

Hladký's conception for Bratislava's development was anchored in the principles of post-war modernist planning. It imagined the city as a radial, centralised and growth-bounded urban structure. A key role was assigned to transport solutions, the building of new residential quarters on the city's outer perimeter and rebuilding of the centre. The professional debates occurring in Czechoslovakia from the mid-1950s onward reflected similar themes, indeed ones resonating in international discussion. And indeed, Czechoslovakia's architects had long been active members of international associations, participating in the activities of the *Congrès Internationaux d'Architecture Moderne* (CIAM) and later after WWII in the *Union internationale des Architectes* (UIA). At the same time, though, even the context of international relations illustrated the lower-ranked position of Slovakia's professional scene. Membership in CIAM from its founding was restricted to Czech (specifically Prague- and Brno-based) architects. Even after the war, when the Czechoslovak representation in CIAM reassembled itself, at first, not one Slovak architect was involved in cooperation. It was only for the congress in Bridgwater that the Czechoslovak group of CIAM was assembled with 41 members, among them two Slovak architects, Emil Belluš and Ján Svetlík. In 1949, the number of the group's members shrank to 32, with Slovakia represented by Belluš, Ladislav Fotlyn and the Czech architect Jan Koula, then working in Bratislava. Considering the severely limited presence of Czechoslovak delegations at CIAM meetings after the Communist seizure of power in 1948, Slovakia's architecture community had an even more restricted chance for direct participation in CIAM. Local debates about the post-war renewal of the city were therefore shaped by Emil Belluš and Ján Svetlík, with active input from the major Czech urban designer Emanuel Hruška. Active in Bratislava after the war as the chief founder of urban-planning education at the Slovak Technical University, Hruška was also one of the most active representatives of Czechoslovakia in international debates on regional and city planning. He was in fact the individual who most decisively contributed to the presentation of Czechoslovakia's urban planning and design on the international stage.

At the end of the 1960s, though, the forces in the Czechoslovak professional scene had grown significantly more balanced in national terms. A new generation of architects and planners was active in Slovakia who had achieved reputations not only on the national but even international levels. Illustrative of their ambitions was the proposal to participate in organising Congress IX of the UIA, which took place in Prague in 1967. On this occasion, a delegation of members from the UIA's standing urbanism commission paid a visit to Bratislava, where architect Milan Hladký, in his authority as head of the City National Committee, received them on 10 July 1967 in the City Hall. An associated development of the time was the proposal for constructing a new congress centre on the right Danube bank, the outcome of a research project prepared in 1966 by the architectural team of Ferdinand Konček, Iľja Skoček and Ľubomír Titl (Andrášiová and Bartošová 2013:22). This plan fittingly encapsulates the era's enthusiasm for bringing about a modern rebuilding of the city and overwriting its peripheral, indeed provincial character. The new pavilion would have been constructed on the site of the Au-café, an early 19th-century building then forming the chief landmark of the western Danube promenade. It was demolished in a fit of "builders-of-socialism" exuberance even before the construction of the pavilion for the UIA congress was approved (Štraus 1992:77). And in the end, the project was never realised: the congress hall would have stood in a floodplain, which would have made its operation more complicated.

The holding of the UIA congress in Czechoslovakia also influenced the organisation of the international urban planning competition for Bratislava's southern edge – the largest satellite town in Slovakia – Petržalka. Arranged by the Council of the MsNV as of 15 June 1966 with a submission deadline of 15 April 1967, it held its jury evaluation in June 1967. The exhibition of 84 competition designs from 19 different countries could, as a result, serve as a backdrop for the Bratislava meeting of the UIA standing urbanism commission and accurately reflected the ambitions of the competition organisers and the municipal authorities.

Subsequently, the results of the international urban design competition needed to be integrated into the directive city master plan, leading to the plan's revision and then replacement by a new directive plan, drawn up after 1970 by a collective headed by Jozef Hauskrecht. This plan was completed in 1973 and then in 1976 approved by the government of the Slovak Socialist Republic in Directive no. 178/76. Again, the authorial team worked with a vision of Bratislava expanding in its area, gradually absorbing into itself other surrounding settlements: eastward to the town of Pezinok and westward (in what they termed the Záhorie Settlement Belt) up to Záhorská Bystrica. Additionally, attention was paid to connecting the city with Petržalka. This "Great Change" plan further assumed in the central area a "gradual concentration of facilities for public service at higher levels at the expense of the housing fund" (Hauskrecht 1978:1–3).

The new generation of architects and urbanists, the first to complete their professional studies at the Bratislava Technical University at the end of the 1950s, could also make full use of the social thaw that followed the end of Stalinism. In their reflections on architecture and urban planning, they drew upon international debates that they followed through (selected) international professional journals, translations of international authors or even direct contacts with the Western world via excursions and visits to professional events. In this connection, it is worthwhile following the trajectory of the discussions on master plan documentation and construction proposals.

While the older generation, represented by such figures as Emil Belluš, cast a critical eye on the radical rebuilding of the city, those then in their 30s, such as Milan Hladký, Dušan Kedro, Milan Beňuška or Štefan Svetko, who had a direct say in the creation of these plans, defended them. And from their position of professional authority, they gave legitimation to their stance in wider social discussion.

The idea of Bratislava as a modern metropolis was to have been brought about through the realisation of three proposals, each of them reflecting themes from international architectural debate. The first was the rebuilding of the city centre; the second was the expansion of the city to the right bank of the Danube; and the third addressed the management of traffic flows, imagining the construction of high-speed roadways and an underground heavy-rail system, i.e., metro. Each of these proposals found support in the city master plans and verification in successive architectonic or planning competitions.

### The new centre for Bratislava: A healthy – and large – hearth of the city

Considerations of the form of Bratislava's new centre were first given a complex presentation in the May issue of the journal *Projekt 1977*. Its editor was urban designer Ján Steller, who had worked on the concept of a new *city-wide centre* from the early 1960s onward. On his own or with other authors, he prepared a long series of proposals for rebuilding the centre of Bratislava, and his ideas had a significant impact on local discussions. Texts were contributed by additional Bratislava planners: Milan Hladký, Tibor Alexy and Milan Gašparec, along with sociologists Ján Pašiak and Dušan Franců. They discussed the city centre as the outcome of a holistic urban-planning stance that viewed it as a "concentration of urban functions, public facilities, or buildings and their complexes", while the sociologists "addressed social relationships, contacts, or social, if necessary social-psychic functions" (Pašiak and Franců 1977:3). And they stressed that "in recent years, the social function of Bratislava as the capital of Slovakia has grown significantly, and equally its social and political centre" (Pašiak and Franců 1977:3). The need for rebuilding the centre's physical fabric was justified through the increased, and partially changed, demands now placed on it. The central core no longer had to meet only elementary functions associated with consumption or housing but now many more complex social functions. To provide them, it was necessary to create both polyfunctional complexes comprising buildings for cultural, educational and artistic facilities but also buildings for administration and governance or commercial complexes. Towards this end, they also discussed that the new city centre would reach across to the western Danube bank in the area of the newly planned residential district. Indeed, the connection of these two sections was regarded as vital so as to "create the continuity and integration of societal functions into a single city-wide urban centre" (Pašiak and Franců 1977:6).

Another influence on the ideas for the new form of central Bratislava was the debate on the "heart of the city" that had been a central theme of CIAM since 1951, following its congress in Hoddesdon that took as its title "The Hearth of the City". It was this congress, or more precisely the congress publication, that Ján Steller invoked in his contribution when describing the city's centre as its physiological "heart", in other words, the "actual fulcrum of social life" (1977:6). His argument then continued to describe the placement of this "fulcrum", its dimensions and its functional content. Serving as a basis for these reflections was the outcome of the functional and spatial analysis of the wider centre of Bratislava (from the historic core up to the outer ring road) covering an area of 2.5 million square metres, realised by the Office of the Chief Architect between 1964 and 1967. Justifying the localisation of Bratislava's centre, he likewise cited José Luis Sert, stating that it should be "a place chosen by the inhabitants themselves, hallowed by their use" (Steller 1977:10). In Steller's view, the conception of the centre should reflect the following requirements: a central location in the city, a focal point for transport communications, links to historic and natural landmarks, ease of servicing and concentration of the most important public facilities. Using these criteria, Bratislava's centre was localised in the area inside the outer ring road, where the historic city core occupied the western edge and the still-undeveloped terrain across the Danube its southern part. The outer city ring road would then assume the form of a high-speed roadway, while the centre would be primarily served for transport through a below-ground urban rail system. This limitation of conception and area was a reaction to the unbounded scope of the previous urban plans and was even reflected in the competition conditions for the urban design of the "City-Wide Centre" of Bratislava, opened in July 1977. Regarded as most significant here was the "placement of the city on the Danube", the expansion of the centre across the river, and the construction of a high-speed urban rail system that would significantly alter the "functioning of the centre and its basic structural composition" (Alexy 1977:12). The new metro stations would serve as focal points for development and sites for intended new construction, whether in the historic core or in the new centre across the river. Mustered in favour of this construction were parallels with new projects in both Western and Eastern Europe: Novi Beograd in Belgrade, the south bank of the Thames in London, Paris's La Défense or Kyiv's expansion to the opposite bank of the Dnipro. Particularly emphasised in this conception was the need to re-evaluate principles "that relied on earlier and asymmetrical development" (Talaš 1977:25).

Bratislava's historic core, which contained the mediaeval town, the adjoining sections around the inner ring road and the Castle complex, an urban heritage zone since 1954, was planned to be "preserved in its full integrity as a material document" but cleansed of "worthless later additions" (Gašparec 1977:19). What was implied by this phrase was, specifically, the "extensive areas with low-rise single-family or functionally and structurally unsuitable construction" (Steller 1979:183). This characteristically modernist approach was based on the Athens Charter and the idea of selective protection of historical building substance.

The centre of Petržalka was to have been grounded – literally – on the main communication corridor combining a high-capacity roadway from the outer city ring, a rail line and a line of the planned Bratislava metro. New public services, spanning retail trade, services, culture or even educational and government institutions, would be concentrated in a gigantic plinth covering this main transport corridor. In time, this ambitious plan was abandoned, with the centre to be spread along this transit artery, now spanned only by footbridges. And this design, though mocked by several architects – "a city for people is becoming a city for cars" (Šlachta 1977:49) – that was gradually realised and is still being realised today, a characteristic instance of the persistent dominance of transport engineering in city planning and construction. Yet at the time of its creation, this idea was largely regarded as a positive development of modernisation. Milan Hladký wrote approvingly of the "entrance of the motorway into the heart of the city" and the "delineation of the central urban area with the southern route of the motorway, the eastern and western bypasses" as a modern solution and a "threshold crossed" (1977:26). The application of radical transport solutions in the city centre was not a unique manifestation of planning in Bratislava. Traffic was handled in a similar way in other European cities. It was a consequence of the growth of motoring after WWII and the prioritisation of the automobile movement as the movement of the future. In the West, this trend had a significant impact, especially on those cities that were most exposed to war devastation. We could mention post-war Britain, where "[a]n especially dramatic introduction was the large-scale imposition of ring roads, reflecting the emerging dominance of planning by traffic solutions" (Larkham 2013:3). In the Eastern Bloc, where city planners and traffic planners had almost unlimited power, it hit almost all cities with higher growth dynamics and car traffic. In the capital of the Czechoslovak Republic, it was the Basic Communication System of Prague approved in 1974, which included three city circuits and a series of radials. The most controversial of them, the so-called "north-south highway" crossed the historic centre of Prague with a pair of three-lane expressways.

Another modern transport solution planned to shape the form of Bratislava's new "city-wide" centre was the underground urban rail plan, with its routes intersecting at key points situated at the edge of the historic core. The planners were convinced that the territory of the centre "needed to be, in this aspect, spatially organised into central districts" (Hladký 1977:28). These would have been the "Historic Core", a cultural-social district on the western edge of the core on the Danube embankment, an administrative-commercial district on the core's eastern edge (Kamenné námestie), a new centre along Obchodná ulica, a political-governmental district on the core's northern edge near the Cabinet Office, a university district (Námestie slobody), another new centre on the southeast edge near the riverbank (today's Pribinova) and a southern section of the centre on the right Danube bank containing four districts: the exhibition grounds, a recreation-sports complex, a new centre and a university campus. Central to the discussion was the theme of "usage of below-ground spaces" and "below-ground urbanism" to link extant and planned commercial or cultural functions with the underground metro stations (Hladký 1977:28). For Bratislava, these were the first conceptions on increasing central urban density not merely through high-rise construction but also through furthering the infrastructure below ground level. To clear space for these admittedly overambitious projects, the central districts would be subjected to wide-scale demolition: entire city blocks would disappear to be replaced with highly complex structures growing equally downward into the earth as much as upward.

Post-war debates on monumentality in architecture did not receive direct theoretical treatment in the Slovak or Czechoslovak context, yet certain principles of the New Monumentalism could well be discerned in the architectural and urban designs from the 1960s and 1970s. The highest-ranked competition designs for urban plans for the city-wide centre of Bratislava likewise reflected the views of city-centre planning as formulated in the wake of WWII by one of the authors of "Nine Points on Monumentality", José Luis Sert. For Sert, commercial and cultural centres "constitute the most important element of a big city, its brain and its governing machine", and thus should contain the "university buildings, the main museums, the central public library, [...] and areas especially planned for public gatherings, the main monuments constituting landmarks in the region, and symbols of popular aspirations" (Mumford 2000:145). That these elements should also be of an exceptional physical scale was also understood as a facet of the "metropolitanization" of Bratislava. An example of such thinking was the proposal for the construction of a new commercial and social centre on one of the city's radials, Obchodná ulica. The megastructure, which was to be built on the place of the original structure from the first half of the 20th century, included all kinds of services and transport infrastructure, spread over six above-ground and two underground floors and had an area of more than 425,000 square metres. In the end, it remained only on paper as an exemplary example of Bratislava's socialist urban utopias (Figure 3.1).

**In the name of growth: The southern city district of Petržalka**

Also linked closely to the new role of Bratislava as the capital of the Slovak Socialist Republic were the efforts to strengthen the city within its own region. Following the passing of Act no. 63/1971, several adjoining municipalities were attached to the city along with their cadastral territories, thus raising the total area of the city by nearly 45% – the last large-scale land

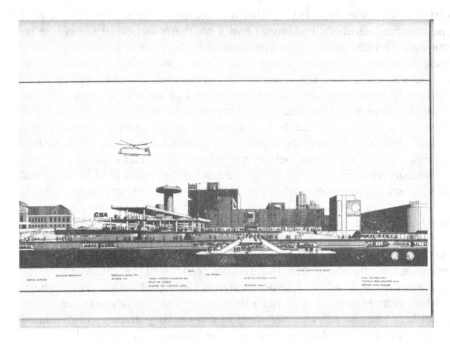

*Figure 3.1* Proposal for a new commercial and social centre on Obchodná ulica, Ivan Matušík, 1970.
*Source:* Department of Architecture HI SAS Archives.

expansion of Bratislava after the end of WWII. Essentially, the surface area of the city and its territorial divisions from 1971 remain unchanged until today. This enormous expansion of 15,000 ha provided the conditions for the construction of new housing complexes near these integrated settlements and also ensured that the number of inhabitants would rise no less sharply. In 1961, the city had 246,845 residents, and 20 years later had reached 381,186. A significant contribution to these figures in the 1970s was, understandably, the construction of the southern urban sector, Petržalka.

Petržalka, originally a village lying on the right bank of the Danube, had been attached to Bratislava already in 1946 and from the outset was discussed as the city's most significant area for development. When construction was launched in 1973 on the "City Sector Petržalka", the plans assumed the realisation of 50,529 flats, housing around 158,000 residents. Moreover, Bratislava's urban planners had the ambition to make the sector an ideal modern city. Following the results of an international competition for the urban plan for this new district, where the jury awarded the highest ranking of third prize to five separate teams, the assumption was clear that international cooperation would continue in the subsequent design stage. The municipal government promised that while the wider project of the South Sector would be created by Slovak urban planners drawing upon the ideas put forward in the competition, selected parts of Petržalka would be assigned to each of the highest-ranked design teams. This proposal would have meant that the final plan of the new Bratislava across the Danube would merge different ideas and visions in an ideal modern city. In addition to two Czechoslovak design teams, the jury awarded town planners from Japan, the United States and Austria.

The selected entries were in fact a showcase of contemporary town planning. On the one hand, it developed ideas of modernist functional city, with its strict spatial organisation on axis and separation to functional units (SK), configurations of megablocks that will allow to create extensive public green areas (CZ) or complex metabolist structures developing ideas of continuity of planning and constructing of the new district (JP). On the other hand, more utopian plans were also awarded. It was the project of Petržalka as a unique island city which enlarged the natural characteristics of the Danube flood plain with its lakes and waterways (USA) as well as consistent artificial symmetrical scheme of 15 identical round housing units (AT).

The 1968 Warsaw Pact invasion of Czechoslovakia, however, not only brought all reformist tendencies to an end but also thwarted any plans for cooperation with the West. The competition results were given the status as "guidelines for preparing a new urbanistic study", yet the authors of the latter were employees of the Office of the Chief Architect under the direction of Ján Steller (Dvorín 1973). The final construction plan was drawn up by designers from the Bratislava state atelier Stavoprojekt under architects Jozef Chovanec and Stanislav Talaš.

The southern urban residential sector of Petržalka was conceived on the principle of a linear "town centre" adjoined by individual "neighbourhood units" (or, following the Soviet terminology, "*mikroraions*") each with local centres containing a primary school and kindergarten, service centre, cultural centre and medical clinic. Petržalka was separated into three neighbourhoods (Háje, Dvory and Lúky), then into nine sections with around 6,000 flats. The system organising transport focused on strict separation of pedestrian and automotive transport, where the main communication axis with the automotive boulevard and urban rail line forming the backbone of the linear centre. Construction of the individual sections was conceived so that the high-rise blocks placed along the transport axis formed a natural noise barrier yet also did not block sunlight from reaching the other areas of the section. This apartment type was termed "envelope construction" and was planned as small flats for young singles; past the envelope construction would have been lower-rise blocks with larger flats for families, here supplemented by school and kindergarten buildings and basic facilities.

It is worth noting that the spatial arrangement of Petržalka along a main communication axis bears a clear resemblance to the "linear forms or tongues extending from the Thames, described as like a herringbone, composed of social units and based around the rail network" in the 1941 plan for London from the MARS Group (Korn, Fry and Sharp 1971:163–73). Even more, the chair of the competition jury for the Petržalka urban plan was Arthur Ling, a member of the MARS Group and one of the authors of the London plan.

Petržalka was intended as a site for realising all the ideas prevalent in European post-war urbanist discussions, from functional zoning through transport segregation up to defined neighbourhood units. Another model for a more open plan of individual built volumes and vertical separation of automotive and foot transport in a linear centre as well as local centres was the La Miraille housing estate in Toulouse designed in 1962 and completed in 1972. As in the French example, Petržalka had conceived the vertical separation of pedestrian routes as open public terraces. Indeed, a specially assembled ferroconcrete frame was developed for this use, UNIVEX, on which the apartment blocks of prefabricated concrete panels would be constructed directly. The architects Chovanec and Talaš published in 1972 their ideas of the growth in volume and technologies of the linear centre, which was expected to expand in stages up to 2000, by which point the communication core would consist of a futuristic automatic high-speed rail system (Zalčík 1972:45) (Figure 3.2).

During the years when the construction of Petržalka was in full swing, international discussion had already shifted to the diametrically opposite standpoint, favouring a return to concentration and traditional city blocks. Czechoslovakia's worsening economic situation and inflexible state governance and management led to the completion of only the most rudimentary infrastructure of the southern sector in the form of apartment blocks and schools. By the end of the 1980s, this fact was the target of criticism even from the unofficial interdisciplinary report on the state of the physical environment of Bratislava, "Bratislava Out Loud" [*Bratislava nahlas*], with the authors openly stating that "Petržalka failed to become the city of the right bank, but only a large monofunctional appendage to Bratislava" (Budaj 1987:40).

**The end of Bratislava's great plans?**

The planned economy of the Czechoslovak Republic, which until the mid-1970s drew on the reserves created in the second half of the 1960s, reached the brink of its possibilities at the end of the 1970s. Despite attempts at certain reforms, economic problems deepened. In the field of construction, it was mainly a large number of unfinished buildings and low-quality housing construction. The public was most sensitive to the lack of apartments. The federal government responded to the situation by making new investment construction conditional on the completion of constructions under construction and by emphasising the capacity and quality of housing construction.

Dreaming the capital 63

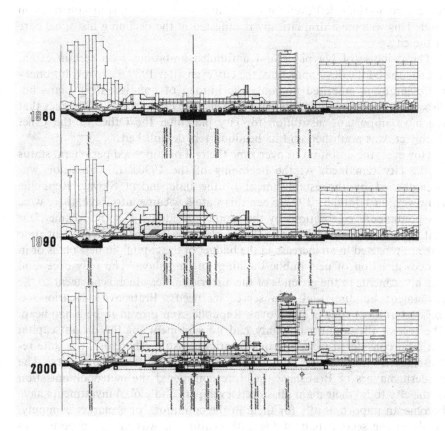

*Figure 3.2* Study of the gradual development of the linear centre of the Petržalka City Sector, Jozef Chovanec and Stanislav Talaš, 1972.

Source: Department of Architecture HI SAS Archives.

This trend dramatically affected Bratislava's ambitious plans. The high degree of incompleteness of large public buildings made it impossible to implement other new plans, such as the new commercial and social centre on Obchodná ulica or the reconstruction of the city centre. Undoubtedly, the stoppage of metro construction was also influenced by the amount of investment and the problems that accompanied the construction of the Prague metro. The Prague metro thus remained the only one in the entire Czechoslovakia. The demand for apartments, in turn, led to the strengthening of housing construction. The structure of the state budget for Bratislava also corresponded to this, where a significant part was state support for the financing of housing construction. At the same time, the city was obliged, in accordance with the state decree, to support housing construction by providing land free of charge and building the infrastructure of new residential districts. There were no more funds left to finance ambitious plans. However, there were no funds left even for the maintenance of the existing construction fund. This was most dramatically manifested in the declining historical core of the city.

The collapse of big plans and unfulfilled ambitions to a certain extent foreshadowed the development of the city even after 1989. Post-revolutionary representations inherited not only the burden of unfinished big plans but also resistance to planning in general. This also corresponds to the fact that the most important institution for city planning, the Office of the Chief Architect, was abolished and its headquarters demolished.

However, the ambition to overcome the real or supposed peripheral status of the city remained. At the beginning of the 1990s, this ambition was strengthened by the establishment of the independent Slovak Republic. However, on 1 January 1993, when this state law came into force, there were no spontaneous celebrations by the capital's residents. It soon became clear that this was a justified restraint. The declaration of a new independent state was not reflected in an increase in the budget of the capital, in new plans or in the construction of new public buildings in Bratislava. The only exception was investments in the grounds of the historic castle, which is adjacent to the parliament building and also serves its needs. Bratislava's location on the westernmost edge of the Slovak Republic even proved to be a handicap. The rest of the country not only had no sentiment for the distant capital but even considered it foreign or downright hostile. Therefore, the state representatives more or less distanced themselves from their own capital. The modern mayors of Bratislava therefore considered the metropolitanisation of the city to be their main mission. Developers and global investments have become an important ally for them in the conditions of a market economy. While in the second half of the 20th century, it was mainly experts who determined the future shape of the city, after 1989, market actors became the determinants, who started bringing their own plans to the city.

### Acknowledgement

This chapter is based on the results of research carried out at the Department of Architecture at the Historical Institute of the Slovak Academy of Sciences (VEGA 1/0286/21; APVV 20–0526), the Institute of History and Theory of Architecture at the Faculty of Architecture and Design of the Slovak Technical University (VEGA 1/0286/21) and at the Institute of Contemporary History at the Academy of Sciences of the Czech Republic (GAČR22–17295S).

### References

Act no. 43/1968 Coll., on the Capital of the Slovak Socialist Republic Bratislava (1968) and Act no. 143/1968 Coll., the Constitutional Act on the Czecho-Slovak Federation. Zbierka zákonov Slovenskej republiky https://www.slov-lex.sk/web/en

Alexy, T. 1977. Ako sa vyvíjala koncepcia bratislavského centra. *Projekt* 19(4):11–15.

Andrášiová, K. and Bartošová, N. (Eds.). 2013. *Konček – Skoček – Titl.* *Exhibition catalogue.* Bratislava: STU.
Budaj, J. (Ed.). 1987. *Bratislava nahlas.* Bratislava: SZOPK.
Dvorín, I. (Ed.). 1973. *Mestský sektor Bratislava – Petržalka.* Bratislava: Investing.
Gašparec, M. 1977. Aby sa zaskvela v plnej kráse. Mestská pamiatková rezervácia v Bratislave. *Projekt* 19(4):16–21.
Gross, K. 1977. *Medzinárodná súťaž Bratislava Petržalka.* Bratislava: Pallas.
Hauskrecht, J. 1978. Úvodom. *Architektura ČSSR* 37(1):1–3.
Hladký, M. 1977. Ako preformujú centrum. Progresívne systémy dopravy. *Projekt* 19 (4):26–8.
Korn, A., Fry, M., and Sharp, D. 1971 The M.A.R.S. Plan for London. *Perspecta* 13/14:163–73.
Larkham, P. (Ed.). 2013. *Infrastructure and the rebuilt city after the Second World War.* Centre for Environment and Society Research Working Paper series no. 22, Birmingham: Birmingham City University.
Moravčíková, H. et al. 2012. *Bratislava Atlas of Mass Housing.* Bratislava: Slovart.
Mumford, E. 2000. *The CIAM Discurse on Urbanism, 1928–1960.* Cambridge: MIT Press
Pašiak, J., and Franců, D. 1977. Stred mesta. *Projekt* 19(4):3–6.
Spurný, M. 2020. Urban Experts in the Building of Post-Stalinist Bratislava. *Architektúra & Urbanizmus* 54(1-2):33–43.
Steller, J. 1977. Za zdravé srdce mesta. Riešenie celomestského centra Bratislavy. *Projekt* 19(4):6–10.
Steller, J. 1979. Súťaž Bratislava – centrum. *Architektúra a Urbanizmus* 13(4):183.
Šlachta, Š. 1977. O asanovaní. *Projekt* 19(4):49.
Štraus, T. 1992. *Slovenský variant moderny.* Bratislava: Pallas.
Talaš, S. 1977. Konečne krok cez rieku. Pravobrežná časť celomestského centra. *Projekt* 19(4):22–5.
Tyrwhitt, J., Sert, J.L., and Rogers, E.N. (Eds.). 1952. *The Hearth of the City: Towards the Humanisation of Urban Life.* London: Lund Humphries.
Zalčík, T. 1972. Petržalka / Nový organizmus. *Projekt* 14(6-7):44–6.
Žatkuliak, J. 2008. Otvorenie problematiky federalizácie štátu roku 1968'. In: Londák, M. (Ed.), *Eto vaše delo.* Bratislava: HÚ SAV.

# 4 The Yugoslav Skopje
## Building the brutalist city, 1970–1990

*Maja Babić*

**Introduction**

In the summer of 1963, the city of Skopje crumbled to the ground. The early morning hours of 26 July brought screams of terror from the rubble of what was a growing modern city only a day before. At 5.17 a.m., an earthquake struck the capital of Macedonia, rendering over 80% of the city unliveable. In the next two decades, a brutalist city of "international solidarity"[1] and Yugoslav "brotherhood and unity"[2] arose from the remnants of Skopje's historical urban layers, entwined with Yugoslav communist ideology and Cold War political negotiations.

Skopje is the capital of the former Yugoslav Republic of Macedonia, now the Republic of North Macedonia, a country in the Western Balkans. It is a city on the periphery of European politics and 20th-century architectural and planning historiography. A city that stood on the fringe of Yugoslav political space and architectural developments after WWII became a beacon of hope for the Yugoslav government and the United Nations during the 1960s when the local and international architects and urban planners together envisaged the modernist city to be built in the aftermath of the 1963 earthquake. Unlike peripheral cities of other state-socialist countries that largely saw architecture as derivative of that of the centre, Skopje stood a unique ground for urban experimentation, only facilitated due to the vast destruction of the earthquake and the Cold War need for diplomacy. Nonetheless, the city of "international solidarity" and its utopian architectural project met the reality of the Cold War Yugoslav politics and its economy of debt: by the late 1960s, the international community deemed its part in the project completed and departed from Skopje; the brutalist capital of the Balkans built from exposed concrete and in minimalist design, emerged from the hands of Yugoslav architects throughout the 1970s and early 1980s.

Following the 1963 earthquake, the international community came together under the patronage of the United Nations and created plans for the reconstruction of Skopje: 1964 and 1965 saw the production of a master plan and the city centre plan, respectively. In a joint effort with Yugoslav architects and Eastern European and Soviet planners and experts, the global architectural community manufactured plans for Skopje in a unique diplomatic effort of the Cold War decades. In Skopje, architects from the global geopolitical West worked with their counterparts from communist Europe, utilising Skopje as a ground ripe for experimentation. As a result, the global architectural community unearthed plans seldom seen in Western Balkans: the brutalist Skopje was to be a city of progressive 20th-century architecture, diplomacy and knowledge transfers. The resulting urban plans were of a modern, brutalist city, one fit for a particular country such as Yugoslavia, a participant in neither of the two embattled sides of the Cold War division; ever since parting ways with the Soviet Union in 1948, Yugoslavia stood isolated from its former state-socialist allies.

As the 1970s unfolded, the progressive and optimistic urban plans for the city centre and the Skopje metropolitan area remained largely on paper. The brutalist city of western architects and their socialist counterparts – the metabolist city centre of Kenzo Tange and the multi-level traffic system – were constructed only partly, and the plans were significantly pared down. The city arose in a more modest Yugoslav iteration of brutalist architecture. As the geographer Stefan Bouzarovski wrote in 2011, the "public finance gradually started to dwindle during the 1970s", and the local and national governments soon faced the necessity for the "downscaling—and in most cases, cessation—of construction activities aimed at implementing the urban development provisions of the 1964 Master Plan" (Bouzarovski 2011:267). The Japanese architect Kenzo Tange's monumental proposal for the city centre met a similar fate. Outside of the segments of City Wall housing structures and the City Gate railway station, the Japanese architect's arresting colossal buildings are found mainly in archives or reproduced in monographs (Figure 4.1). Ultimately, ideological and financial anxieties of the 1970s and 1980s in Skopje and Yugoslavia unearthed the city known for its brutalist architecture created by local, national and few international architects.

The city of Skopje and its architecture of the second half of the 20th century exist on a dual periphery. The collaborative production of the 1960s Cold War reconstruction plans for Skopje and the particular brutalist architecture of the 1970s remained on the architectural and geopolitical periphery of the global events of the era. While Yugoslavia held a place of importance as a country effectively straddling the Iron Curtain, an interest in the country's urban developments – similar to the one paid to Yugoslav politics and the economy at large – was only extended to Belgrade, the capital, and even then, not in a far-reaching manner. Skopje, the capital of the southernmost Yugoslav republic, held no such place in the political constellation of the second half of the 20th century. Further, the architectural and urban historiographies of the era have paid little attention to Skopje; this has only come to change in the past two decades and most often by architectural historians from and in the region. While Skopje's relative

*Figure 4.1* Skopje City Wall, 1970s. Postcard.

geopolitical inconsequentiality can be understood in the context of the Cold War politics of division between the global centres of power, indifference towards the brutalist architecture erected in the city during the 1970s is less comprehensible.

The architectural historian Carmen Popescu traces the marginalisation of Eastern European architectural history – within which the architecture of the Western Balkans ostensibly falls – and argues that this marginality was mainly due to its "alterity – both cultural and political". The othering of Eastern Europe and the Balkans can be traced back to the European Enlightenment when the region served as the "internal other", and following the post-1989 globalisation, much has remained the same in this context. Popescu further argues that this marginalisation is partly methodological, as the canonical discourse seeks to address the architecture of Eastern Europe from the perspective of pre-established "grand narratives", which do not apply to the region in their original context. The rhetoric of "'creative' centres" and "'following' peripheries" only furthered the otherness of the architectural particularities of the region's urban spaces during the 20th century, and it persists to a large extent nowadays, demanding extensive contextualisation in its attempt to explore the region (Popescu 2014:9–11). In this context, we seek to historicise the Cold War peripherality of brutalist architecture of Skopje: on the fringes of ideological spaces of Yugoslav economy and architecture, the Macedonian capital of the 1960s stood as a fertile ground for vast architectural

*The Yugoslav Skopje* 69

experimentation missing in the more prominent urban environments in the country and further.

This chapter examines the period of the 1970s construction of Skopje and the peripheral place the city holds as a repository of brutalist architecture in the contemporary architectural discourse. The chapter further explores the peripheral place Skopje and Yugoslavia held in the post-war architectural space and in the context of the intricate links of global knowledge exchanges. The overwhelming focus of the contemporary scholarly community on the involvement of the UN, international architects and the master and city-centre plans produced in the 1960s obfuscates the history of the effectively Yugoslav construction of brutalist Skopje. Local and regional architects designed the structures erected in the 1970s and early 1980s; they were based on the modernist-era plans and under the influences of Tange's brutalist tendencies and constructed by local companies. As such, they tell a story of the construction of a Yugoslav city and its complex urban identity.

**The Yugoslav politics, economy and knowledge transfers: 1970–1990**

By 1970, significant shifts in the ideologically bipolar world destabilised the established political patterns of the clear division between the global West and its Soviet-led counterpart. New concerns arose in the Cold War political arena: the two decades of the Vietnam War had led to a global economic decline, and the 1973 oil crisis further disrupted the balance of power between the ideological East and West. The development of the Non-Aligned Movement – in which Yugoslavia held a dominant position – further contributed to the imbalance of power as the formerly colonised Third World countries established and asserted their independence. Following the crush of the Prague Spring in 1968, the fragile stability of the post-war years was gone; reawaken hostilities engrossed the globe in the late 1970s and early 1980s. These events profoundly altered global politics and the economy, and although the economic reform of the 1960s briefly "made possible the economic boom of the late 1970s" (Ramet 2006:228), they consequently swayed an already unstable Yugoslav economy of self-management.

During the 1970s in Yugoslavia, there was a continuous increase in architectural production: the Yugoslavs held their steadfast focus on the modernisation of the country, and "the opportunities grew even more after Yugoslavia's policy of non-alignment opened the door to Third World markets". As the architectural historians Vladimir Kulić and Maroje Mrduljaš argue, to practice architecture in Yugoslavia was a lucrative endeavour, and "until the early nineteen-eighties, the booming urbanisation made sure that jobs were aplenty" (Mrduljaš and Kulić 2012:29). Yugoslav architects studied in the country's newly founded schools of architecture – influenced by the country's interwar modernist traditions – and with liberties to travel for education and professional training. They produced an urban environment that merged local traditions and local architectural modernisms with international inspirations and impacts. Throughout Yugoslavia, the architecture of the 1970s mainly featured works of late-stage modernism bound to the vanishing tenets of the International Style, and the architects only sporadically engaged in experimentation with brutalist architectural language.

The brutalist Skopje was the outcome of the architectural expertise of Yugoslav architects and the continuous transfers of knowledge, both national and international. Knowledge transfers and exchanges amongst architects from Yugoslav republics, as well as with their counterparts and educational institutions in the United States and Western and Eastern Europe, have been instrumental in the reconstruction of Skopje since the project's inception. In 1970, the historian Derek Senior wrote that international and local experts – "over a hundred consultants from more than twenty countries" – worked together to create urban plans for Skopje, "constantly exchanging ideas and experiences". The Soviet government donated a factory of prefabricated elements and dispatched Soviet specialists to provide "expertise in training local professionals" (Mariotti and Hess 2021). Czechoslovak, Bulgarian and Romanian governments donated complete buildings, and the UN facilitated the allocation of academic fellowships: the Skopje reconstruction project manager, the Polish planner Adolf Ciborowski, had the "task of selecting, in consultation with the local authorities, the 14 professionally qualified candidates to be awarded Special Fund fellowships" (Senior 1970:123–4).

UN officials envisaged these fellowships to be awarded to experts in the fields pertinent to earthquake destruction and reconstruction: the fourteen awards were given for "post-graduate work in seismology, town planning, architecture, traffic engineering and water engineering", with experts selected from different universities around the globe, including the Federal Republic of Germany, Finland, the Netherlands, Poland, Sweden, the Soviet Union, the United Kingdom and the United States. Host countries were typically those that played a significant role in the 1960s UN work in Skopje or those with vital expertise in seismology as paramount for the safe future of the city. Ciborowski recommended that the fellowships run "from the end of the Project's planning period so that the best use might subsequently be made of the successful candidates' services in working out the detailed implementation of a Master Plan they had themselves helped to prepare". The UN awarded the fellowships to professionals already involved in the reconstruction of the city, many of whom would come to define the urban fabric of 1970s Skopje (Senior 1970:124).

Seven Macedonian architects left Yugoslavia to participate in the United States-sponsored master's degrees at American universities. After spending time at the American public and private universities and interning in American architecture studios, "they all returned to Skopje to design some of the most prominent structures" in the new city (Mrduljaš and Kulić 2012:46). The architects who returned from the United States left an architectural mark on Skopje evident to this day and influenced the new generations of architects, either as teachers or through their works that came to form the cityscape of the North Macedonian capital. Through these means, Skopje "served as an open-air classroom for a younger generation of Yugoslav architects" (Stierli and Kulić 2018:22), a space for experimentation where global know-how merged with local architectural traditions, both modernist and historical.

The canonical supposition that the creation of brutalist Skopje rests on the influences exclusively assigned by the UN and the United States exhibits a simplistic understanding of the events that transpired: the architecture of the Macedonian capital was created in a multifaceted manner that overarches this assumption. The construction of Skopje's built environment took place through an amalgamation of interwar modernism, regional particularities and centuries-long heritage, along with the transfers of knowledge from various parts of the ideologically divided globe that built the city's multi-layered urban fabric. The following examples illustrate the processes that constructed Skopje and the links between the local, national and international influences and actors.

**Building the brutalist Skopje**

On 26 July 1970, on the seventh anniversary of the earthquake, the Macedonian daily newspaper *Nova Makedonija* recalled the destruction of 1963. Journalists praised the construction completed in the years prior. The unknown author of the short front-page article emphasised the perseverance of Macedonians and the Yugoslav "brotherhood and unity" that had rebuilt the city (*Nova Makedonija* 1970:1). The Skopje City Council reportedly took pride in the "rational execution" of new buildings and the repair of damaged ones, further highlighting that the rebuilding of the city was not yet over and that it can only be done through the camaraderie and compassion of all Macedonians (*Nova Makedonija* 1970:1). The article concluded with the statement that the memory of the earthquake and the reconstruction project had transformed the city into a living monument and a vehicle for progress.

In the 26 July 1969 issue of *Nova Makedonija*, only a year prior, journalists regarded Skopje as a construction site. The city was a transformed modern capital, novel architecture plentiful throughout (*Nova Makedonija* 1969:6). In 1966, the local and national construction firms – *Granit* from Skopje, for example – erected the first structures of the post-earthquake city. Still, the ambitious city centre plan was only partly executed: The City Wall residential complex and the new train station were the only segments of Tange's proposal that stand today. Tange's City Wall was planned in a format of "massive residential blocks circling the central area in a wall-like formation" (Grčeva 2013:3). The Japanese architect envisioned the residential complex as an "expression of permanency" (Tange 1967:38) and designed it in imposing overlapping segments. The lower elements of the buildings were constructed for seismic stability and housed commercial amenities, while the architect planned for residential spaces on the upper segments of structures. Although Tange envisioned the City Wall to encircle the city centre perimeter and, perhaps, to serve as a psychological anti-seismic defence mechanism, the architect himself only designed one segment of the Wall. While Tange produced the initial proposal for the City Wall, the Macedonian architects completed the designs for different towers that constituted the large complex. Due to the financial obstacles, these followed Tange's model only partly and exemplified the problematic of an indiscriminate application of global architectural trends in a country in a precarious geopolitical and financial position.

The optimistic notions that accompanied the erstwhile construction of the City Wall were dulled by the time of the completion of the structure. In its 5 July 1970 issue, *Nova Makedonija* reported that "after much anticipation", the first tenants had finally moved into their new homes (*Nova Makedonija* 1970:10). The new dwellings failed to measure up to the high expectations set by the government and the ever-present rhetoric of urban progress; while the tenants conveyed to newspapers that the apartments were comfortable and spacious, they noted that the quality of construction was poor; the new inhabitants regularly complained about the faulty electricity, and draughty doors and windows. Those in pressing need of housing – Skopje's citizens first lived in tents in parks and then in makeshift structures following the earthquake – brought up yet another concern: the city-administered allocation of units was extremely slow, and, by 1969, the city assigned the tenants to only thirty out of hundreds of future apartments (*Nova Makedonija* 1970:10). The construction process had been delayed, and many citizens of Skopje had to wait for long periods since the city allocated the apartments according to employment seniority and families' needs. Architectural concerns accompanied these issues: the architect Živko Popovski wrote in 1981 that, while the towers were "healthy architectural productions" when taken on their own, as a complex, they were missing "visual motivations [...] and urban character" (Popovski 1981:14). Still, at the time of the building's completion and as the tenants were moving in, the design of the buildings was not the primary concern if it was at all, and the city government's near-dogmatic treatment of the 1965 plan and its execution saw the local and national newspapers suppress any criticism that may have arisen.

In Skopje, the subdued high-rises of the City Wall stood as signifiers of the 1960s city centre plan and the role the UN and Kenzo Tange had played in the city's reconstruction. However, the buildings that came to define the post-earthquake brutalist city were produced by local architects and, under the influences of local traditions, merged with international architectural developments. As the decade of the 1960s reached its end, Georgi Konstantinovski designed a building complex that would initiate a trend of brutalist architecture throughout the city. The Goce Delčev Student Dormitories were completed in two segments: the first phase between 1969 and 1971 and the second between 1973 and 1977. Supported by one of the UN fellowships established in the aftermath of the earthquake, Konstantinovski first studied at Yale University under the supervision of Paul Rudolph, a visionary modernist and brutalist architect, and later interned in the studio of I. M. Pei, yet another modernist architect with a proclivity for combining traditional architectural influences with thoroughly modernist architectural principles. Konstantinovski's Dormitories exemplify his professional development and merge the "sculptural, textured *béton brut* characteristic of Rudolph, with Pei's geometrically rigorous forms" (Mrduljaš and Kulić 2012:46), illustrating the foundational elements of the 1970s architectural language in Skopje.

The Macedonian and Yugoslav public and architectural professionals deemed the Dormitories complex a marvel of brutalist architecture. Composed of four buildings of different heights connected by "flying bridges", Konstantinovski designed Goce Delčev in exposed concrete, a key element in the Dormitories' architectural expression. Architectural historians Martino Stierli and Vladimir Kulić argue that the complex allowed for an "exclusive use of that brutalist material par excellence (to) subvert the conventional modernist distinction between structure and enclosure, resulting an aesthetic reduction in terms of materiality and colors" (Stierli and Kulić 2018:161). Konstantinovski utilised national motifs and elements of traditional Macedonian embroidery – albeit minimally – as an inspiration for the Dormitories' facades, further merging his local iteration of global brutalism with Macedonian heritage (Bogoeva 2018).

In his 2013 monograph, Georgi Konstantinovski summarised the design inspiration and fundamental architectural principles employed in Goce Delčev: the architect defined architecture as pure art that requires the architect to "inevitably be acquainted with architecture of past civilisations, so that he would be able to locate himself with his work in the period of time he lives and creates [sic]" (Konstantinovski 2013:11). The architect argued that to produce quality works of architecture, one must always study, further emphasising the notion of an architect as a social being, one required to acknowledge his or her place and role in society and the role society plays in the development of any architect's design. Konstantinovski defined the basic principle of his architecture as "creating a space for living or working that will be worth for man [sic]" (Konstantinovski 2013:11). Arguably, this can be traced to Konstantinovski's studies at Yale; the work with Rudolph and Pei influenced his architectural path regarding the use of materials and space. However, the local idiosyncrasies of Skopje and the architect's attuned stance towards the city's historical lessons pointedly characterised his architectural trajectory and the overall feasibility of his projects as much as his foreign education.

The Goce Delčev Student Dormitories – like Konstantinovski's earlier work on the nearby Skopje City Archive constructed between 1966 and 1968 – illustrate the brutalist architecture of the 1970s. The large complex constructed in exposed béton brut makes a mark in the urban fabric of the Macedonian capital and serves as a signifier of urban development: the Dormitories were constructed westward from the historic city centre and in what was to become the neighbourhood of Karpoš, interspersed with clean-lined modernist housing, hospitals and schools. The *béton brut* used extensively by Konstantinovski connotes a sense of progress and urban expansion sought in the aftermath of the war and the earthquake.

The Slovenian architect Marko Mušič designed the complex of the Ss. Cyril and Methodius University in Skopje (Bogoeva 2018). Between 1970 and 1974, Mušič's structures were erected in *béton brut* (Figure 4.2). They stand imposing, all elements of the composite urban unit seemingly alike. Nevertheless, distinctions between the architectural segments do exist. The Slovenian architect designed buildings of the University's different faculties and departments with subtle distinctions mainly exhibited in the designs of the facades. The architectural historian Mirjana Lozanovska describes the University as "Brutalism in speed" and references Mušič's work in comparison to Paul Rudolph's design of the University of Dartmouth, further arguing that the Slovenian architect was "interested in other, parallel developments of Brutalism" (Lozanovska 2015:158), with more dynamic forms.

Mušič's work is not only significant for the qualities of the architect's design of the vast complex but also for its affirmation of inner-Yugoslav knowledge exchange processes and resulting projects: Mušič, a renowned

*Figure 4.2* Ss. Cyril and Methodius University, 1970s. Postcard.

Slovenian architect, falls into a group of highly successful architects from the north-western Yugoslav republic. These architects were "exceptionally successful at architectural competitions around Yugoslavia, spreading their taste for expressive structural figures to other republics" (Mrduljaš and Kulić 2012:87). Like other brutalist structures of the time, the University complex serves as a signifier of space and the architectural manifestation of the monumentality of design and the use of *béton brut*. Mušič's design of the University complex is not distinct from the rest of the brutalist structures in the city due to his different utilisation of *béton brut*; the architect's design is different in its spatial explorations within the site and the surrounding urban fabric of Skopje and in his urban compositions of open and closed spaces, traditions extensively explored in the Slovenian architectural landscape.

Although structures clad in *béton brut* would come to permeate Skopje and essentially create its new urban identity, the modified traditions of European modernism still found their place in the city. The Museum of Contemporary Art, which overlooks the city from atop the Kale fortress just up the street from the Ottoman Bazaar, is strikingly dissimilar to the brutalist architecture of Konstantinovski. The building was a donation from the Polish government as a part of a collaboration of socialist countries: designed by the Polish *Grupa Tigri* between 1969 and 1970, the museum is a repository of an impressive collection of contemporary art.[3] The structure is an archetypical modernist building with an open floor plan enclosed in glass with external columns supporting the upper floor. Constructed in reinforced concrete with coffered ceilings and completely painted in white, the Museum is a significant building in regard to its design, its prominent site and its donation from the Polish government as part of a multi-national socialist partnership. The construction of the museum and the donation of artworks exemplify the dual nature of socialist countries' exchange: art and architecture were utilised as a tool of diplomacy and support as well as ideological exchange. At the same time, the art donation further facilitated the continuous process of knowledge exchange.

The architectural designs produced in the late 1960s and 1970s show few signs of uniformity. They gave birth to a new city, one that exceeded earlier Yugoslav architectural experimentations, similar examples seldom evident in the rest of the federation.[4] The creation of the brutalist urban narrative of Skopje exemplified knowledge transfers of the era, both national and international. At the same time, it illustrated the place Skopje and Yugoslavia held in the global bipolar division and the problem of the historicisation of the architecture in the so-called periphery. The brutalist structures transformed the city into a locus of cutting-edge design based on urban plans created by international and Yugoslav planners further modified to fit the local histories and vernacular motifs. Konstantinovski's Student Dormitories initiated and exemplified a new design path; the massive complex of the Dormitories illustrated the unique amalgamation of global trends, local executions and engineering feats of the Yugoslav industry.

Still, the architecture of Skopje did not receive uniform approval, and Macedonians were some of the brutalist city's harshest critics. In his 1981 article in Zagreb's *Arhitektura*, the Macedonian architect Živko Popovski outlined the development of the brutalist architectural style in Macedonia during the 1970s. Popovski – the architect of the 1973 modernist, streamlined and open-air Gradski Trgovski Centar shopping centre in Skopje (GTC) – opened his multipage treatise by acknowledging that the "results are not always in line with the wishes". He both praised the new architecture of Skopje and offered a rare critique of the lauded brutalist structures, deriding the architecture of Konstantinovski as derivative of global architectural trends of the period. Popovski's critique is centred on the fact that the architect's employment of vernacular motifs was minimal, and the massive complex is almost wholly brutalist in its design and execution. Popovski further argued that the lack of an established school of architecture in Macedonia resulted in the creation of "parallelisms in architectural expression" – the first university-level studies of architecture in Macedonia started in 1949 at the Technical Faculty – perhaps best seen in the works of Macedonian architects who have studied in the United States and were exposed to Western influences (Popovski 1981:8–14).

## Conclusion

The 1970s in Macedonia were a period of economic growth. After decades of receiving indispensable aid from wealthier Yugoslav republics, the economic tide changed, and in 1977 Macedonia "showed exceptional growth, especially in heavy industry". Although not long lasting, the unexpected financial boom allowed for the construction endeavours that characterised Skopje during this period. The Yugoslav national budget for 1978 "nearly doubled the amount of money being turned over to the three underdeveloped republics and Kosovo [Bosnia and Herzegovina, Montenegro, and Macedonia]" (Ramet 2006:268). Still, while beneficial for the construction industry, these financial peaks and investments were exceptions, and "between 1975 and 1986 Macedonia's economic position relative to the Yugoslav average declined steadily". These economic developments only further emphasise the uniqueness of the built environment of Skopje: the city was designed and constructed despite the issues that engulfed the rest of Macedonia during its time as a Yugoslav republic with extensive financial support from the Yugoslav centres of power (Ramet 2006:271). By the late 1980s, Skopje was a distinctly different city than it had been 20 years earlier. Its population grew by almost half a million, and the city spread significantly. Brutalist structures permeated the Macedonian capital, and large parts of the city had been reconstructed based on the 1965 blueprints. The buildings clad in *béton brut* exemplified the urban identity of Skopje yet were only a part of the narrative: the urbanisation and technological advancement of the Yugoslav and Macedonian construction industry and the architectural know-how were on full display in the city by the early 1990s.

The 1970s and the 1980s ushered in widespread changes in global architecture. The era of modernism, dominated by the International Style, gave way to post-modernist explorations of high-tech and organic architecture; Yugoslav architects engaged with this transition only in the 1980s. By the mid-1980s, Yugoslav post-modernists became more prominent within the larger architectural field. Just as the 1960s in Yugoslavia "brought a taste for structurally advanced design with a pervasive focus on honesty of materials and of structure" that can be seen in regional brutalist explorations, the "taste for structure gradually lost its appeal with the onset of the 1980s" (Skansi 2018:66, 71). A new architectural period unfolded congruently with a political one, and the early 1990s brought upon violent ends in both architectural explorations and in the existence of Yugoslavia.

\* \* \* \* \*

Today, the brutalist architecture of Skopje has been applied yet another layer of peripherality: a local one. The neoclassical city centre reconstruction project, *Skopje 2014*, has seen the cladding of previously modernist structures in *faux* classical elements, the city labelled the "capital of kitsch". While only a few brutalist structures were covered in the neoclassical façade of the new millennium, they were rendered an element of the past and left without the state's financial support. Students in Konstantinovski's Dormitories argue that the city "abandoned them;" the installations in the building are no longer, or barely, functioning. On its path towards the proverbial Europe – cultural, financial and geopolitical – Skopje's brutalist heritage stands as a reminder of the urban periphery of the past, best left behind, as argued by local politicians. Architectural historians, socio-cultural anthropologists, urban geographers and sociologists reject this narrative and seek to examine the brutalist and modernist history of the city. The recent proliferation in publications dealing with the topic shows this interest is only growing: the question remains as to where Skopje's urban heritage belongs in the unrelenting Western and westernised canon, as well as in the architectural space of contemporary Europe and the globalised world.

**Notes**

1 In the aftermath of the 26 July earthquake and after 85 countries from all over the globe sent aid to the demolished Macedonian capital, Skopje became known as the "city of international solitary".
2 "Brotherhood and unity", a slogan developed during the Liberation War in Yugoslavia (1941–1945) and employed by the Yugoslav communists throughout the country's existence. The slogan designated the official policy towards Yugoslav nations and national minorities and granted them equal standing before the law.
3 The *Warsaw Tigers* was comprised of modernist Polish architects Wacław Kłyszewski, Jerzy Mokrzyński and Eugeniusz Wierzbicki.
4 A notable exception is Mihajlo Mitrović's Western City Gate in Belgrade. Commonly known as Genex Tower, the building was designed in 1977.

## References

Bogoeva, E. 2018. *Skopje: Architecture as a Photographic Sculpture (1963–1990)*. Skopje: Ars Libris.

Bouzarovski, S. 2011. Skopje. *Cities* 28(3):265–77. 10.1016/j.cities.2010.05.002

Grčeva, I. 2013. The growth of Skopje and the spatial development of Macedonia 1965–2012. *International Conference on Earthquake Engineering*. Skopje, North Macedonia.

Konstantinovski, G. 2013. *Патот на еден архитект 1958–2013 / Path of an architect, 1958–2013*. Skopje: Kultura.

Lozanovska, M. 2015. Brutalism, metabolism and its American parallel: encounters in Skopje and in the architecture of Georgi Konstantinovski. *Fabrications* 25(2):152–75. 10.1080/10331867.2015.1032482

Mariotti, J., and Hess, D.B. 2021. Enlargement of apartments in socialist housing estates in Skopje under transition: the tension between individual preferences and collective action. *Journal of Housing and the Built Environment* 38:39–59. 10.1007/s10901-021-09875-4

Mrduljaš, M., and Kulić, V. 2012. *Modernism In-between: The Mediatory Architectures of Socialist Yugoslavia*. Berlin: Jovis.

Popescu, C. 2014. At the periphery of architectural history – looking at Eastern Europe. *Artl@s Bulletin* 3(1):9–17. https://docs.lib.purdue.edu/artlas/vol3/iss1/2

Popovski, Ž. 1981. O mladoj makedonskoj arhitekturi – on young Macedonian architecture. *Arhitektura* 176(7):8–16.

Ramet, S. 2006. *The Three Yugoslavias: State-building and Legitimation, 1918—2005*. Bloomington, IN: Indiana University Press.

Senior, D. 1970. *Skopje Resurgent: The Story of a United Nations Special Fund Town Planning Project*. New York: United Nations.

Skansi, L. 2018. Unity in heterogeneity: building with a taste for structure. In: Stierli, M., and Kulić, V. (Eds.), *Toward a Concrete Utopia: Architecture in Yugoslavia 1948–1980*, pp. 64–71. New York: The Museum of Modern Art.

Stierli, M., and Kulić, V. 2018. *Toward a Concrete Utopia: Architecture in Yugoslavia, 1948–1980*. New York: The Museum of Modern Art.

Tange, K. 1967. Skopje urban plan. *The Japan Architect* 130:30–69.

# 5 From reverse colonial trade to antiurbanism
## Frustrated urban renewal in Budapest, 1950–1990

*Daniel Kiss*

**Theories of modernisation, global capitalism and the Soviet world order**

In their attempt to describe the process of modernisation within societies, classical theories of modernisation of the 1950s and 1960s drew on sociological theses of Karl Marx, Emile Durkheim and Max Weber. Their dominant paradigm suggested that traditional societies will follow the development path of industrialised countries in adopting modern practices – made possible by their gradual involvement in the global market. By the same token, proponents of modernisation theory also claimed that modernised states are wealthier, more powerful, with their citizens being more likely to have access to a higher standard of living.

The economist Walt Whitman Rostow, one of modernisation theory's chief architects, argued that economic modernisation occurs in five basic stages of varying length: from a traditional society, through meeting the preconditions for take-off, the take-off itself and the drive to maturity, to the stage of high mass consumption. His "Stages of Economic Growth" (Rostow 1960) was without doubt one of the 20th century's most influential development theories. However, it was published at the height of the Cold War, and with the subtitle "A Non-Communist Manifesto". Rostow himself was fiercely anti-communist and modelled his theory solely after western capitalist countries that were in an advanced stage of their industrialisation and urbanisation. Thus, "[h]is model illustrated a desire not only to assist lower-income countries in their development process but also to assert the United States' influence over that of [Soviet] Russia" (Rogers and Gentry 2021:193).

Not long after its consolidation, the modernisation paradigm was challenged by radical economists, who stated that the development of capitalism into a world system has a geographical structure. According to their dependency theory, which emerged out of attempts to explain persistent levels of under-development in Latin America, the territorial distribution of development organises the world into centre and periphery. Thereby, traditional capitalist democracies – that were first in replacing their feudalist arrangements with urbanised and industrialised societies, modern law and bureaucracy, as well as a free market – are the most advanced in their process of modernisation and, thus, constitute the centre of global economic and political power, while less developed societies are considered peripheries that are economically dependent on the centre. Notable among scholars of dependency theory is André Gunder Frank, who studied the development of Latin American countries in the 1960s (see e.g., Frank 1967). He asserted that underdeveloped countries' joining the global market will not lead to their comprehensive development. Instead, they will become attendants to foreign capital by serving it with raw materials, cheap labour and out-of-date industrial production. Frank and the like-minded economist Raúl Prebisch (1962:1–22) argued that Rostow's model assumes a false dichotomy between traditional and modern societies and challenged his presumption that internal barriers to development would be responsible for Latin America's underdevelopment. In their view (Frank 1966:18), the region's underdevelopment was much rather the historical product of its economic relations with developed countries that have systematically kept it in a state of dependency.

The economic historian Immanuel Wallerstein went on to organise criticisms of global capitalism into a comprehensive theory, published between 1974 and 1988 as three volumes of "The Modern World System", complemented in 2011 by a fourth volume. Wallerstein asserts in these that the modern world system is based on the expansion of the global market and is distinguished from empires by its reliance on economic control of the world order by a dominating capitalist core in systemic economic and political relation to peripheral and semi-peripheral areas (Lemert 1993:426–32). Thereby, according to Wallerstein (2004:57), an endless accumulation of capital by competing agents accounts for frictions, while the core's continued practice of "unequal exchange" with the peripheries preserves the world system's asymmetry (ibid.:60).

Wallerstein (ibid.:24) adds that the capitalist regimes at the core of the global economy can maintain their domination over the periphery without direct colonial rule through the efficacy of their division of labour alone. By contrast, the Russian Empire and its successor, the Soviet Union, relied on classical colonialism (see e.g., Annus 2019:43–8). After WWII, the Soviet Union extended its imperial domain to East Central Europe, a region that was economically, culturally and politically more developed than the imperial centre. Thus, the less developed core was unable to exert any attraction on the more advanced periphery. Countries of East Central Europe, and even the Baltic states, had a wider middle class, more modern industry and more balanced urbanisation than Soviet Russia, the latter securing its domination over them by superiority of size and military force. Thus, Moscow's centre-function was not based on economic supremacy and attractivity but on dictatorial command, whereby the Soviets, in part by way of empowering local communist proxies, forced their internal structure on the annexed regions. However, in order to sustain their regime, they did not only oppress but, until they could afford, also subsidised their attached parts, for example with cheap oil and gas.[1]

The core's lack of economic superiority and the resulting "reverse colonial trade", in which the Soviet Union supplied East Central Europe with raw materials, energy, food and low-processed goods, while the latter sent back industrial products, is a substantial anomaly of the Soviet world order and will serve as the interpretive framework of this chapter. Due to this deviation from the relationship between core and periphery as known in the capitalist world, forced industrialisation was inherent in the socialist system's arrangement and resulted in shortcomings of services and consumption but also of infrastructure and housing. These, in their turn, were responsible for underurbanisation, irrational land use, shortages of space and other structural distortions characteristic to the socialist city's development (see more in Kiss 2018:20–35). Consequently, anomalies to the territorial distinction between centre and periphery surfaced in socialist urbanisation, whereby urban peripheries emerged within geographical centres. This chapter argues that while this alternative, urbanist use of the centre-periphery dichotomy is conceptually not identical to the terminology's use in Wallerstein's, Frank's and others' explanation of the world being split into centres of power and peripheral regions economically and politically dependent on these, an apparent causality exists between the two.

In order to demonstrate this nexus, this chapter studies the case of the Hungarian capital, Budapest's post-war urbanisation, focusing on the planning and execution of its urban renewal between 1950 and 1990. In doing so, it identifies four anomalies that offer partial explanations to what can be described as urban renewal's frustration in the socialist period. The featured anomalies concern the controversial territorial distribution of development in Budapest, the gradual liberalisation of property in response to the Soviet-type development path's unsustainability, tensions caused by the lack of means for executing normative plans provided by the Soviet core and impacts of knowledge transfer across the Iron Curtain, despite the core's continued efforts for intellectual isolation. This chapter introduces these internal anomalies of the Hungarian socialist system as effects of the general contradiction of core-periphery relations within the Soviet world order and, thus, also interprets the observed "antiurbanism" as a logical consequence of a less developed core, Soviet Russia, dominating a more advanced periphery, East Central Europe.

## From "antiurbanism" to "camelback urbanisation": The emergence of urban periphery in central areas of socialist Budapest

After WWII and with the consolidation of state socialism in the 1950s – like so many other areas of life – the pattern of urban planning and development in Hungary had also become significantly distinct from western practices of the time. The abolition of real self-governance and of private property in the tenement sector, the expulsion of market actors from urban development and the all-encompassing influence of the party-state have streamlined the post-war renewal of Hungarian cities. The case of Budapest was no exception. Shortages of available resources and the prioritisation of Soviet-style forced industrialisation have resulted in its war-torn central areas being only partly reconstructed, with renewal plans abandoned in favour of erecting new industrial compounds and, further on, mass housing satellites. The far-reaching consequences of this paradigm were noticeable even beyond the regime change of 1989, with a survey of Budapest's eclectic downtown neighbourhoods still registering some 450 empty plots, that is former bomb sites, in the 1990s (Kovács and Wiessner 1996). What is more, by the 1960s severe housing shortages emerged as neither the pace of renewal nor the development of new housing estates was able to keep up with the rate of industrialisation and subsequent housing demand on the part of the new proletariat.

These phenomena are linked to a general trend that the sociologists Pearse Murray and Iván Szelényi (1984) describe in their account on the development of cities in transit towards a socialist mode of production as socialism's "antiurbanism". In their interpretation, while socialism in theory is definitely urbanist, its practices appear to be rather "antiurbanist". They base their statement on the observation that in socialist states the "rate of urban growth, and especially the rate of [...] metropolitan growth, appears to be slower than the rates one can observe under similar circumstances in market capitalist economies" (Murray and Szelényi 1984:91). The economic geographer György Enyedi (1984) explains this phenomenon with Eastern Europe's belated urbanisation cycles and the asynchronic relation between the socialist regimes' centralising planning and deconcentrated spontaneous development. Szelényi and the novelist György Konrád (1971) conducted empirical research in Hungary in the 1960s and found that with the society's socialist transformation the growth of the urban population appeared to have fallen behind the expansion of the industrial population. They claimed that the proletarianisation of the formerly agricultural workforce was not followed by a comparably fast migration of this population from rural areas to urbanised centres.[2] On the contrary, administrative measures were often put in place to restrict metropolitan growth by preventing would-be migrants from settling in the major cities. In the case of Budapest this resulted in the exponential growth of villages in poorer areas of the city's agglomeration. It was relatively easy to get a new job in the growing socialist industry but almost impossible to become a legal resident of the capital. György Berkovits (1976) outlines this phenomenon and the resulting aggravated life circumstances of the worker-turned-farmers in his local history publication of Budapest's agglomeration. He describes therein the new proletariat settling under miserable living conditions beyond the fringes, being forced to commute long hours every workday, deprived of the advantages that came with being an inhabitant of a metropolitan centre.

In response to this devastating situation, from the 1960s onwards the issue of urban development in Hungary had increasingly been linked to the state's central housing programme, which commenced in 1961. Its framework was set by a 15-year housing construction scheme, with the aim of building one million homes nationwide by 1975, of which 250,000 were planned in Budapest (Kovács 2005:160). With the first precast panel housing estate erected in 1965, promising a speedy and economical way of building in large masses, the focus of the development shifted exponentially towards "green fields" in peripheral locations. Meanwhile, overwhelming areas of land in central urban areas were left as reserves for industry-related use and most of the housing stock in the city's second urban belt remained neglected and deteriorated further, resulting in a structurally weak zone between the historical centre and peripheral new towns. Szelényi (1983:148) coined the term "camelback urbanisation" for this type of slum formation and identified it as a model characteristic of the socialist city. A study by the historian Elisabeth Lichtenberger (1994:94–8) also confirms that, while slum formation in neighbourhoods along Budapest's second ring road already made its appearance in the inter-war era, the area's deterioration accelerated during socialism as the city council's Real Estate Administration Company lacked both the sufficient funds and the political will to launch the large-scale renewals that had been necessary to reverse the trends.

This observation points to what this chapter identifies as the first anomaly of socialism and its urbanisation. As the case of Budapest demonstrates, the territorialisation of slum formation to the second urban belt and, furthermore, this area's procrastinated renewal under socialism testify to "urban periphery" not coinciding with the territorial outskirts of the socialist city. The underlying controversial territorial distribution of development in Budapest is linked to the Soviet system's fundamental principles: the prioritisation of industrial production over consumption, housing and services, and the absence of the concept of land values in the evaluation of real estate.

### Reforming the "premature welfare state": Private property and market-type relations in Hungary's command economy

Despite the increasingly massive and concentrated public housing construction of the 1960s, the rapidly growing housing demand of the time remained far from being satisfied. This deficit can be contextualised in the general trend that the economist János Kornai (1997) described as the deficiency of those socialist states, amongst them Hungary, that introduced generous welfare services, for example in the form of public housing and health care systems, while the state of their economies did not allow for the allocation of sufficient resources for these measures. Kornai coined the term "premature welfare state" to characterise these unsustainable regimes.

A theoretical model of the socialist housing system also worked based on the economic model of the "premature welfare state". The exclusion of a real market, the omission of housing costs from incomes and the centralisation of all important investment decisions characterised a system in which relevant aspects of housing were meant to be under the control of state institutions. The economist József Hegedüs and the sociologist Iván Tosics (1996) claim in their account on the East European housing model that housing was intended to be a form of public service in which the private sector should not have a role either in production or in distribution. Rents were kept artificially low with the immediate consequence that the state was neither able to provide new housing at an acceptable level nor was it capable of properly maintaining tenements already in its portfolio.

The anthropologist Kathrine Verdery (1996) adds a further explanation for the dysfunctionality of state socialism's system of paternalistic redistribution. She summarises this regime as the Party's efforts to secure legitimacy and popular support by taking care of people's needs through centralising the social product and redistributing it in the form of jobs, affordable goods, free health care and education, subsidised housing and so on. "Herein lay the Party's paternalism: it acted like a father who gives handouts to the children as he sees fit" – argues Verdery (1996:25). Accordingly, the socialist mode of operation sacrificed demand and consumption in favour of production and the control of supply. In his acclaimed theory, the "Economics of Shortage", Kornai (1980, Volume A:100–4) explains this phenomenon with the constant need to hoard the means of production in order to enhance redistributive power, a "tendency further strengthened by uncertainty in the sphere of production and trade" (ibid.:101). This "expansion drive" also resulted in heavy industry having been the party-state's dominant preference at the expense of consumer industry. The aim of maintaining strong central power was better served by producing things the regime could continue to control than by giving away goods. In other words, while the socialist state claimed to satisfy people's needs, redistributing products and assets to the masses remained secondary to accumulating things at the centre.

In response to the resulting tension, the Hungarian party-state cautiously installed economic reforms in 1968, aiming at a partial marketisation of the production and distribution of goods. The "New Economic Mechanism", an all-around restructuring of the planning and commanding of the socialist economy, reduced the role of central planning, increased corporate autonomy and installed a limited price competition.[3] Kornai emphasises that the provisions were supposed to maintain or regain support for the socialist regime and had controversial results (interview with Blanchard, 1999:441). On the macroeconomic level, the welfare reforms caused high inflation, budget deficits and a growing over-demand for loans, unfavourable trade balance and surging debt and, as such, had a negative impact. However, they had positive consequences on the micro level: real property, a well-functioning legal infrastructure and a management elite and working class whose members knew more about how the market economy works. All this had contributed to making reformist states, like Hungary, more attractive for foreign investment – argues Kornai (ibid.:442). Nevertheless, while some functioning of demand and supply and of price mechanisms was acknowledged in the limited markets that came into existence as a consequence of the reforms, the behaviour of individuals and institutions was not wholly determined by these market-type relations.

With the shift towards "market socialism", the decentralisation of housing has also taken its first steps. Most importantly, private property made its reappearance in the construction of new dwellings. A 1969 Government Decree[4] awarded the right of designating properties for alienation to the territorially competent councils, and the construction of privately owned apartments and single-family houses was also made possible. Approximately 100,000 such dwelling units had been erected in Budapest between 1961 and 1980, almost twice the initial target number of 54,000.[5] Most of these apartments were built as housing co-operatives, a few in the form of condominiums realised with mortgages from the party-state's savings bank[6] or by converting income earned in the second economy into real estate, and some as single-family houses, often aided by the voluntary co-operative work of friends and family members (see e.g., Horváth 2012:108–42; Molnár 2013:71–5). Developer projects aiming at private investment instead of providing for one's own dwelling were not made possible, while the real estate market remained very limited and state-controlled, resulting in the emergence of informal practices for the exchange of property (Horváth 2012:129). Hegedüs and Tosics (1988:21) call this system, which is partly based on the informal economy, allows self-help provision in development, and introduces a limited market, the peculiar "Hungarian model of housing".

The Hungarian New Economic Mechanism's effort to respond to the severe housing shortages through measures of decentralisation, that is, by replacing parts of the welfare programme via facilitating private property, reveals a second anomaly of socialism. Despite the command economy being by and large based on the idea of collective ownership, private property played a central role in patching the holes within the socialist state's public welfare system. The share of private property and of the limited and in part informal "real estate market" successively grew in Hungary in the 1970s, coinciding with Soviet Russia increasingly shifting from subsidised energy and raw materials towards market pricing of commodities in their trade with COMECON countries.[7] Thus, the private property anomaly also finds partial explanation in the basic economic unsustainability of the Soviet world order, presented in this chapter's introduction.

**Ambitious plans from the centre meet shortage economies on the periphery: The frustration of tabula rasa type renewal in Budapest**

Within the framework of the socialist party-state, the central will prevailed in the fields of urban planning and governance. Budapest's District Councils were subordinated to the Metropolitan Council which, in its turn, was both under the supervision of the state and of the Communist Party's Budapest Committee. The Party's influence was guaranteed, among other things, by the fact that the leading officials of both the Metropolitan and District Councils were either appointed or nominated as candidates for election by the Party. The District Councils operated social infrastructures, maintained public spaces and played an important role in the management of public housing, but they had hardly any authority in the field of urban planning and policy. Instead, the Metropolitan Council's Urban Planning Department and the Budapest Urban Planning Company (BUVÁTI),[8] a planning bureau owned by the council, played the central role in urban planning. The latter prepared the city's General Urban Plans and the districts' local development regulations, as well as the preliminary studies and surveys, which served as the basis for urban development concepts in the socialist era – obviously in line with five-year plans, Party and government resolutions and other central directives concerning the development of Budapest and its subsystems. Development strategies of the 1960s and 1970s prioritised the quantitative response to the massive housing shortages of the time and only planned to follow thereafter with qualitative upgrades to existing tenements. While the socialist leadership interpreted the run-down eclectic blocks as undesirable remnants of the "bourgeois past" and, thus, cherished the idea of replacing them by modern dwellings, the envisioned large-scale demolition in a dense urban area with predominantly small households would have been rather unfavourable for the regime's already poor housing statistics. This dilemma greatly contributed to the freezing of renewal in central urban areas in the decades to follow, turning attention to extensive new construction on the city's fringes instead. Accordingly, the General Urban Plan of 1960, albeit addressing the necessity of neighbourhood renewal in the second urban belt, did not consider this a priority.

The case of Budapest's eighth district exhibits a prime example of this concern. Being the city's poorest area, the working class, the Roma and other groups from society's lower strata were overrepresented in its post-war population (Ladányi 2010:340). Moreover, it also displayed the highest density of old buildings in poor condition and of unused parcels in Budapest, as suggested by statistical data from the 1980s (Lichtenberger 1994:152). The combined effect of these was that the middle class – especially young and better-earning families, as well as intellectuals – moved out en masse in the decades of socialism. The most privileged moved to villas and large apartments on the Buda hillsides, nationalised after WWII and allocated to them by the socialist regime. Those less privileged by the bureaucratic allocation system received apartments in new mass housing neighbourhoods, while those without any privileges but possessing the means and skills to build a small house engaged in private construction on Budapest's outskirts (Csanádi and Csizmady 2010:27–32). The district's population declined sharply, by nearly a third, between 1970 and 1980, although Budapest's overall population grew by 3.7% in the same period. The community was also characterised by high ageing; at the time about 30% of its inhabitants were over 55, and the proportion of children under 15 remained well below the city-wide average (Lampel and Lampel 1998:36). This tendency was followed by the influx of poor, mostly Roma families from the 1970s on, increasing the district's social segregation even further.

It was against this challenging background that, while in practice the district's 19th-century tenements remained largely neglected throughout the post-war socialist era, plans had been repeatedly developed for their renewal. These schemes also testify to multiple significant paradigm shifts in Budapest's socialist urban planning which will be presented in the following sections. The district's detailed redevelopment plan from 1965 (see Figure 5.1) was yet to conceive a complete demolition of its most underprivileged area.

*Figure 5.1* Streetscape from the 1965 integrative renewal scheme of Budapest's eighth district (János Brenner et al., BUVÁTI).

The proposal's chief planner, János Brenner (1965:22), proposed replacing the 19th-century urban blocks with a radically transformed land-use pattern, characterised by the composition of residential towers and slabs, albeit maintaining and integrating the historical blocks in comparably good condition and preserving the street network in its original form. Altogether, BUVÁTI's scheme suggested diversifying 55 acres of dense urban tissue in a development that was planned to become the Hungarian pilot model for the socialist renewal of deteriorated downtown areas (Tomay 2007:337). In the development's first phase, planned to be carried out between 1965 and 1970, a total of 4,000 tenement flats were supposed to be torn down and replaced by 6,000 modern apartments, exclusively in prefabricated slabs of ten stories or higher.[9] Instead, a mere 192 apartments were provided by 1971 in the few newly erected buildings.

However, the 1970s have witnessed the project's rejuvenation, following a new plan by Árpád Mester, another chief planner of the city council's planning bureau (see Figure 5.2). The updating of Brenner's previous plan became necessary for three main reasons. First, new panel housing systems had been introduced, leading to a paradigm shift in the planning of housing and urban renewal. Second, by this time the building stock had decayed to an extent that the city council was moved to consider a full tabula rasa

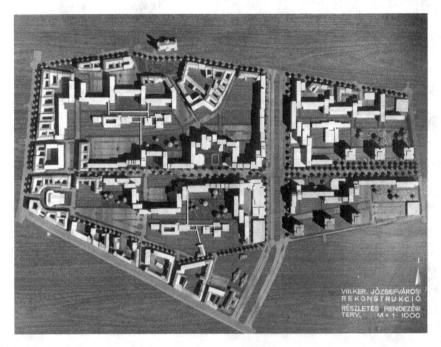

*Figure 5.2* The eighth district's tabula rasa plan from 1971 (Árpád Mester et al., BUVÁTI).

redevelopment based solely on a prefabricated mass housing scheme. Third, the requirements concerning public and social infrastructure had also been enhanced. Accordingly, the revised plan proposed larger green areas, centralised car parking, a zone dedicated to public institutions, the placement of tertiary functions in the ground floors and organised all this along a new east-west boulevard. Mester's scheme was based on the neighbourhood's almost complete demolition, conforming Soviet renewal policies of the time. A total of 6,700 apartments were suggested to be torn down and the aimed number of newly built apartments was increased to 8,000. Despite the soaring ambitions, hardly any new slabs were erected and, thus, the renewal carried on to miss its quantitative targets by wide margins.[10] The scale of the project and the proposed massive and simultaneous demolition of apartments were feasible neither financially nor logistically under the conditions of budgetary and housing shortages of the time. The clean slate renewal's consequent procrastination had a double-negative effect on the neighbourhood. Beyond resulting in the absence of newly built dwellings, it also contributed to the historic buildings' accelerating decay, as no refurbishment allowances were allocated from the central budget to tenements that occurred in the respective five-year plans as designated for demolition.

As the case of Budapest's eighth district demonstrates, tabula rasa renewal based on the normative concepts of modernist planning and prefabricated mass housing systems – both advocated throughout the Soviet sphere of interest by the political core – proved unaccomplishable under the economy of shortage. It is conspicuous that the planned developments did not fail due to the periphery's resistance of any kind but because of logistical and financial insufficiencies alone. Let's call this phenomenon the third anomaly of the socialist system. The Soviet core provided its periphery with normative plans but neither with the means, nor with the institutions required for their execution.

## Western influence on the Soviet peripheries: Paradigm shifts in heritage protection and urban renewal in Budapest

The international discourse and practice of heritage protection experienced a general change of attitudes amid the post-war reconstruction of Europe, with a shift towards the conservation of urban ensembles and even entire neighbourhoods of cultural significance instead of the previously dominant protection of single monuments (Horler 1984:55–7). Yet this paradigm shift remained largely unnoticeable in Budapest until the mid-1970s, even if the first Hungarian attempt to list culturally relevant architectural assemblages – for example, along major urban avenues and squares – dates back to the inter-war era (Sipos 2010). However, the retentive attitude was never extended to the eclectic urban blocks, neither by development programmes of the 1930s nor by respective post-war policies. This parallelism also testifies to the fact that the highly selective attitude towards the 19th-century tenement districts was not a feature exclusive to socialist planning. It was rather an approach that had already been demonstrated prior to the post-war communist takeover by architects committed to modernism. Their dominant mindset is well illustrated by an urban design survey conducted in 1953, in which 12 architects expressed their ideas about the development of Budapest. The subsequent publication (Preisich et al. 1954) interpreted the tenement blocks as mediocre records of bygone eras, not worth the resource-intensive renovation their devastating condition had necessitated. Nor was the period between 1949 and 1970 in general favourable to heritage protection. Although the preservation of national heritage was declared a priority and an institutional system for the protection of monuments existed, in reality the number of listed buildings decreased in these two decades. It was not uncommon for even monuments considered highly valuable by prominent architects to become endangered of demolition (Szívós 2010:386).

From the beginning of the 1970s, the "anti-monument" atmosphere dissolved. As a sign of this, the authority over heritage protection was decentralised in 1974, making it possible to place buildings deemed valuable under council protection, that is, to declare them protected under local jurisdiction (Román 1996:34). The change of attitudes towards the built heritage had also affected the perception of the 19th-century tenement neighbourhoods and, consequently, their preservation and rejuvenation have become a frequently discussed issue in the fields of architecture, urban planning and economy – a new credo also approved by Party officials (Tomay 2007:328–9).

With the turn towards a more retentive approach, anxiety over the condition of the building stock also grew stronger. Its deterioration had reached a stage by the mid-1970s where it was feared that if comprehensive renewal does not begin shortly, it will become impossible to save most of the dilapidated buildings later. The continuing decay in Budapest's second urban belt, aligned by the alarming demographic processes illustrated in this chapter with the case of the eighth district's population, was another major boost to the council's urban renewal plans (Cséfalvay 1994:20–3). Sociological surveys carried out by BUVÁTI on behalf of the city council (Szűcs and Varga 1975) played an important role in raising awareness of the emerging crisis and confirmed the urgency of the tenements' renewal. BUVÁTI's planners also played a key role in developing the new principles and modus operandi of urban renewal. Once it had become general opinion that the renewal of historic neighbourhoods by means of precast mass housing systems, typical for instance to East German cities, is highly unlikely to become a common practice in Budapest, they proposed the renovation of a number of blocks in the sixth and seventh districts as pilot projects. This new, block-scale strategy was adopted by the Metropolitan Council in 1978 in the form of a decree ordering the experimental renewal of a single block in the seventh district (Kovács 2005:161–2). A plan for the comprehensive renewal of Budapest's second urban belt was formulated thereafter, with BUVÁTI first preparing a preliminary study in 1984 entitled "The Rehabilitation Concept of the Inner Districts of the Capital". This was followed by the urban renewal programme being put forward by the Metropolitan Council and approved by the central government in 1986, declaring the goal for the first time that "the historical structure of the districts, together with their urban values and [...] character, must be preserved" (Cséfalvay 1994:22).

There is an apparent coincidence of this paradigm shift in the central directives concerning urban renewal with similar trends in the West, most notably with the methods of "soft renewal" and "critical reconstruction", developed during the International Building Exhibition (IBA) Berlin between 1979 and 1987. Hungarian architects and planners followed with great interest the comprised projects and professional discourses, involving some of the most appreciated international architects of the time.[11] Such appearances of Western intellectual influence in the Hungarian architecture scene were not isolated cases at the time. As the architecture historian Ákos Moravánszky (2017a:8) points out in his introduction to the trilogy "East West Central, Re-Building Europe 1950–1990", travels by architects and professional organisations between East and West had intensified from the 1970s onwards, and these encounters have also contributed to dissolving the dichotomy of the prevailing East-West bloc-thinking, bringing new perspectives into the respective discourses. Moravánszky (2017b:33, 39–40) elaborates in more detail on the emerging knowledge transfer, taking the example of study trips to Budapest by French, Finnish, Estonian, Swiss and Polish students of architecture and engineering he had helped organise. He further illustrates the exchanges with Charles Polónyi's memorable summer schools in the 1980s with Hungarian and international students, culminating in 1987 in a workshop with Alison and Peter Smithson aboard a boat on the Danube and with a 1976 special issue on Robert Venturi's work in the Budapest Architecture School's journal.

Notwithstanding the significant differences among Eastern Bloc countries in terms of their openness to Western influences, these Hungarian examples testify to knowledge transfer over the Iron Curtain having been increasingly influential in shaping policies on the Soviet peripheries, challenging the so-far hegemonic core-periphery interactions within the bloc itself. While isolation was an essential element in maintaining the Soviet world order, East Central Europe's more advanced level of modernisation and pre-war integration into the pan-European flow of goods and ideas resulted in its segregation becoming increasingly unsustainable, especially given the Soviet power centre's falling ability to compensate the dependence with economic benefits. This constitutes another, the fourth, important anomaly of socialism.

**The Soviet world order's final hours: A postmodern renewal attempt's fiasco in late socialism and thereafter**

In line with the change of attitudes in the planning of urban renewal in Hungary and under the influence of the international paradigm shift towards rehabilitation that aimed at keeping former inhabitants, the 1980s have also witnessed the evolution of an alternative renewal strategy for Budapest's eighth district. The protection of the neighbourhood's social and economic networks, as well as its built substance, was the focus of the scheme first published in 1987, however, turned into a legally binding local decree only after the regime change, in 1994.[12] This was the first plan to claim that, in the absence of sufficient renewal measures, the district was facing the formation of a metropolitan ghetto rather than mere housing tensions (Perczel 1992:90–1). Its leading authors at the Urban Research and Planning Institute (VÁTI), architect Anna Perczel and sociologist János Ladányi,[13] praised the diversity of the neighbourhood's building types as a reminiscence of former social life there, with cafés, cinemas, workers' clubs, workshops and so forth. Thus, maintaining as much as possible of the old milieu has become a central aim of the project. The planned renewal was based on a detailed value assessment and suggested retaining and refurbishing the majority of buildings. The planners' goal was to only replace the most dilapidated units, to slightly loosen up the urban structure and to implement a network of meandering green open spaces throughout the area (see Figure 5.3). In the end, however, such block-scale renewal was

*Figure 5.3* Façade sketches and masterplan from the 1987 gradualist, block-scale renewal scheme of Budapest's eighth district (Anna Perczel et al., VÁTI).

never actuated in the district, mainly due to discrepancies between physical planning and fiscal policies, the latter still being developed within the framework of five-year plans. No economic feasibility study was carried out for the proposal in the 1980s, which was not unusual at the time. The work of state-employed urban planners was limited to proposing spatial plans, whereas the task of assigning budgets to these was reserved for political authorities. This often resulted in huge gaps between normative visions and financial realities, and thus in development projects regularly getting frustrated – as visible in the case of both the 1965 and 1987 schemes.

In 1989 the Berlin Wall and the Iron Curtain fell, with the Soviet Union's disintegration following a mere two years later, also bringing the Soviet world order's existence to an abrupt end. The renewal misery in Budapest's eighth district persisted long beyond the regime change, as the plans still developed under socialism were unfit to respond to the introduction of land values, a real estate market and the following fragmentation of property in the 1990s, while the neighbourhood's bad reputation discouraged spontaneous private investment there. Thus, the renewal of Budapest's most battered urban quarter remained unaccomplished for generations, resulting in its further decay and adversely affecting its inhabitants, despite decades of its continuous and diverse planning under the centralised decision-making and ownership of socialist Hungary (see Figure 5.4).[14]

**Epilogue: Working class on the periphery**

This chapter's initial observation was that the relationship between the Soviet core and its peripheries was anomalous, and this, in turn, was responsible for further structural anomalies inherent in the socialist system's functioning. The next consideration was that the division into centre and periphery, a conception of geography and urbanism to territorialise development, is different from the term's use within economic world system models – however, the two are related. No matter if it is the controversial territorial distribution of development in Budapest, the cautious introduction of private property in Hungary's "premature welfare state", the failure of its tabula rasa urban renewal or the emerging knowledge transfer over the Iron Curtain despite the regime's efforts for isolation, all these anomalies find partial explanation in the peculiar centre-periphery relations imposed on East Central Europe by Soviet Russia, an imperial power less developed and modernised than its dependencies in the region.

In addition, the case of thwarted urban renewal in Budapest, and the adverse impacts this has had on the Hungarian working class, testifies to the fact that the anomalies described here have also prevented the socialist regime from achieving some of its most fundamental societal goals. In spite of the working class having been overrepresented in the deteriorated neighbourhoods of Budapest's second urban belt, it can be asserted that the historical tenements' continuing decay in the socialist era contributed to the segregation of working-class communities – against the key promise of Marxist doctrine,

*Figure 5.4* Frustration spanning ages: scarce remnants of socialism's ambitious but largely thwarted urban renewal programme hovering above a dilapidated 19th-century tenement in Budapest's eighth district. Photograph by the author.

the establishment of a classless, egalitarian society. What is more, in the few blocks that were affected by the renewal schemes after all, the original population was largely replaced by higher-status, older residents, representing the Party's prime social base. In their response to this phenomenon, József Hegedüs and Iván Tosics (1991) coined the term "socialist gentrification" to describe forced population exchange carried out amid urban renewal with the support of higher-level state authorities. György Konrád's and Iván Szelényi's (1969) empirical research confirmed already two decades into socialism that – despite the official propaganda on the proletariat's social rise to the position of the ruling class – their disadvantageous treatment within the administrative allocation of housing had been a general trend under socialist rule. According to Konrád and Szelényi, the new mass housing estates were predominantly inhabited by middle-class people, such as professionals, intellectuals and bureaucrats, while the bulk of the working class was nowhere to be seen there. Thus, the working class, while being central to state propaganda, remained on the social peripheries throughout socialism – a fundamental paradox of the Soviet world order.

## Notes

1 The accounting of raw materials and energy amongst countries of the Council of Mutual Economic Assistance (COMECON) began to be adjusted to world market prices in the 1970s – directly triggered by repercussions of the global oil crisis of 1973 but also as a result of the Soviet economy's general difficulties in further subsidising countries in its sphere of interest (see e.g., Kramer, 1990; Marrese and Vanous, 1983).

2 Konrád and Szelényi asserted that urban expansion fell behind industrial growth in the 1960s within the Eastern Bloc. According to them, this period of "actually existing socialism" could be best described as a process of "under-urbanisation". By this, they were referring to the regional system in which an increasing proportion of the newly proletarianised population maintained its rural residence and started commuting from villages to urban workplaces.

3 The so-called "New Economic Mechanism" was elaborated in the mid-1960s and enacted on 1 January 1968. It contained the following major changes: (1) it reduced the role of central planning and increased corporate autonomy in production and investment; (2) it liberalised pricing, allowing the price of certain products to be set in accordance with market demand; and (3) it replaced the centrally determined wage system by a flexible regime, in which companies could determine wages, albeit within certain limitations (M. Rainer 2010:40–4).

4 Government Decree 32/1969 (September 30).

5 19,504 units between 1961 and 1965; 22,507 between 1966 and 1970; 28,922 between 1971 and 1976; and 27,654 between 1976 and 1980 (Kondor and Szabó 2007:242).

Between 1966 and 1970 – a period characterised by the opposing trends of an expanding public housing sector due to the introduction of prefabricated housing systems on the one hand and growing private home ownership following the economic reforms on the other – state construction amounted to 37.5% of newly built housing, building co-ops to 10.7%, non-building organisations to 4.8%, and the private sector to 47% – the highest share of private property in housing within the Eastern Bloc (Központi Statisztikai Hivatal 1981:200).

6 National Savings Bank: Országos Takarékpénztár, abbreviated as OTP in Hungarian. Socialist Hungary's biggest financial institution, which initially dealt with the collection of deposits and the provision of loans for private individuals. From the 1970s on it additionally managed the finances of the councils, while in the 1980s also provided financial services to state and co-operative enterprises.

7 The socialist industry's unsustainability and decreasing Soviet subsidies within the COMECON resulted by the end of the 1970s in severe current account deficits and the depletion of foreign exchange reserves that prompted Hungary to join the World Bank and the International Monetary Fund (IMF) and to immediately borrow from the latter. Hungary submitted its application in 1981, which was approved by both institutions in 1982. Beyond its positive economic impacts, the move also resulted in Hungary's decreased isolation from the West. Trade and cultural relations improved especially with West Germany, also triggered by their "Wandel durch Handel" (change through trade) policy.

8 Budapesti Városépítési Tervező Iroda, abbreviated as BUVÁTI in Hungarian, also known as Budapesti Városépítési Tervező Vállalat (BVTV).

9 Target numbers of the City Council's Executive Committee until the end of the third five-year plan, 1963.

10 Between 1960 and 1970, merely 665 apartments were newly built in the eighth district overall, instead of the planned 18,000 (Kondor and Szabó 2007:253).

11 Magyar Építőművészet [Hungarian Architecture], the Journal of the Association of Hungarian Architects, reviewed projects from the IBA Berlin in the early 1980s, for example, Robert Krier's Ritterstrasse block in its fourth issue of 1981.
12 Decree 32/1994 (5 July) of the Municipality of the Eighth District regarding the Detailed Development Plan for the Northern Quarter of Central Józsefváros.
13 Perczel and Ladányi were colleagues and followed a then-novel transdisciplinary planning method at the Urban Research and Planning Institute (Városépítési Tudományos és Tervező Intézet, abbreviated as VÁTI in Hungarian), a progressive think-tank that was also Iván Szelényi's intellectual hinterland prior to his emigration to the United States in 1975.
14 Finally, in the year 2000, the eighth district's municipality officially abandoned its former gradualist renewal plan and announced the "Corvin Quarter", the biggest tabula rasa urban development in post-socialist Budapest's history, involving 22 acres of redevelopment area. Its execution commenced in public-private partnership in 2005, bringing the century-long procrastination of the neighbourhood's renewal to an end.

**References**

Annus, E. 2019. *Soviet Postcolonial Studies. A View from the Western Borderlands*. London and New York: Routledge.
Berkovits, Gy. 1976. *Világváros határában* [On the Margins of a Metropolis]. Budapest: Szépirodalmi Könyvkiadó.
Blanchard, O. 1999. An interview with János Kornai. *Macroeconomic Dynamics* 3(3):427–250.
Brenner, J. 1965. A Józsefváros egy részének szanálási terve [Plans for the partial rehabilitation of Józsefváros]. *Városépítés* 2(2):22–27.
Csanádi, G., and Csizmady, A. 2010. Budapest térbeni-társadalmi szerkezetének változásai [Changes in the Spatial and Social Structure of Budapest]. In: Csanádi, G. et al. (Eds.), *Város, tervező, társadalom*, pp. 11–37. Budapest: Sík Kiadó.
Cséfalvay, Z. 1994. Stadtverfall und Stadterneuerung im Staatskapitalismus [Urban Decay and Renewal in State Capitalism]. In: Lichtenberger, E., Cséfalvay, Z., and Paal, M. (Eds.), *Stadtverfall und Stadterneuerung in Budapest*, pp. 20–30. Vienna: Verlag der Österreichischen Akademie der Wissenschaften.
Enyedi, Gy. 1984. *Az urbanizációs ciklus és a magyar településhálózat átalakulása* [Urbanisation Cycles and the Transformation of the Hungarian Settlement Network]. Budapest: Akadémiai Kiadó.
Frank, A.G. 1966. The development of underdevelopment. *Monthly Review* 18(4): 17–31.
Frank, A.G. 1967. *Capitalism and Underdevelopment in Latin America: Historical Studies of Chile and Brasil*. New York: Monthly Review Press.
Hegedüs, J., and Tosics, I. 1988. Is there a Hungarian model of housing system? Changes in the Hungarian housing system in the past three decades, paper presented at the 7th Reunion of CIB W 69 in Santo Kiriko, Bulgaria.
Hegedüs, J., and Tosics, I. 1991. Gentrification in Eastern Europe: The case of Budapest. In: van Weesep, J., and Musterd, S. (Eds.), *Urban Housing for the Better-Off: Gentrification in Europe*, pp. 124–136. Utrecht: Stedelijke Netwerken.
Hegedüs, J., and Tosics, I. 1996. Disintegration of the East-European housing model. In: Clapham, D. et al. (Eds.), *Housing Privatization in Eastern Europe*. Westport, Connecticut and London: Greenwood Press.

Horler, M. 1984. *A műemlékvédelmi gondolat kialakulása Európában* [The Evolution of the Idea of Monument Protection in Europe]. Budapest: Budapesti Műszaki Egyetem Mérnöki Továbbképző Intézet, pp. 55–57.

Horváth, S. 2012. *Két emelet boldogság. Mindennapi szociálpolitika Budapesten a Kádár-korban* [Two Floors of Happiness. Everyday Social Policy in Budapest during the Kádár Era]. Budapest: Napvilág Kiadó.

Kiss, D. 2018. The socialist urban legacy narrative: tensions of socialist urbanization. In Kiss, D., *Modeling Post-Socialist Urbanization. The Case of Budapest.* pp. 19–54. Basel: Birkhäuser Verlag.

Kondor, A. Cs., and Szabó, B. 2007. A lakáspolitika hatása Budapest városszerkezetére az 1960-as és az 1970-es években [The impact of housing policy on the urban structure of Budapest in the 1960s and 1970s]. *Földrajzi Értesítő* 56(3-4):237–269.

Konrád, Gy., and Szelényi, I. 1969. *Az új lakótelepek szociológiai problémái* [Sociological Problems of New Mass Housing Estates]. Budapest: Akadémiai Kiadó.

Konrád, Gy., and Szelényi, I. 1971. A késleltetett városfejlődés társadalmi konfliktusai [Social conflicts of delayed urbanisation]. *Valóság* 14(12):19–35.

Kornai, J. 1980. *Economics of Shortage.* Amsterdam, New York, Oxford: North-Holland Publishing Company.

Kornai, J. 1997. The reform of the welfare state and public opinion. *American Economic Review* 87(2):339–343.

Kovács, Z. 2005. A városrehabilitáció eredményei és korlátai Budapesten [Achievements and limitations of urban renewal in Budapest]. In: Egedy, T. (Ed.), *Városrehabilitáció és Társadalom,* pp. 159–174. Budapest: MTA Földrajztudományi Kutatóintézet.

Kovács, Z., and Wiessner, R. 1996. A lakáspiac átalakulásának főbb jellemzői és városszerkezeti következményei Budapest belső városrészeiben [The main characteristics of the housing market's transformation and its consequences to urban structure in Budapest's central districts]. In: Dövényi, Z. (Ed.), *Tér-Gazdaság-Társadalom,* pp. 29–48. Budapest: MTA Földrajztudományi Kutatóintézet.

Központi Statisztikai Hivatal [Central Statistical Office of Hungary] 1981. *Statistical Pocketbook of Hungary 1981.* Budapest: Publishing House for Economics and Law.

Kramer, J.M. 1990. *The Energy Gap in Eastern Europe.* Toronto: Lexington Books.

Ladányi, J. 2010. Gondolatok a Középső-Józsefváros rehabilitációjának társadalmi összefüggéseiről [Thoughts on the Societal Context of Renewal in Middle-Józsefváros]. In Ladányi, J. (Ed.), *Egyenlőtlenségek, redisztribúció, szociálpolitika. Válogatott tanulmányok 1974–2010,* pp. 334–348. Budapest: Új Mandátum Kiadó.

Lampel, É., and Lampel, M. 1998. *Pesti bérházsors. Várospolitika, városrehabilitáció* [The Fate of Tenements in Pest. Urban Policy and Renewal]. Budapest: Argumentum Kiadó.

Lemert, C. 1993. Immanuel Wallerstein. In: Lemert, C. (Ed.), *Social Theory: The Multicultural and Classical Readings,* pp. 398–405. Boulder: Westview Press.

Lichtenberger E., Cséfalvay, Z., and Paal, M. 1994. *Stadtverfall und Stadterneuerung in Budapest* [Urban Decay and Renewal in Budapest]. Vienna: Verlag der Österreichischen Akademie der Wissenschaften.

M. Rainer, J. 2010. *Magyarország története. A Kádár-korszak 1956–1989* [Hungary's History. The Kádár Era 1956–1989]. Budapest: Kossuth Kiadó.

Marrese, M., and Vanous, J. 1983. *Soviet Subsidisation of Trade with Eastern Europe: A Soviet Perspective.* Berkeley: Institute of International Studies, University of California.

Molnár, V. 2013. *Building the State: Architecture, Politics, and State Formation in Post-War Central Europe*. London and New York: Routledge.

Moravánszky, Á. 2017a. Foreword. East west central: re-building Europe. In: Moravánszky, Á. et al. (Eds.). *East West Central. Re-Building Europe 1950–1990*. pp. 7–11. Basel: Birkhäuser Verlag.

Moravánszky, Á. 2017b. Piercing the wall: east-west encounters in architecture, 1970–1990. In: Moravánszky, Á., and Lange, T. (Eds.), *Re-Framing Identites. Architecture's Turn to History, 1970–1990. East West Central, Re-Building Europe 1950–1990*, Volume 3. pp. 27–43. Basel: Birkhäuser Verlag.

Murray, P., and Szelényi, I. 1984. The city in the transition to socialism. *International Journal of Urban and Regional Research* 8(1):90–107.

Perczel, A. 1992. A Közép-Józsefváros északi területére készülő részletes rendezési terv programja [Program of the Detailed Development Plan of Central-Józsefváros' northern part]. *Tér és Társadalom* 6(3-4):89–162.

Prebisch, R. 1962. The economic development of Latin America and its principal problems. *Economic Bulletin for Latin America* 7(1):1–22.

Preisich, G., Sós, A., and Brenner, J. 1954. *Budapest városépítészeti kérdései* [Urban Design Issues of Budapest]. Budapest: Építésügyi Kiadó.

Rogers, S., and Gentry, S. 2021. *Economic Development and Planning*. London: ED-Tech Press.

Román, A. 1996. *Műemlék, építészeti örökség, város. Válogatás (cikkek és előadások), 1970–1995* [Monuments, Architectural Heritage, and the City. Selection (articles and lectures), 1970–1995]. Budapest: Országos Műemlékvédelmi Hivatal.

Rostow, W.W. 1960. *The Stages of Economic Growth: A Non-Communist Manifesto*. London, New York: Cambridge University Press.

Sipos, A. 2010. "Örökségvédelmi" szempontok a budapesti várostervezésben a 20. sz. első felében [Aspects of Heritage Protection in Budapest's Urban Planning in the First Half of the 20th century]. In: Bódy, Zs., Horváth, S., and Valuch, T. (Eds.), *Megtalálható-e a múlt? Tanulmányok Gyáni Gábor tiszteletére*, pp. 520–529. Budapest: Argumentum Kiadó.

Szelényi, I. 1983. *Urban Inequalities under State Socialism*. Oxford: Oxford University Press.

Szívós, E. 2010. Terhes örökség: Budapest történelmi lakónegyedeinek problémája a Kádár-korszakban a Klauzál tér és környéke példáján [Burdensome Legacy: The Problem of Budapest's Historic Residential Neighbourhoods in the Kádár Era, on the Example of Klauzál Square and Its Surroundings], In: Varga, L. Á. et al. (Eds.), *Urbs, Magyar várostörténeti évkönyv*, Vol. 5. pp. 381–402. Budapest: BFL.

Szűcs, I., and Balázsné Varga, M. 1975. *Budapesti településszociológiai vizsgálatok – Belső-Erzsébetváros vizsgálata* [Studies of Budapest's Urban Sociology – The Case of Inner-Erzsébetváros]. Budapest: BUVÁTI.

Tomay, K. 2007. Józsefváros és Ferencváros – két rehabilitációs kísérlet a fővárosban [Józsefváros and Ferencváros – Two Renewal Attempts in the Capital]. In: Varga, L. Á. et al. (Eds.), *Urbs, Magyar várostörténeti évkönyv*, Vol. 2. pp. 323–357. Budapest: BFL.

Verdery, K. 1996. What was socialism, and why did it fall? In: Verdery, K. (Ed.), *What Was Socialism, and What Comes Next?*, pp. 19–38. Princeton: Princeton University Press.

Wallerstein, I.M. 2004. *World-Systems Analysis: An Introduction*. Durham and London: Duke University Press.

# Part II
# Architects and urban planners in the socialist city
## Roles and positions in the periphery

# 6 Passive agents or genuine facilitators of citizen participation?
## The role of urban planners under the Yugoslav self-management socialism

*Ana Perić and Mina Blagojević*

## Introduction

Socialist Yugoslavia refrained from the polarisation provoked by the Cold War. Besides turning back to the war ally of the Union of Soviet Socialist Republics (USSR) in 1948 to take the leading position in the non-aligned movement in 1961, Yugoslavia followed a distinct path of socialism known as self-management, an emancipatory project in pursuit of a democratic socialist society. As such, Yugoslavia was assigned different roles and attributes: for sure, it was a melting pot of criticism (from both East and West); more positive prospects saw it as a hybrid between East and West; inevitably, Yugoslavia was condemned to be somewhat distanced from both power centres, thus being a periphery to both East and West. Though the periphery is challenging to define due to the heterogeneity of the countries forming it (Becker et al. 2010), during the Cold War, southeast Europe (SEE) has mainly been considered a periphery to the western world (Göler 2005). Despite the existence of the so-called Western European peripheral countries, the absence of capitalism was considered the most influential parameter for diversifying SEE from the West (Bohle and Greskovits 2012; Bohle 2018). However, due to unstable political relations between the Soviet Union and Yugoslavia, the latter was considered detached from the communist ideology, too. This was particularly seen in the architectural and planning discourse, which after 1948 was informed almost exclusively by Western sources, while references to the communist bloc became exceedingly rare (Kulić 2009). Although nowadays the so-called Western Balkans region (that largely coincides with the former Yugoslavia) is considered to be a "super-periphery" (Bartlett and Prica 2013), in the Cold War period, Yugoslavia "was softening the contrast between socialism and capitalism, between the planned economy and the free market, and between liberal democracy and the 'dictatorship of the proletariat'" (Kulić 2009:129).

Against such a background, Yugoslavia emerged as a testbed where the "third way" was searched for. Development of the Yugoslav "third way" officially started after the political expulsion of Yugoslavia from the Cominform in 1948, which inevitably led to a distinctive economic restructuring, too.

DOI: 10.4324/9781003327592-9
This chapter has been made available under a CC-BY-NC-SA 4.0 license.

For example, while the centrally planned economy (with the five-year plans) was a specificity of the communist Eastern Bloc for decades after WWII, e.g., the Soviet Union experienced the changes of the socio-political-economic system just after the *"glasnost"* and *"perestroika"* initiatives in 1985 (Grava 1993; Golubchikov 2004), the centrally planned economy in Yugoslavia lasted only until 1950 (Nedović-Budić and Cavrić 2006). Shortly after, in 1953, Yugoslavia introduced the self-management socialism as the main tool of economic liberalisation (Dawson 1987; Liotta 2001). As a distinctive Yugoslav feature in comparison to other countries behind the Iron Curtain, self-management meant societal ownership over the means of production aimed to prevent the concentration of power in the hands of state bureaucracy and distribute it to the "working people" (Lydall 1989). From a practical point of view, such an "industrial democracy" (Ramet 1995) introduced a number of instruments (e.g., self-management arrangements) aimed at coordinating the interaction among numerous administrative bodies and individual enterprises. From a more abstract perspective, the goal of the socialist evolution was to eliminate the very existence of the state as a condition of ultimate democracy, making the self-management a tool against bureaucratic dogmatic communism and uncontrolled speculative capitalism (Ignjatović 2012). As a result, the "market socialism", i.e., the free-market principles introduced into a state-controlled economy, facilitated massive housing construction and the proliferation of educational, scientific and cultural activities (Zukin 1975).

To support the self-management model and triggered by the internal tensions among the Yugoslav republics over the federal administrative level as the key decision-making body, the political and administrative decentralisation started in 1965 and continued over the next two decades (Vujošević and Nedović-Budić 2006). Hence, the Yugoslav socio-economic planning included not only the previously mentioned self-management approach but also the so-called societal planning (Dabović et al. 2019; Blagojević and Perić 2023). The main units in charge of the self-management planning were basic organisations of associated labour (BOALs) (*osnovne organizacije udruženog rada*) and self-managed interest-driven communities (SICs) (*samoupravne interesne zajednice*), while various socio-political communities (*društveno-političke zajednice*) – from the federation to municipalities/communes (*opštine*) and local communes (*mesne zajednice*) as constitutive elements of a commune – were crucial for societal development.

Although self-management certainly failed in eliminating social inequalities, many members of the middle strata could prosper by virtue of their competence (Zukin 1975). Intellectuals particularly enjoyed a relatively high level of cultural autonomy and international mobility (Jovanović and Kulić 2018; Mrduljaš 2018). In the Yugoslav socialist experiment, urban and spatial planning played a key role, contributing significantly to societal emancipation, modernisation and welfare. Rather than a mere tool of economic growth and industrialisation (as under the centrally planned economy), urbanisation was instrumentalised in pursuit of a higher interest:

establishing a self-management socialism. However, as self-management was relatively short-lived for a genuinely democratic political culture to be developed, there is a significant gap between the profound self-management narrative, revolving around the ideas of political decentralisation and citizen participation, and their practical implementation. Accordingly, when translating the self-management narrative into the urban and spatial planning discourse, a body of literature focuses on the prominent urban planners and their core ideas and principles aligned to the self-management model (Blagojević 2007; Mrduljaš and Kulić 2012; Le Normand 2014; Kulić 2014). However, evidence on their implementation into planning processes is rare.

To address such a gap, this research critically examines the planners' pursuit for citizen engagement under the self-management socialism. This is considered valid as, on the one hand, planners enjoyed freedom in self-management conditions in terms of organisational aspects and the content of their work (Mrduljaš 2018). On the other hand, since self-management was imposed from the highest political tiers, some authors question the role of planners as independent mediators in a seemingly conflict-free, socialist society revolving around the common interest as the fundamental societal value (Blagojević and Perić 2023). To tackle such a dichotomy through the lens of citizen participation in urban planning, this chapter elucidates planners' role under the self-management socialism. In other words, did planners act only as technical executors (of the high-level political goals and visions) or as active agents in pursuing citizen participation (and fostering local community needs)?

The chapter is structured as follows: after a brief introductory section, the critical features of Yugoslav socio-economic, physical and urban planning are briefly discussed to elucidate the norms within which urban planners operated. To situate the narrative beyond the national borders and official instruments, the next section briefly reflects upon the core international influences and domestic planning discourse. Both sections serve to set the scene, i.e., to depict the socio-spatial circumstances and main planning topics, approaches and mechanisms the Yugoslav planners dealt with. After a brief methodological part, the central section presents the results of the analysis of planners' role in affecting citizen participation as the core mechanism of socialist planning. The conclusion puts the planners' pursuit for citizen participation into the context of socialist self-management, also blurring the definitions of East and West.

## Setting the scene: Yugoslav socio-economic planning, physical planning and urban planning

To understand the nature of urban and spatial planning in socialist Yugoslavia, it is helpful first to observe the broader social development of the state led by a specific political ideology different from both the mainstreams of West and East. The primary legislative documents that resulted from the political paradigm during the Cold War and their core substantive and procedural features are given in Table 6.1.

Table 6.1 Timeline of key federal (Yugoslav) and national (Serbian) legislative documents and their main substantive and procedural features

| Year | Legislative documents | Main substantive (S) and procedural (P) features |
|---|---|---|
| 1953 | Constitution of the Federal People's Republic of Yugoslavia | S: self-management socialism<br>P: societal agreements, self-management arrangements |
| 1961 | Act on Urban and Regional Spatial Planning of the People's Republic of Serbia | S: citizen participation as societal support and plan verification<br>P: public discussion |
| 1963 | Constitution of the Socialist Federal Republic of Yugoslavia | S: commune as a territorial and political unit<br>P: bottom-up participatory approach to policy- and decision-making |
| 1974 | Constitution of the Socialist Federal Republic of Yugoslavia | S: societal ownership; more democratic re-distribution of power (decentralisation) in the process of policy-making; strengthening of the role of the local commune<br>P: advanced mechanisms of obligatory public participation |
| 1974 | Act on Planning and Spatial Arrangement of the Socialist Republic of Serbia | S: early involvement of public (comment possible throughout the entire phase and not only in the final phase of policy-making)<br>P: public viewing, public consultation |
| 1976 | Act on the Foundations of the System of Societal Planning and the Societal Plan of Yugoslavia | S: integration of physical planning into socio-economic planning<br>P: agreement on the plan's foundations |
| 1985 | Act on Planning and Spatial Arrangement of the Socialist Republic of Serbia | S: coordination and integration of plans and policies<br>P: expert debate on a draft plan<br>P: expert debate on a draft plan |

Source: Authors.

Postulated as the pillar of social governance by the Yugoslav Constitution of 1953 (OG FPRY 3/1953), the original intention of self-management was to replace the state bureaucracy with empowered workers at the helm of Yugoslav enterprises, thus establishing workplace democracy focusing on leadership development and continuous learning among all employees (Lynn et al. 2012). Gradually, self-management was meant to spread over all segments of society, transitioning from workers' self-management to societal self-management or self-government (Zukin 1975). Problems related to diversity and social and economic heterogeneity among the federal republics were tackled through administrative decentralisation of the federal state, in which the commune (municipality), as the essential socio-political unit, played a critical role (Fisher 1964). Like an enterprise, communes were supposed to raise their funds, set their budget and provide their residents with various social services (Zukin 1975).

Yugoslav socio-economic planning included societal planning and self-management planning to forecast social and economic developments and their interdependence. More precisely, socio-economic planning was a social relationship between, on the one hand, socio-political communities at various administrative levels (from municipality to federation) in charge of societal development and, on the other hand, BOALs in different sectors and governmental levels, responsible for overall production and consumption. The main instruments of each institution were societal agreements and self-management arrangements, respectively, and they were mutually coordinated by the principle of "cross-acceptance" (Dabović et al. 2019; Blagojević and Perić 2023). Under such circumstances, urban planning was perceived only as physical planning, i.e., a tool to support socio-economic development and ensure the rational use of resources through "top-down" allocation (Perić 2020).

Towards the end of the 1950s, the role of physical planning in societal development was challenged. At a conference in Arandjelovac in 1957, professionals (architects, geographers, engineers and sociologists) gathered from all Yugoslav republics agreed upon a need for a new discipline that should become an integral part of the socio-economic planning system (Nedović-Budić and Cavrić 2006). The idea was to enable cross-sectoral coordination in the spatial development and establish the profession of an urban and regional planner. Accordingly, the nature of planning shifted from the physical planning towards the so-called integrated and comprehensive planning, attending not only to the multidisciplinarity as the fundamental norm but also to the collaboration of planners with citizens (Dabović et al. 2019). To implement such visions, the Serbian Act on Urban and Regional Spatial Planning (OG PRS 47/1961) introduced the instrument of public participation as societal support in the process of verifying the planning documents. Furthermore, the Yugoslav Constitution of 1963 (OG SFRY 14/1963) identified the commune not only as the basic territorial but also socio-political unit in which self-interests and common interests were to be aligned with the public interest.

During the 1970s, several regulatory instruments addressed the way of spatial development decision-making. The 1974 Constitution (OG SFRY 9/1974) facilitated administrative and political decentralisation to enable workers, in narrow terms, and citizens, more generally, to achieve some common interests and needs. Through local communes, the role of technocratic and administrative structures was diminished in favour of growing citizens' impact on their immediate environment. Furthermore, local communes were encouraged to collaborate with BOALs and SICs, as the main self-management units, as well as with the socio-political communities at higher administrative tiers to, hence, become the conveyors of the broader developmental goals both horizontally and vertically (Blagojević and Perić 2023).

Adoption of the 1974 Constitution was followed by another set of legal acts concerning spatial and urban planning in all republics, where all the relevant components of socio-economic, environmental and physical development were considered. Local communes became the leading planning and implementation authorities that enabled the inclusion of the civil sector in the decision-making process using negotiation and consensus-building (Maričić et al. 2018; Perić 2020). To strengthen the exchange between a local community and planners, the Serbian Act on Planning and Spatial Arrangement (OG SRS 19/1974) introduced regular public inspection and public consultation on draft plans. In doing so, planning was considered a right and obligation of the working class, and local communes were envisioned as communities of people (Kardelj 1979). Furthermore, the Act on the Foundations of the System of Societal Planning and the Societal Plan of Yugoslavia (OG SFRY 46/1976) suggested the integration of physical planning into socio-economic planning by establishing the instrument of "agreement on the plan's foundation" as a tool to improve collaboration among professionals, local political representatives and the public, and cooperation among bodies at various administrative levels. Finally, according to the Serbian Act on Planning and Spatial Arrangement (OG SRS 27/1985), the operationalisation of the idea of horizontal, vertical and multi-sectorial cooperation should be achieved through the integration of plans and policies, as well as by introducing the instrument of expert debate on a draft plan (Vujošević and Nedović-Budić 2006).

**Setting the scene: international planning ideas and domestic planning discourse**

The previous overview gives valuable insight into implementing the main ideological narrative into the Yugoslav constitutions and urban and spatial planning legal frameworks. Nevertheless, as Yugoslavia differed from the countries behind the Iron Curtain in terms that it was more exposed to international influences, to properly grasp the nature of Yugoslav socialist urban and spatial planning, it is important to note how and to what extent the dominant foreign ideas and principles were accepted in the domestic planning discourse. The most significant international and national planning events and policies, and their central ideas classified into substantive and procedural categories, are briefly indicated in Table 6.2.

Early after WWII, Yugoslavia (re)started its engagement in some of the most influential international networks: in 1950, Yugoslavia joined the International Union of Architects (UIA), in 1953, it re-joined the International Congress of Modern Architecture (CIAM) and in 1960, it joined International Federation of Housing and Town Planning (IFHTP). Due to all this networking, the planning system in Yugoslavia during the socialist era evolved

Table 6.2 The overview of main substantive and procedural aspects of the planning process in leading international and national events and policies

| Year | International and national events and policies | Main substantive (S) and procedural (P) features |
|---|---|---|
| 1961 | IFHTP Congress, Paris | S: neighbourhood unit<br>P: scientific-based conceptual foundations |
| 1962 | 10th Conference of the Association of Urban Planners of Yugoslavia | S: urbanism as a societal agency<br>P: citizens as informed agents in public debates |
| 1971 | IFHTP Congress, Belgrade | P: multistakeholder cooperation; decentralised government |
| 1972 | Belgrade Master Plan | S: public consultation<br>P: interdisciplinarity, formal and informal collaboration, transparency; sociological survey; public discussion of a draft plan; extensive public informing (exhibition, visual presentations, specialised publications, information in daily newspapers) |
| 1973 | IFHTP Congress, Copenhagen | P: involvement of multiple actors; symbiosis between planners and local administration |
| 1974 | International Planning Seminar ("U 73"), Ljubljana | P: citizen participation as an alternative to urban design; integration of rational and irrational input |
| 1976 | Vancouver Declaration (UN) | S: dynamic incorporation of people in the social life<br>P: a cooperative effort of people and their governments; providing information in clear and meaningful language; two-way flow of information |
| 1980 | Third Meeting of Planners and Urbanists of Yugoslavia | S: protection of municipalities; community cohesion<br>P: genuine citizens' inputs; rising political awareness |
| 1981 | UIA Congress, Warsaw Warsaw Declaration of Architects | S: overcoming professional blind-mindedness<br>P: planners as equal participants in collective endeavours; genuine democratisation of urban development |
| 1982 | Conference of the University of Belgrade and the Centre for Marxism of the League of Communists of Yugoslavia | S: planning as a process<br>P: inclusive decision-making (self-interests of heterogenous public beyond technical rationality); absence of technical jargon; design competitions (alternative proposals) |

Source: Authors based on Blagojević and Perić 2023.

through synthetic innovation and selective borrowing, primarily from the West (Nedović-Budić and Cavrić 2006). Western planning principles, listed in the Athens Charter, were dominantly influential in the 1950s due to the professional relations that some leading Yugoslav architects established with CIAM (Jovanović and Kulić 2018). Notably, CIAM X was held in Dubrovnik in 1956. For example, Athens Charter's "functional city model" was widely adopted among planning authorities as a suitable tool for catching up with the rapid urbanisation process and ever-increasing demands for housing. However, throughout the 1960s, criticism against rationalist "big schemes" started to evolve from social science and architecture perspectives, leading to the emergence of alternative urbanistic concepts (Kulić 2014; Le Normand 2014) and, hence, placing more emphasis on planning as a social practice.

Similarly to the international experiences, pluralism and diversity of critical and theoretical thought were institutionally promoted, aiming at a continuous advancement of Yugoslav planning practice (Kulić 2014). Interdisciplinary collaboration, as well as connecting research and practice, was standard in many planning and design institutions (Mrduljaš 2018), while professional organisations and associations at various scales (from federal to local) flourished (Nedović-Budić and Cavrić 2006). Since the 1960s, the Association of Urban Planners of Yugoslavia (AUPY) served as an instrument to develop international connections, as Yugoslav delegates regularly participated in international architectural and planning congresses (Perišić 1965; Stupar 2015). Internally, AUPY was oriented towards revising certain theoretical foundations and planning practices, considering the general social development of Yugoslavia (Bjelikov 1962b). In general, professional conferences and symposia served as channels for rethinking the role of socialism within various scientific and professional fields (Martinović 2020).

However, the main difference between international ideas and domestic discourse was in different viewpoints and, hence, priorities given to the importance of the planning procedures on the one hand and the methods for improving the planning practice on the other (Blagojević and Perić 2023). For example, the focus of the discourse in the national reports was on public discussions and consultations as a tool to increase citizen participation and diminish the dominant role of planners as professionals. International declarations, on the contrary, focused more on scientifically proven methods (e.g., surveys) that foster true feedback from the locals in creating planning solutions. The flaws in the loosely defined planning procedures were inevitably seen in the practice of creating planning instruments (as shown in the central part of this chapter).

**Methodological approach**

To contribute to the discussion on how self-management socialism influenced professional thinking and the practice of citizen participation, the research attends to the roles, viewpoints and actions of participants in planning processes. As professional journals are considered a tool to disseminate key information among the authorities, professionals and a wider public (Blagojević and Perić 2023), the data were collected through the archival research of the two most influential Yugoslav professional journals – *Urbanizam Beograda*, published by the Belgrade Urban Planning Institute, and *Arhitektura-Urbanizam*, a publication of the Serbian Urban Planners Association. In selecting the articles, the focus was on those prepared as critical reflections on the socialist self-management urban planning theory and practice, and, hence, addressing the broadly used concepts such as local commune – considered both the object and subject of planning, participation – considered the key mechanism of socialist planning, and interests – public, common and self-interest as the triggers of any planning activity, thus elucidating the roles of different actors in planning processes. The articles cover the period between 1961 and 1982 to secure the representativeness of planners' perspectives as the professional feedback in this period was often inspired by significant international networking events as well as formal decisions regarding the planning system and urban development, e.g., adoption of legal reforms or major plans. The professional backgrounds of the authors were notably diverse: architects, engineers, geographers, sociologists, economists, lawyers, archaeologists, etc.

**Planners in the pursuit of citizen participation: critical reflections**

At the 11th AUPY Conference (1963), a broad consensus was reached that the commune, being a unit of socio-economic planning closest to urban settings, should be accepted as an object of spatial planning (Bjelikov 1963). Zooming into the urban level, instead of large-scale urban schemes dogmatically dedicated to functionalist principles, a more sensitive, human-centred and small-scale approach was sought, the one that nurtures local specificities, memory and atmospheres (Janković 1969; Mutnjaković 1964; Radović 1964). Seen as a fundamental organisational and spatial module for a meaningful co-existence of citizens, where the sense of emotional security and belonging should be developed, the local commune was an essential topic of urban planning (Figure 6.1).

Furthermore, the calls for a more interdisciplinary effort to conceive a unified, systematic approach and establish scientific methodology in tackling the challenges of the local commune were typical for the period of the early 1960s. Namely, the deficiency of adequate studies led to arbitrary and inconsistent approaches and uncritical replications of international concepts and practices (Bjeličić 1962; Maksimović 1963). The gaps between the "static" visions of spatial planners (which were, at the time, mainly architects) and the objective possibilities of the society were to be bridged by the elaboration of dynamic studies regarding socio-economic trends (Bjelikov 1962a).

*Figure 6.1* Centre of the first local commune in New Belgrade.
Source: Arhitektura-Urbanizam 72–73 (1975).

The constitutional reform of 1963 saw the commune not only as a basic territorial unit but also as a socio-political community where common and self-interests should be aligned with the public interest. Accordingly, the planning community reached a consensus on understanding urbanism as a social activity that involves the broad public in decision-making processes (Bjelikov 1962a). Nevertheless, it was challenging to implement the principles of "planning as a societal practice" due to the relatively low public awareness about the possibility of actively changing the environment they were directly living in (Bjelikov 1963). Furthermore, urban development issues were not sufficiently and adequately communicated to the masses, giving way to the misuse of power by individuals. Hence, it was stressed that the popularisation of urbanism ought to take place using all forms of public informing (Bjelikov 1962a) and by introducing public debate (Perišić 1965).

Since the enactment of the 1974 Constitution, as the primary cells of the self-management society, local communes were increasingly regarded as crucial to enabling decentralised, organically developing urban structure instead of the alienation and dehumanisation of the rapidly growing urban environment (Jakšić 1978; Krstanovski 1977; Tomić 1980). Accordingly, discussions regarding the social role and tasks of urban planning intensified. The fundamental goal of the reforms was seen as ensuring the redistribution of power in decision-making on human environment in a more democratic way, i.e., avoiding power concentration in techno-bureaucratic structures. As constitutional reforms elaborated the rights and responsibilities of different actors in the self-management planning system, planners felt responsible to rethink their roles and revising their methodologies (Bojović 1976; Đorđević 1974; Vasić 1976). As often stated, the very activity of planning was not clearly defined, and different sectors and levels of planning were not appropriately mutually coordinated (Đorđević 1974; Milenković 1981; Vasić 1976).

A robust planning methodology was needed to eliminate arbitrary decision-making (Bojović 1976), against a common bias that urban planners held power over people's lives (Đorđević 1974), as well as to provide a clear division of responsibilities and coordination between expert research and self-management decision-making processes. In other words, the role of planners was to collect and organise relevant data and propose multiple development alternatives regarding the commonly agreed development goals and criteria. In sum, planning agencies should act as neutral professional services to inform the self-management decision-makers (Vasić 1976). Planners should focus on research regarding integrated planning models and their evaluation as the basis for coordinating individual and collective interests to promote proper information transfer and closer collaboration with the primary planning actors (Vasić 1976). As for the local communes, the essential precondition to pursuing their upgraded role as the decision-makers was to develop their organisational and individual staff capacities. The role of local (communal) authorities was marked as crucial to ensuring citizens' participation in planning processes (Veljković 1975).

Criticism regarding the misbalance between clear social orientation towards self-management socialism and the inability to translate it into spatial policies and coordinated planning activities prevailed in the years that followed (Vasić 1976). Liberalism and bureaucratic dogmatism were seen as major systemic threats to the self-management society. The former referred to autonomous economic structures that used the semi-market system to maximise their interests, while the latter was embodied in the authorities' oligarchy that tended to misuse spatial rights in the name of "higher interests" (Milenković 1981). Concerning detailed urban plans for Belgrade's reconstruction from the late 1970s, it was underlined that citizen participation had been reduced to the final stages of a planning process when the plan could not be practically changed anymore (Jakšić 1978). Moreover, surveys demonstrated that the deficiency of time and resources, coupled with insufficient levels of neighbourhood cohesion, hindered the local communities from sustaining the torrential decision-making functions on their shoulders (Milenković 1981).

In such a context, self-critical stances were common and directed towards the planners' technocracy, i.e., ignorance of political awareness (Milenković 1981) and their elitism and distorted perceptions about their professional role, power and personal responsibility (Srdanović 1981). In other words, planners were criticised for not being interested in anything beyond their conceptions of demands, needs and standards. Instead of being obsessed with procedures and efficiency, planners should have been more responsive to the needs of the social community (Srdanović 1981). To utilise the "advantageous position" of Yugoslav self-management and enable a genuine democratisation of urban development, a whole new system of urbanism was to be contemplated (Radović 1982). Citizens needed to be provided with comprehensive information and included in all planning stages (Jakšić 1978). Hence, information on public display should be suited to the competencies and interests of the non-experts. This meant that each development alternative's environmental and practical consequences should be presented broadly and comprehensively. Furthermore, decision-making should not be done only on formal, special occasions but often in a continuous planning process (Krešić 1982; Radović 1982). The local community was stressed as the point where individual and common interests should consolidate in a united, general interest (Jakšić 1978). Finally, there were calls for a profound and comprehensive urbanist critique that could catalyse progressive change in planning practice by facilitating debate, alternative approaches, continuous planning process and impetus for research rather than blueprint solutions (Krešić 1982).

*Participatory planning in practice: the Master Plan of Belgrade 1972*

To test the previously mentioned participatory planning narrative in practice, this section briefly elucidates the process of making the Master Plan of Belgrade of 1972, which lasted between 1967 and 1972 and featured an extensive range of collaborative and participatory activities across number of phases (Figure 6.2). More precisely, first we give an overview of different procedures for public engagement, political instruments and expert knowledge and skills, to then critically reflect on effectiveness of implementing such measures.

As previously given, national planning instruments often emerged from consultation with international bodies. Interestingly enough, the Belgrade Master Plan 1972 was born out of the collaboration between the Urban Planning Institute of Belgrade and Wayne State University (from the United States), bringing the cooperation beyond European borders. As a result, one of the pillars ingrained in the plan-making process was the tendency towards public consultation (Le Normand 2014), supported by interdisciplinarity and transparency (Đorđević 1973). Regarding the first, around 150 studies, including a sociological survey, were conducted to strengthen exchange (both formal and informal) among scientific and public institutions, including

*Passive agents or genuine facilitators of citizen participation?* 113

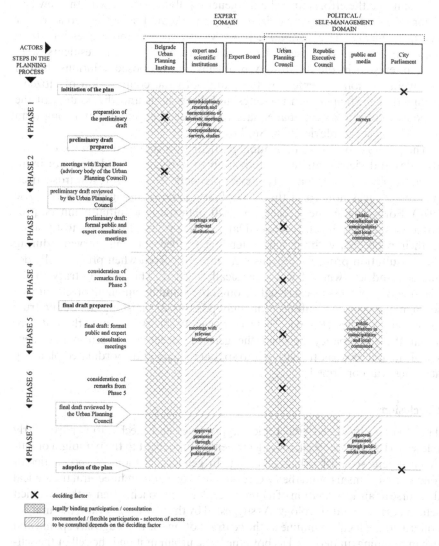

*Figure 6.2* Main phases and participants in making the Belgrade Master Plan of 1972.
*Source:* Authors.

international participants (Đurović and Marinko 1969). As for the latter, in addition to the public consultation on the final plan proposal, which was binding by law, additional triggers for public engagement in communes and local communes – seen in exhibitions, public presentations, specialised publications, supplements in daily newspapers and written information to all households – served to boost public feedback on the planning proposals (Stojkov 1972; Đorđević 1973).

Despite all the efforts to make the general public familiar with the new city vision, the extent of genuine citizen engagement could have been greater. For example, during the consultative process, citizens had more questions than objections to the proposed plan versions (Stojkov 1972); residents were more interested in the day-to-day problems and proposed solutions than in the overall vision presented by the plan; both the composition of citizens' groups that participated in meetings and discussions and the content of the responses collected were inadequate to represent a genuine and long-term public interest (Đorđević 1974; Stojkov 1972).

The previous obstacles to genuine participation stemmed from the citizens' attitudes and viewpoints (e.g., motivation to participate) and were not influenced by planners' approach. However, planners' decisive role towards citizen engagement was noticed in the early phase of the planning process (Bojović 1976). For example, the preliminary draft plan (i.e., the initial planners' proposal) was rarely critically discussed later in the process. Planners mostly stuck to their proposals without fully attending to ideas brought forward during the consultation process – citizens were more involved when procedurally demanded and less when it was intrinsically needed. This was contrary to the plan-making process as exercised through the genuine public involvement as a self-management convention, ultimately possibly diminishing the public trust in planning bodies (Bojović 1976; Đorđević 1974). Briefly put, the criticism about the discrepancy between the general narrative on self-management socialism and its weak transition into spatial policies and coordinated planning activities was confirmed.

## Conclusions

The chapter elucidated vital topics, approaches and mechanisms particularly related to the process of citizen engagement as immanent to the paradigm of self-management socialist urban planning. In general, ideological specificities of Yugoslav socialism found their expression in urban and architectural theoretical discourse as an appreciation of a human-centred approach, often with no explicit references to the state ideology. As suggested by the literature review, the analysis confirmed the local commune as the central notion for the Yugoslav spatial and urban planning, understood as both the basic urban unit and the cell of the self-management society. In other words, the conceptualisation of the local commune as an object of urban planning was seen as key to creating conditions that would enable and facilitate self-managed social relations. However, the efforts to develop a unified methodological approach in planning local communes persisted throughout the entire analysed period, with no broad and clear consensus reached. Until the mid-1970s, there was almost no word on the citizens' role in planning processes. As the 1974 Constitution designated the local commune as the principal agent (decision-maker) in the planning processes, discussions concerning participatory aspects of planning intensified. Also, the role of planners concerning other actors in the planning processes was increasingly debated.

Yugoslav planners saw their role as a neutral professional service of the society by providing comprehensive information basis to facilitate a deliberate exchange between planning actors. Planners performed interdisciplinary, methodologically sound research, proposing as many alternatives as possible, including clear, practical implications for future urban development scenarios. Accordingly, planners developed a comprehensive platform and a toolkit for the self-management authorities, enabling them to balance different interests and shape their living and working environment. In short, both initially proposed viewpoints on planners' position – technocrats vs. enablers of societal change – are partially supported. Nevertheless, planners' position towards citizens is not to be observed in an isolated manner, i.e., without attending to the specificities of the socio-political context. Though self-management included various forms of vertical and horizontal cooperation and involved various actors (political councils, professional bodies, local communities), due to the strong political ideology pursued through different socio-political units, both the planners' decisions and citizen needs could have been eroded. In other words, hardly any decision could have been made without the previous consent of the local and central governments (Perić 2020). Consequently, the constant tension between the unitary, i.e., politically imposed public interest, and self- and collective interests blocked the planners from fully utilising the instrument of citizen participation. In a nutshell, the personal and collective identities of Yugoslav citizens coloured with the right and duty to shape their immediate environment, and the planners' image of emancipated intellectuals marked by the constant exchange with the internationals were curbed by the uncontested communist political regime. Hence, the Yugoslav self-management seemed to emerge from two opposed but rather complementary interpretations of East and West.

**References**

Bartlett, W., and Prica, I. 2013. The deepening crisis in the European super-periphery. *Journal of Balkan and Near Eastern Studies* 15(4):367–382. 10.1080/19448953.2013.844587

Becker, J., Jäger, J., Leubolt B., and Weissenbacher, R. 2010. Peripheral financialization and vulnerability to crisis: a regulationist perspective. *Competition & Change* 14(3–4):225–247. 10.1179/102452910X12837703 6153

Bjeličić, S. 1962. Stambena zajednica – osnovna planska urbanistička celina grada. *Arhitektura-Urbanizam* 14:41–42.

Bjelikov, V. 1962a. Lik našeg urbaniste. *Arhitektura-Urbanizam* 14:46–47.

Bjelikov, V. 1962b. Sa desetog Savetovanja Saveza urbanističkih društava Jugoslavije. *Arhitektura-Urbanizam* 14:48–49.

Bjelikov, V. 1963. Komuna kao prostorna pojava, Sa jedanaestog Savetovanja Saveza urbanističkih društava Jugoslavije. *Arhitektura-Urbanizam* 22–23:5–7.

Blagojević, Lj. 2007. *Novi Beograd: Osporeni modernizam*. Belgrade: Minerva.

Blagojević, M., and Perić, A. 2023. The diffusion of participatory planning ideas and practices: the case of socialist Yugoslavia, 1961–1982. *Journal of Urban History* 49(4):797–820. 10.1177/00961442211044501

Bojović, M. 1976. Metodologija i algoritmika savremene izrade prostornih planova i njihova usaglašenost sa potrebama. *Urbanizam Beograda* 37:24–26.

Bohle, D. 2018. Mortgaging Europe's periphery. *Studies in Comparative International Development* 53(2):196–217. 10.1007/s12116-018-9260-7

Bohle, D., and Greskovits, B. 2012. *Capitalist Diversity on Europe's Periphery*. Ithaca, NY: Cornell University Press.

Dabović, T., Nedović-Budić, Z., and Djordjević, D. 2019. Pursuit of integration in the former Yugoslavia's planning. *Planning Perspectives* 34(2):215–241. 10.1080/02665433.2017.1393628

Dawson, A.H. 1987. Yugoslavia. In: Dawson, A.H. (Ed.), *Planning in Eastern Europe*, pp. 275–291. New York: St. Martin's Press.

Đorđević, A. 1973. Uvodno izlaganje o Generalnom urbanističkom planu Beograda 2000. *Arhitektura-Urbanizam*, 72–73:53–55.

Đorđević, A. 1974. Sistem i organizacija prostornog planiranja u Beogradu. *Urbanizam Beograda* 26:1–27.

Đurović, Đ., and Marinko, V. 1969. Osnovni oblici saradnje zavoda sa drugim institucijama i javnošću. *Urbanizam Beograda* 4:48–49.

Fisher, J.C. 1964. The Yugoslav Commune. *World Politics* 16(3):418–441.

Göler, D. 2005. South-east Europe as European periphery? Empirical and theoretical aspects. In: *Serbia and modern processes in Europe and the world*, pp. 137–142. Belgrade: University of Belgrade, Faculty of Geography.

Golubchikov, O. 2004. Urban planning in Russia: towards the market. *European Planning Studies* 12(2):229–247. 10.1080/0965431042000183950

Grava, S. 1993. The urban heritage of the Soviet regime: the case of Riga, Latvia. *Journal of the American Planning Association* 59(1):9–30. 10.1080/01944369308975842

Ignjatović, A. 2012. Tranzicija i reforme: Arhitektura u Srbiji 1952-1980. In: Šuvaković, M., Daković, N., Ignjatović, A., Mikić, V., Novak, J., and Vujanović, A. (Eds.), *Istorija umetnosti u Srbiji XX vek. Realizmi i modernizmi oko Hladnog rata*, pp. 689–710. Belgrade: Orion Art i Katedra za muzikologiju Fakulteta muzičke umetnosti.

Jakšić, M. 1978. Karakter i dometi učešća građana u planskom usmeravanju razvoja postojećeg gradskog tkiva kroz detaljne urbanističke planove rekonstrukcije. *Urbanizam Beograda* 50:41–45.

Janković, M. 1969. Otuđenje u savremenom gradu. *Urbanizam Beograda* 4:27–28.

Jovanović, J., and Kulić, V. 2018. City Building in Yugoslavia. In: Stierli, M., and Kulić, V. (Eds.), *Toward a Concrete Utopia: Architecture in Yugoslavia 1948–1980*, pp. 58–63. New York: The Museum of Modern Art.

Kardelj, E. 1979. *O sistemu samoupravnog planiranja. Brionske diskusije (drugo izdanje)*. Belgrade: Radnička štampa.

Krešić, M. 1982. Gradovi i proces odlučivanja. *Arhitektura-Urbanizam* 88–89:23–25.

Krstanovski, M. 1977. Elementi egzistencijalnog i arhitektonskog prostora u okviru mesne zajednice. *Urbanizam Beograda* 41:22–24.

Kulić, V. 2009. 'East? West? Or Both?' Foreign perceptions of architecture in Socialist Yugoslavia. *The Journal of Architecture* 14(1):129–147. 10.1080/13602360802705106

Kulić, V. 2014. The Scope of Socialist Modernism: Architecture and State Representation in Postwar Yugoslavia. In: Kulić, V., Penick, M., and Parker, T. (Eds.), *Sanctioning Modernism: Architecture and the Making of Post-War Identities*, pp. 37–65. Austin: University of Texas Press.

Le Normand, B. 2014. *Designing Tito's Capital: Urban Planning, Modernism, and Socialism*. Pittsburgh: University of Pittsburgh Press.

Liotta, P.H. 2001. Paradigm lost: Yugoslav self-management and the economics of disaster. *Balkanologie* V(1–2):37–57.

Lydall, H. 1989. *Yugoslavia in Crisis*. Oxford: Claredon Press.

Lynn, M.L., Mulej, M., and Jurse, K. 2012. Democracy Without Empowerment: the grand vision and demise of Yugoslav self-management. *Management Decisions* 40(7/8):797–806. 10.1108/00251740210437752

Maksimović, B. 1963. Povodom XXVI Svetskog kongresa Internacionalne federacije za stanovanje, urbanizam i regionalno planiranje. *Arhitektura-Urbanizam* 18:18–19.

Maričić, T., Cvetinović, M., and Bolay, J.-C. 2018. Participatory Planning in the Urban Development of Post-socialist Serbia. In: Bolay, J.-C., Maričić, T., and Zeković, S. (Eds.), *A Support to Urban Development Process*, pp. 1–28. Lausanne/Belgrade: CODEV EPFL/IAUS.

Martinović, M. 2020. *Yugoslav Self-management in Architecture of Local Community Centres in Belgrade 1950–1978* (doctoral dissertation). Belgrade: University of Belgrade.

Milenković, A. 1981. Treći susret planera i urbanista Jugoslavije: Aktuelne teme prostornog planiranja. *Urbanizam Beograda* 61:84–89.

Mrduljaš, M. 2018. Architecture for a Self-managing Socialism. In Stierli, M., and Kulić, V. (Eds.), *Toward a Concrete Utopia: Architecture in Yugoslavia 1948–1980*, pp. 40–57. New York: The Museum of Modern Art.

Mrduljaš, M., and Kulić, V. (Eds.). 2012. *Unfinished Modernizations—Between Utopia and Pragmatism: Architecture and Urban Planning in Socialist Yugoslavia and Its Successor States*. Zagreb: Croatian Architects' Association.

Mutnjaković, A. 1964. Grad našeg radosnog sutra. *Arhitektura-Urbanizam* 25:14–16.

Nedović-Budić, Z., and Cavrić, B. 2006. Waves of planning: a framework for studying the evolution of planning systems and empirical insights from Serbia and Montenegro. *Planning Perspectives* 21(4):393–425. 10.1080/02665430600892146

OG FPRY 3/1953. *Constitution of the Federal People's Republic of Yugoslavia [Ustav Federativne Narodne Republike Jugoslavije]*.

OG PRS 47/1961, 14/1965, 30/1965. *Act on Urban and Regional Spatial Planning of the People's Republic of Serbia [Zakon o urbanističkom i regionalnom prostornom planiranju Narodne Republike Srbije]*.

OG SFRY 14/1963. *Constitution of the Socialist Federal Republic of Yugoslavia [Ustav Socijalističke Federativne Republike Jugoslavije]*.

OG SFRY 9/1974. *Constitution of the Socialist Federal Republic of Yugoslavia [Ustav Socijalističke Federativne Republike Jugoslavije]*.

OG SRS 19/1974. *Act on Planning and Spatial Arrangement of the Socialist Republic of Serbia [Zakon o planiranju i uređenju prostora Socijalističke Republike Srbije]*.

OG SFRY 46/1976. *Act on the Foundations of the System of Societal Planning and the Societal Plan of Yugoslavia [Zakon o osnovama sistema društvenog planiranja i o društvenom planu Jugoslavije]*.

OG SRS 27/1985, 5/1986, 6/1989. *Act on Planning and Spatial Arrangement of the Socialist Republic of Serbia [Zakon o planiranju i uredjenju prostora Socijalističke Republike Srbije]*.

Perić, A. (2020). Citizen Participation in Transitional Society: The Evolution of Participatory Planning in Serbia. In: Lauria, M., and Schively Slotterback, C. (Eds.), *Learning from Arnstein's Ladder*, pp. 91–109. New York: Routledge.

Perišić, D. 1965. Godišnja skupština Urbanističkog saveza Jugoslavije. *Arhitektura-Urbanizam* 33–34:43–45.

Radović, R. 1964. Sinteza ili zbir. *Arhitektura-Urbanizam* 26:54–55.

Radović, R. 1982. Od kritike modernog urbanizma do sopstvene radne filozofije i prakse ljudskih naselja naše sredine. *Urbanizam Beograda* 61:40–43.

Ramet, S.P. 1995. *Social Currents in Eastern Europe: The Sources and Meaning of the Great Transformation*. Durham: Duke University Press.

Srdanović, B. 1981. Plan protiv čoveka. *Arhitektura-Urbanizam* 86–87:60–62.

Stojkov, B. 1972. Neke karakteristike učešća javnosti u izradi GUPa Beograda. *Urbanizam Beograda* 20:5–8.

Stupar, A. 2015. Cold War vs. architectural exchange: Belgrade beyond the confines? *Urban History* 42(4):623–645. 10.1017/S0963926815000528

Tomić, V. 1980. Humanizacija života u gradu. *Urbanizam Beograda* 59–60:14–16.

Vasić, Z. 1976. Samoupravno planiranje. *Urbanizam Beograda* 37:20–23.

Veljković, V. 1975. Nacionalno i regionalno planiranje kao okvir u lokalnim upravama. *Urbanizam Beograda* 31:23–25.

Vujošević, M., and Nedović-Budić, Z. 2006. Planning and Societal Context – The case of Belgrade, Serbia. In: Tsenkova, S., and Nedovic-Budic, Z. (Eds.), *The Urban Mosaic of Post-Socialist Europe. Space, Institutions and Policy*, pp. 275–294. Heidelberg: Physica-Verlag.

Zukin, S. 1975. *Beyond Marx and Tito: Theory and Practice in Yugoslav Socialism*. Cambridge: Cambridge University Press.

# 7 The influence of nuclear deterrence during the Cold War on the growth and decline of the peripheral town of Valga/Valka

*Kadri Leetmaa, Jiří Tintěra, Taavi Pae, and Daniel Baldwin Hess*

**Introduction**

Estonia's location in the geopolitically turbulent frontier area between the Russian and Western worlds has always affected the spatial development of Estonian cities and regions. Frontier lands inevitably experience militarisation at times when international tensions build and demilitarisation and disarmament when tensions ease (Woodward 2005). In the 1990s, when Europe transformed at the end of the Cold War, one of the fastest growing research areas in the field of urban and regional studies was demilitarisation. Thirty years later, we are again living in times of rapidly growing military capabilities in Russian-Western border territories, including in Estonia where the deployment of military installations is back on the agenda (e.g., deployment of NATO troops, residences of military personnel, and expansion of training grounds).

It is during these times that it is especially important to look back upon the previous period of intense militarisation – following the conclusion of WWII until the departure of the Soviet military from Estonia in the early 1990s – and to examine how the presence of the military affects spatial development during and after periods of international tension. This chapter explores the fortunes of Valga/Valka (Valga, Estonia with 17,709 and adjacent Valka, Latvia with 7,911 inhabitants in 1990), a small town on the Estonian-Latvian border (cf. Tintěra 2019; Lundén 2009), amidst the development of Cold War-related military capabilities in the former Soviet Union. The Baltic States represented the periphery of the Soviet Union in terms of geographical location, but at the same time were critical from a military point of view – the Iron Curtain traversed through the Baltic Sea and formed the external border of the Soviet Union. We consequently consider Valga/Valka a double periphery – a geographical periphery on the scale of the Soviet Union and for their two host countries (Estonia and Latvia), the cities served as peripheral county ("districts" in Soviet parlance) seats. Yet this small and seemingly peripheral place played a very important role in global relations at the time.

In this chapter, we first provide an overview of the role Estonia played in the military system of the Soviet Union between 1940 and 1991 (representing

DOI: 10.4324/9781003327592-10
This chapter has been made available under a CC-BY-NC-SA 4.0 license.

the period of Soviet occupation), the facilities and military forces stationed here and the effect of military presence on urban and rural landscapes and societies. Next, we unravel the story of Valga/Valka. We present the details of how military needs affected the town and its surroundings and how militarisation led to the rapid growth of the town and subsequent demilitarisation led, on the contrary, to severe shrinkage. Case of Valga/Valka illustrates how "small places and large issues" (Baumann et al. 2020; Seljamaa et al. 2017) are intrinsically connected, or in other words, how globally relevant international relations – e.g., the Cold War arms race and deterrence tactics – transform small and seemingly peripheral places into sites of importance. The chapter demonstrates how large issues can give small places the impetus for development and growth, or vice versa, for diminishment and shrinkage. In addition, we explore how the spatial needs of the military were related to regular spatial planning under state socialism; in other words, we demonstrate how investments in defence and preparation for war had a higher priority (cf. Sjöberg 1999) than civilian aims and served as priorities for spatial planning.

As key sources for this research, we use maps of historical plans stored in the National Archives of Estonia and, as secondary sources, reviews by Estonian military historians on the location of the Soviet army in Estonia and reports by environmental scientists of residual contamination from the military presence. In addition, conversations with local experts familiar with the conditions of the region during the Soviet era helped to enhance this study. Population data are based on regular population statistics of the Statistical Offices of Estonia and Latvia.

**Militarisation of Estonia during the Soviet occupation (1940–1991)**

Estonia has been a strategic frontier land throughout its history, and therefore military facilities from earlier historical periods can be found here (Peil 2005; Jauhiainen 1997). For example, Peter the Great's zone of sea fortifications, which was established on the eve of WWI on the shores of the Gulf of Finland to protect St. Petersburg, has never directly been used in battle. This demonstrates that building military readiness and deterrence tactics can change landscapes, not merely wars themselves. Following WWI and the Estonian War of Independence (1918–1920), Estonia became an independent nation state. Its independence was interrupted before WWII by the Soviet-Nazi secret agreement. The Soviet occupation of Estonia began in 1940, was replaced by a short period of German dominance (1941–1944), and continued consecutively from 1944 until 1991 when the Soviet Union disintegrated and Estonia restored its independence. Military historian Jüri Pärn (2020) estimates that in 1944, during periods of battle, up to one million military personnel passed through Estonia (the population of Estonia at the time was approximately one million people); immediately after the war, however, the number of military forces significantly reduced (to approximately 150,000 in 1946–1947), and instead of troops directly involved in military conflict, units with defence and deterrence purposes were placed in border areas of the USSR.

To protect its sphere of influence, the Soviet Union displayed its military prowess by deploying troops throughout the Eastern Bloc as a means of keeping the satellite states in check (Seljamaa et al. 2017). The Baltic States were directly occupied by the Soviet Union and became a strategic border area in the defence and deterrence tactics of the USSR. Estimates of how many military personnel were present in Estonia during the decades of occupation vary, but the number changed over time depending on military objectives during specific periods of the Cold War. By the beginning of the 1990s, just before the final withdrawal of troops from Estonia, there were between 32,000 and 36,000 military personnel (or 2% of Estonia's population) (Pärn 2020). Soviet forces throughout Estonia numbered approximately 80,000 during the 1950s (or 7% of Estonia's population), and immediately after the end of WWII military personnel of the occupation forces amounted to 150,000. Still, in the final years of the Soviet occupation, the share of military personnel in many cities was remarkable; in some cities and settlements, military personnel constituted the majority (including military towns Paldiski and Tapa) (Jauhiainen 1997:122). The population connected to the military was usually larger, since it included not only those in direct military service but also family members of military personnel and civilian employees of military units.

Between the 1940s and 1980s, almost 2% of Estonia's territory was entirely closed for military reasons (Jauhiainen 1997:118). The extent of direct and indirect military control (including closed towns, sensitive scientific and industrial facilities, coastal zones, etc.) reached as high as 14–25% of the Estonian territory (Miller 2019). The entire country of Estonia was of key importance militarily to the USSR, and therefore strategic military facilities and various types of troops were placed in all regions, both in cities and rural places. Many coastal areas were entirely off-limits for non-military people, while other places required special permission for access by non-residents. For example, an Estonian could travel to the Estonian islands only for the purposes of recreation or to visit relatives living there, and a special border zone permit was required. Many cities had special military functions. Since the town of Paldiski, along the northwestern coast of Estonia, contained a nuclear submarine training centre and other secret military facilities, it became a closed city in 1964 (Miller 2019; Peil 2005). Another closed city in the period 1947–1991 was Sillamäe in northeast Estonia, where a uranium enrichment plant was critical to the USSR defence industry (Miller 2019). Even Estonia's second-largest city, the traditional university city Tartu, was semi-closed (for foreigners) due to one of the largest Eastern Bloc military airfields on the edge of the city. The airfield near Tartu was a secret site excluded from Soviet maps but paradoxically included on contemporaneous United States intelligence maps

(Hess and Pae 2020; cf. Veldi and Bell 2019 on secret cartography). There were many closed territories all over Estonia (missile bases, training sites, airfields, radar stations, etc.) within and outside cities, often preventing access to beloved former heritage sites and landscapes for locals.

Since Estonia lost its independence *de facto* in 1940, the Soviet army was perceived there as an occupation army. Its location in the occupied territory also meant that military bases in Estonia needed to fortify themselves. Various structures were created in the landscape to protect military installations, including observation posts and trenches (Miller 2019). Even though young Estonian men were required to serve time in the Soviet army, it was typical, due to loyalty factors, that locals were sent to military service in far-flung places in the USSR while young men from distant Soviet republics served their time in Estonia (Rakowska-Harmstone 1986). Consequently, the Soviet military bases in Estonia mostly functioned as isolated communities with relatively little contact with local life (cf. Grava 1993). However, to the extent that these contacts functioned, the military presence contributed to the Russification of Estonia, which was already in full swing due to rapid industrialisation and related immigration (Hess et al. 2012; Tammaru and Kulu 2003). Unconnected to the local community and with little feeling of responsibility for the locale led to a predatory attitude among military personnel towards the natural resources of Estonia. Within and near its installations the military caused extensive soil and water contamination. The cleanup of these spills was both expensive and complicated, and continued for decades after the military left (Valga municipality 2018; Mander et al. 2004; Auer and Raukas 2002; Raukas 1999). When the last soldiers departed Estonia in 1994, they left behind extensive environmental damage, unsafe empty buildings and various disused facilities. For Estonians, military heritage from earlier stages of history is more easily and neutrally associated with the country's history, while Soviet-era military remnants in its cities and landscapes are often associated with (too recent) injustice and occupation (Peil 2005).

**Strategic missile forces expansion in Estonia during the Cold War**

Estonia gained an even more significant role in the military system of the Soviet Union during the build-up of Cold War tensions and the nuclear arms race. This occurred between the late 1950s and the mid-1980s, concluding when nuclear arms reduction agreements were finally signed between the United States and the USSR. Strategic missile forces were stationed throughout the Baltic countries. The 23rd Guard Missile Division installed missile regiments in Estonia and Northern Latvia along with supporting technical units responsible for the assembly and repair of nuclear warheads (Kivi 2020; Pärn 2020; Raukas 1999; Tähiste and Mõniste 2018). The command of the missile division and the communication battalion was headquartered in Latvia (Valka town and district) near the

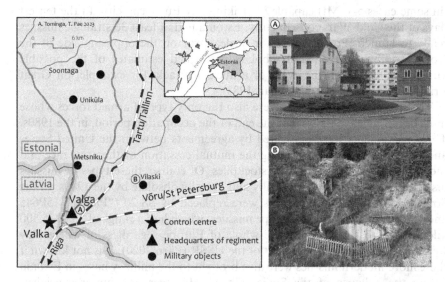

*Figure 7.1* Effects of military installations on Valga/Valka and surrounding landscapes: locations of the command centre of the missile division and the headquarters and missile launch bases of Valga regiment, (A) socialist-era housing complexes in the town centre (photo by Valga municipality 2023), (B) underground missile shaft (photo by Valga municipality 2018).

Estonian border. Two of the Estonian missile regiments were in northern and northwestern Estonia (in the then Haapsalu and Rakvere Districts) and two in southern Estonia (near Võru, another small district centre in South Estonia, and in Valga and its surroundings; see Figure 7.1). The 23rd Guard Missile Division installed the fifth regiment in northern Latvia (Alūksne and Gulbene Districts).

In the military heyday (1961-1978), the composition of the missile forces was between 7,000 and 9,000 military personnel in Estonia, including 1,000 officers (the proportion of senior military personnel was particularly high in military units associated with nuclear weapons). A training school for junior military specialists of the missile forces of all-Union importance was located in the southern Estonia Võru regiment. Administration centres for the regiments were located in small peripheral towns and influenced these towns mainly through an expanding housing stock and related civic infrastructure development (schools, local transport and urban growth). However, the ready-to-launch nuclear missiles themselves were located in rural settings, hidden among forests, mostly omitted from official maps (Tähiste and Mõniste 2018), and of course, protected by top-level security regulations.

There were nine missile launch facility locations in all regiments located in Estonia, and each facility had four medium-range missiles (missile types with the designation R-12, R-12U or R-12N) with a flight distance of more than 2,000 kilometres and with thermonuclear warheads on the missiles 1–2.3 Mt (in some cases 5–6 Mt) capable of striking any European city. In the tensest days of the Cold War, consequently, Estonia had at least 36 strategic missiles ready to launch. Preparedness for nuclear war on the opposing side of the Iron Curtain was real, and scholars performed estimates of the possible number of victims in the event of a nuclear attack, for example, on Great Britain (Openshaw and Steadman 1983).

The strategic missile forces were also the only type of armed forces whose units began to leave Estonia at the end of the occupation period, in the 1980s. These movements were motivated by agreements between the United States and the Soviet Union concerning the mutual cessation of the arms race and the reduction of nuclear missile stockpiles. Of course, these types of missiles installed in Estonia were physically and morally obsolete by the 1980s. In the Soviet Union R-12 type missiles were replaced by SS-20 ("Pioner" 15P645 and "Pioner-UTTH" 15P653) type missiles, which had a range of 5,000–5,500 kilometres and thus covered the whole of Europe regardless of where they were placed in the European part of the USSR (Pärn 2020). It is not known if these more modern missiles were also brought to the military bases in Estonia before the collapse of the Soviet Union. However, the junior specialist-training centre in Võru was still training specialists in the mid-1980s for these newest missile complexes.

### The transformation of the peripheral town Valga/Valka in the midst of the Cold War

We now focus on Valga/Valka, the small Estonian-Latvian border town and transformations wrought by the militarisation of the Baltic States. We extend our inquiry by exploring the nuclear arms race as a manifestation of the Cold War and how military escalation affected the development and planning of this small peripheral location. It is no surprise that in the 1960s and 1970s Valga/Valka experienced a rapid increase in the number of inhabitants and corresponding spatial changes in the urban area. This was, however, the second wave of growth for this peripheral border town.

Valga/Valka (then called "Walk") became a strategic industrial city under the Russian Empire in connection with the construction of a railway in the region (Tintěra 2019; Tintěra et al. 2018). The city has served since 1889 as a junction where railway lines from Estonia (Tartu and Võru), Latvia (Riga) and Russia (Pskov and St. Petersburg) came together. The railway expansion launched the first wave of population growth of Walk: important Tsarist-era industries were established in Walk, including textile and metals factories and railway manufacturing (Kant et al. 1932), and the German-speaking town (until then Baltic Germans dominated urban centres in Estonia and Latvia) became an Estonian- and Latvian-speaking town. The growth lasted until Estonia and Latvia gained independence after WWI. For Valga/Valka, the independence of Estonia and Latvia resulted in an administrative separation between the two countries.

Although both towns now functioned as administrative centres, a significant stagnation in their economies was experienced, as the markets for industrial goods in Tsarist Russia were now lost. The population of Walk grew from 8,400 in 1893 to 16,000 in 1913 and then declined to 14,110 by 1934–1935 (including Valga with 10,842 inhabitants and Valka with 3,268 inhabitants according to the Estonian (1934) and Latvian (1935) censuses, respectively).

The second wave of growth in the region began after WWII (Tintěra 2019; Tintěra et al. 2018) when both Estonia and Latvia were again incorporated into the Russian empire, this time as republics of the Soviet Union. During WWII, the Valga/Valka region was an important frontline of battle (Kivi 2020). The city was bombed by the Germans in 1941 and by Soviet forces in 1944, and the Red Army returned in September 1944. The city centre and the railway station environs were the main sites of damage. After the war, even when split between two Soviet republics administratively, Valga/Valka soon restored its industrial base and functioned as an important railway node. According to the first Soviet Census in 1959 the population of Valga had increased to 13,354 and that of Valka to 4,872 (18,226, combined population for the two cities). In the first decades after WWII, Valga/Valka was developed again as a well-connected small industrial town (e.g., a wagon repair factory of all-USSR importance, a meat processing plant, textile and furniture factories) and an administrative district centre. In 1949, a new Stalinist-inspired railway station building was built in place of the main building of the war-destroyed train station, which in its scale and dimensions foretold plans for the town's rapid industrial growth. The town's role as a military centre was added only during the late 1950s when Cold War tensions gathered momentum globally.

In 1960, the 23rd Red Flag Oryol-Berlin Guard Missile Division with the Order of Lenin was formed (a successor to the former bomber aviation division) with a command centre in Valka (in Latvia, only metres from the Estonian border) (Kivi 2020; Pärn 2020; Tähiste and Mõniste 2018). The command centre was also supported by a communications centre, a military prosecutor's office, a special department of the KGB and other military supports. All five Estonian and Northern Latvian regiments were controlled from here. One of the regiments was headquartered in Valga along with a repair shop and technology centre. Three missile launch bases in the Valga regiment were hidden in forests in the Valga District: Soontaga, Uniküla and Vilaski (see Figure 7.1). Among them, the Soontaga and Uniküla locations were armed with ground-based medium-range missiles (R-12 or R-12N), while Vilaski was armed with group launch complexes (R-12U, also designated as R-12V) with missiles stored in underground shafts (four shafts per complex) (see Figure 7.1B). The shaft structures were sunken below ground (with six underground floors and a total depth of 35 metres). From the upper floor, passages led to the centre of the facility where there was a command post and service facilities. USSR experts in subway building from Leningrad (later, St. Petersburg) oversaw the construction of shafts with one-metre thick walls. Extensive military supports – including technical bases, assembly brigades, communications and radio offices and other military services were scattered throughout Valga and Valka districts on both sides of the Estonian-Latvian border.

The locations of such complexes were deliberately chosen to be removed from urban density and within nature settings, making it possible to conceal from view armaments and related facilities. Figure 7.1 shows the locations of the Valga district missile regiment (headquarters in town and launch bases in Soontaga, Uniküla and Vilaski in surrounding territory) and other military installations. Small unimportant towns attracting little or no international attention, with ample surrounding greenspace, were ideal sites for such secret facilities and were thus subsumed into the high stakes of the Cold War. A concrete driveway was built to provide access to the shooting complex, shown on contemporary maps only as an insignificant forest path (Tähiste and Mõniste 2018). Typical of the security environment at the time it was established, the entire complex was hidden from view and thoroughly secured and of course not shown on contemporaneous maps (cf. Veldi and Bell 2019).

The secret missile complexes hidden in the forest – and related administration and service functions in nearby towns and settlements – required a remarkable number of skilled military personnel. Up to 1,760 military personnel were designated for each missile regiment containing underground shafts (such as the Valga regiment), including 280 officers (Pärn 2020). Additional personnel were required for repair shops and technology centres; in regiments with three launch complexes this support could amount to approximately 460 personnel (including 150 officers). Thus, a Cold War missile base in the region affected a peripherally located town like Valga/Valka in the first place with the sheer presence of military personnel. When their family members were included, military-related persons constituted a significant part of the population of small towns. During a 20-year period between the early 1960s and the early 1980s, the population of Valga/Valka increased by a factor of 1.5. By 1979, the population of Valga had increased to 18,474 and that of Valka to 8,023 according to the Soviet Census in both republics (26,497 combined population for the two cities) (cf. Tintěra 2019; Tintěra et al. 2018).

Such an increase in population put enormous pressure on the housing supply and consequently created demand for new housing construction. Housing shortages were a hallmark of the Soviet system experienced throughout the USSR. During the years of military urban growth, a great deal of new housing was built of necessity in Valga, primarily large apartment buildings typical of the centrally planned mass housing construction of the time in the Soviet Union (Hess and Tammaru 2019). Building demolitions following WWII destruction in city centres resulted in a great deal of open space in central locations, and historical buildings were not at the time considered architecturally or culturally valuable.

Large modernist complexes of prefabricated apartment buildings were built in the centre of the town, sometimes requiring the demolition of remaining historic buildings (see Figure 7.1A). In addition to vast residential space, an underground shelter was built in Valka for evacuation in the event of a possible nuclear attack. The reuse and adaptation of this shelter into a modern bomb shelter is again on the agenda today as a response to the Russian invasion of Ukraine (Valka municipality 2023). More population also meant the need for updating social infrastructure. Since the military population was Russian-speaking, a double social infrastructure existed in Valga/Valka (like in other major cities in Estonia and Latvia), i.e., there were Russian-language schools and kindergartens in addition to Estonian- and Latvian-language educational institutions.

Despite the secrecy of military facilities located in and around Valga/Valka, the militarisation of the region inevitably affected the daily life of residents of the community. For example, the street named "Punaarmee tänav" (Red Army Street) was a rather rare name among Soviet places. In 1978, a monument-tank was erected on the border of the town. One of five such monuments in Estonia, it was intended to legitimise the presence of the Red Army and engender respect for and acceptance of the army. The presence of the military was also clearly visible in the city on a daily basis; the military provided its own shops, and a significant share of the Valga/Valka local economy was related to serving the military (e.g. local food production that supplied the local military).

## A joint general plan guides border town development during the 1970s

What is particularly significant is that during the period of military-related urban growth during the 1970s, a combined general plan for Valga and Valka was prepared and officially adopted. This achievement in comprehensive planning has not been repeated, even during the three decades following independence in Estonia and Latvia when town planners have intentionally attempted to connect these two adjacent towns. The plan was commissioned by executive committees in both Valga and Valka. The 1970 Valga-Valka General Plan (see Figure 7.2) was composed by the National Design Institute "EstonProject", a central government agency responsible for urban planning and civil engineering. As there were no private design companies in the Soviet Union, all planners, designers, and architects were engaged in state-controlled institutes.

The new 1970 plan followed the trajectory established in the previous post-WWII plan of Valga (1954, found publicly and digitally in the Estonian National Archives under code "EAM.3.1.467 leht 1", weblinks in reference list) but was the first combined spatial plan for the town divided between the two republics. Due to the necessity to accommodate newcomers, both industrial workers and military personnel, densification of the built-up area of the town was the main goal of the general plan. An east-west corridor

128  *Kadri Leetmaa et al.*

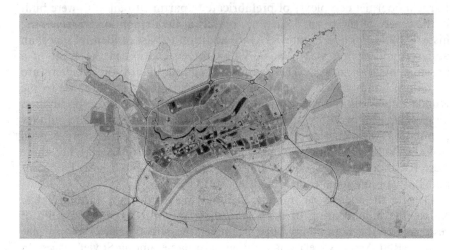

*Figure 7.2* Valga/Valka combined general plan, 1970 (Source: Estonian National Archives) (The original colour drawing of the 1970 Valga/Valka General Plan (code in archive: ERA.T-14.6.14., weblink in reference list) and its 1988 Revision (code in archive: VAMA.379.1 s.163) is available in the Estonian National Archives.

traversing the old town centre contained new standardised apartment buildings (depicted by dark properties on the plan) planned between the railway in the south and the Pedeli River in the north.

Since the areas destroyed during WWII in the town centre and around the railway station had already been built at a high density (Koop 2012), new apartment buildings were now planned to replace older mainly wooden low-rise buildings dating from the end of Tsarist period. This earlier housing, containing structures built before WWI, was generally oriented along the property boundary at the street line with gardens and yards at the rear of plots. The new socialist-era buildings were first built in the former yards between older buildings. Residents of the older housing units were eventually forced to relocate, and the outdated housing stock was demolished when the new construction was completed. Oftentimes, however, the planned demolition did not follow, and new large-scale apartment buildings remained in close proximity to the former low-rise structures, creating the seemingly unplanned but expressive urban structure that remains today throughout Valga (See Figure 7.1A).

In accordance with the spirit of the time, older housing units were considered unhealthy and outdated and not worth preserving. The Valga/ Valka plan called for removal of nearly all pre-WWI structures with the exception of a few important public buildings. As Valka leaders had already approved the demolition of old urban structures in the town centre in the 1960s, there was public pressure to do the same in Valga. In fact, similar processes were ongoing in all cities and towns of Soviet Estonia. However, attempts to enhance heritage protection emerged as a counter-reaction, and in 1973 nine conservation areas were declared in small Estonian towns, a movement that was the first of its kind in the Soviet Union. Valga, however, was not among them (an architectural conservation area was formed here in 1995), and consequently rampant demolition proceeded here through the 1980s.

As the housing shortage continued, new apartment buildings were planned for the northern and eastern peripheries of the town. New residential districts were accompanied by public amenities and greenspace between the housing blocks. Following the planning rules of the time, public facilities (e.g., schools and kindergartens) were placed in proximity to apartment buildings. Apart from the large residential space to the north of the city, all of the planned residential areas and a significant share of the amenities were completed during the Soviet period. Due to secrecy in the central administration, military facilities were not indicated on the general plan, and the map even contained intentionally inaccurate information about development in the military space. Apartments for prominent Soviet military figures were, however, allocated in the new housing areas in the centre or on the eastern edge of town.

The 1970 General Plan was amended in 1988, and the new version took into account the urban development that had occurred since 1970. Since pressure for new housing was not satisfied, additional apartment buildings were built during the 1980s on the south bank of the Pedeli River on sites initially intended to be preserved as greenspace. Throughout the Soviet period, the town was clearly planned according to population forecasts; the 1988 Revision of the General Plan established a projection of 40,000 inhabitants for the two towns combined.

## Conclusion

After nuclear disarmament agreements were signed by global superpowers (US and USSR) in the 1980s, the military importance of Valga/Valka immediately decreased. By 1981, the Valga regiment had halted its combat readiness. Under international treaties enacted in the late 1980s, many missiles were subject to destruction by both parties in the Cold War. Valga/Valka began to experience shrinkage – amid the global circumstances that led to the dissolution of the Cold War – which quickened in pace after the complete withdrawal of the Soviet army from Estonia in the early 1990s. The population decreased during the 1980s from 26,517 in 1979 (in the two towns combined) to 25,620 in 1990 (Valga, 17,709; Valka, 7,911). However, during the period of independence in Estonia and Latvia during the last three decades Valga/Valka has become a symbol of urban shrinkage. The two towns together have lost more than one-third of their former population; as of 01 January 2022, the combined towns had only 16,531 inhabitants (Valga, 12,009; Valka, 4,522).

Today, urban planners in Valga/Valka (two distinct municipalities in two countries) are attempting to overcome the effects of urban shrinkage and to reallocate urban space accordingly. The aim is to adapt the urban fabric to the needs of a smaller number of residents through various strategies, such as implementing a new master plan, demolishing unused buildings, concentrating density in and near the town centre, restoring historic buildings and sensibly revitalising and connecting urban space of the two towns. Valga/Valka is planned to be intentionally smaller with a more vibrant town centre and greener suburbia. The goal of the transformation is to boost residents' pride and satisfaction in their hometown and its urban space (Tintěra et al. 2018). Disused missile bases in the surrounding hinterland have been mostly eliminated, but filling the dangerous underground shafts in Vilaski (Figure 7.1B) occurred as recently as 2018 (Valga municipality 2018). Valga/Valka still features a remarkable industrial base (textile, furniture, food, logistic, etc.), but its potential as an international railway junction (between Estonia, Latvia, and Russia) is perpetually unrealised because the eastern markets are inaccessible by railway.

Estonia and Latvia also function as frontier lands in the context of current international tensions; however, the Iron Curtain is now situated to the east of these countries, not along the Baltic Sea as it did in the mid-20th century. Some small towns and rural places in Estonia (Tapa, Võru County) carry military functions today since Estonia and Latvia are situated along NATO's eastern border, but Valga/Valka has not resumed military development. Consequently, our research suggests how large global issues relate to peripheral small places: global nuclear deterrence tactics occurred within peripheral border areas of the Soviet Union depicted in seemingly irrelevant small towns (such as Valga/Valka) and their surroundings. Military installations and facilities functioning as large spatial priorities affected the cities and regions of many countries during the Cold War period (Woodward 2005; Westing 1988) and are similarly now a part of urban and rural space as international tensions have escalated since 2022. Many localities also need to make concessions today, for example, sacrificing strategic areas – meaningful to local communities for socio-cultural reasons – to military purposes. Likewise, the military sector can in turn provide jobs in such regions, which can revive housing markets and attract resources for infrastructure development.

The example of Valga/Valka sheds light on realistic trajectories for places experiencing urban shrinkage. An emphasis in the scholarly literature exploring urban shrinking tends to be placed on peripheralisation and functions that gradually disappear from the declining regions. However, besides the shrinkage phenomenon, we argue that it is important to place more emphasis on the nature of growth factors at play before shrinkage began and whether this growth was sustainable when "large issues" suddenly disappeared. Many small towns fall into decline following the gradual disappearance of key industries. However, the priorities of the military sector (cf. Gentile and Sjöberg 2010; Sjöberg 1999) can cause even greater ripples in the development of small places. If the development of a place is built only on an external force that eventually recedes or falls away completely, then severe shrinkage is expected and inevitable. However, it is unreasonable to assume that the prioritised spatial needs of the military sector can be aligned with regular spatial planning, regardless of the political regime; consequently, it is also difficult for planners to prepare strategically for future stages of demilitarisation and their socio-spatial impacts. Yet, the waves of militarisation and demilitarisation tend to alternate in frontier regions, and this inevitably is reflected throughout the development of cities and landscapes.

**Acknowledgements**

This research was supported by the Estonian Research Council grant (PRG1919) "Rethinking smartification from the margins: co-creating smart rurality with and for an aging population". The authors also thank Meelis Kivi, Kalev Härk and Kairid Leks for sharing written and unwritten information about the history of Valga with us and Ago Tominga for his help in designing the map.

**References**

Auer, M.R., and Raukas, A. 2002. Determinants of environmental clean-up in Estonia. *Environment and Planning C Government and Policy* 20(5):679–698. DOI:10.1068/c0011j

Baumann, B., Kretschmer, D., Von Plato, J., Pomerance, J., and Rössig, T. 2020. "Small places, large issues" revisited: Reflections on an ethnographically founded vision of new area studies. *International Quarterly for Asian Studies* 51(3–4):99–129. DOI: 10.11588/iqas.2020.3-4.13362

Gentile, M., and Sjöberg, Ö. 2010. Soviet housing: who built what and when? The case of Daugavpils, Latvia. *Journal of Historical Geography* 36(4):453–465. DOI: 10.1016/j.jhg.2010.01.001

Grava, S. 1993. The urban heritage of the Soviet regime the case of Riga, Latvia. *Journal of the American Planning Association* 59(1):9–30. DOI: 10.1080/01944369308975842

Hess, D.B., and Pae, T. 2020. Competing militarisation and urban development during the Cold War: how a Soviet air base came to dominate Tartu, Estonia. In: Brook, R., Dodge, M., and Hogg, J., (Eds.), *Cold War Cities: Politics, Culture and Atomic Urbanism, 1945–1965*. London: Routledge, pp. 148–166. DOI: 10.4324/9780203701478

Hess, D.B., and Tammaru, T. 2019. Modernist housing estates in the Baltic countries: formation, current challenges and future prospects. In: Hess, D.B., and Tammaru, T. (Eds.), *Housing Estates in the Baltic Countries: The Legacy of Central Planning in Estonia, Latvia and Lithuania*. Dordrecht, Netherlands: SpringerNature, pp. 3–27. DOI: 10.1007/978-3-030-23392-1_13

Hess, D.B., Tammaru, T., and Leetmaa, K. 2012. Ethnic differences in housing in post-Soviet Tartu, Estonia. *Cities* 29(5):327–333. DOI: 10.1016/j.cities.2011.10.005

Jauhiainen, J. 1997. Militarisation, demilitarisation and re-use of military areas: The case of Estonia. *Geography* 82(355):118−126. ISBN: 9783642571275
Kant, E., Luha, A., and Tammekann, A. 1932. *Valgamaa: maadeteadulik, tulunduslik ja ajalooline kirjeldus* (Valga County: Geospatial, Economic and Historical Description). Tartu: Eesti Kirjanduse Selts.
Kivi, M. 2020. *Valga ja militaaria* (Valga and the Military). Brochure of the Valga War Museum.
Koop, M. 2012. *Valga linna ruumiline areng alates 18. sajandist* (Spatial development of the city of Valga since the 18th century). Bachelor's thesis, Estonian University of Life Sciences.
Lundén, T. 2009. Valga-Valka, Narva-Ivangorod Estonia's divided border cities—Cooperation and conflict within and beyond the EU. In: Jaroslaw, J., (Ed.), *Conflict and cooperation in divided towns and cities*. Berlin: Logos, pp. 133–149. ISBN: 9783832523541
Mander, Ü., Kull, A., and Frey, J. 2004. Residual cadmium and lead pollution at a former Soviet military airfield in Tartu, Estonia. In: Wieder, R.K., Novák, M., and Vile, M.A. (Eds.), *Biogeochemical Investigations of Terrestrial, Freshwater, and Wetland Ecosystems across the Globe*. New York: Springer Dordrecht, pp. 591−607. DOI: 10.1007/978-94-007-0952-2_40
Miller, A.-L., and Bell, S. 2019. Keep out! No entry! Exploring the Soviet military landscape of the coast of Estonia. In: Bell, S., Fisher, A., Maia, M.H., Pallini, C., and Capresi, V., (Eds.), *SHS Web of Conferences 63: Modernism, Modernisation and the Rural Landscape Proceedings of the MODSCAPES_Conference2018 & Baltic Landscape Forum, Tartu, Estonia, June 11–13, 2018*. Published online (EDP Sciences), 11001. DOI: 10.1051/shsconf/20196311001
Openshaw, S., and Steadman, P. 1983. The geography of two hypothetical nuclear attacks on Britain. *Area* 15(3):193−201. DOI: 10.1016/0260-9827(82)90014-3
Pärn, J. 2020. *Nõukogude Liidu relvajõud Eestis 1944–1995* (The Armed Forces of the Soviet Union in Estonia 1944–1995). In: Eesti Sõjaajaloo Teejuht (Guide to Estonian military history). Available from: https://teejuht.esap.ee/eesti-ringreis/noukogude-liidu-relvajoud-eestis-1944–1995/ [accessed 09 May 2023]
Peil, T. 2005. Estonian heritage connections—people, past and place: the Pakri Peninsula. *International Journal of Heritage Studies* 11(1):53−65. DOI: 10.1080/13527250500037021
Rakowska-Harmstone, T. 1986. Baltic nationalism and the Soviet armed forces. *Journal of Baltic Studies* 17(3):179−193. DOI: 10.1080/01629778600000091
Raukas, A. 1999. *Endise Nõukogude Liidu sõjaväejääkreostus ja selle likvideerimine* (Residual Pollution from the Military of the Former Soviet Union and Its Elimination). Tallinn: Estonian Ministry of Environmental Affairs.
Seljamaa, E.H., Czarnecka, D., and Demski, D. 2017. "Small places, large issues": between military space and post-military place. *Folklore-Electronic Journal of Folklore* 70(7–18). DOI: 0.7592/FEJF2017.70.introduction
Sjöberg, Ö. 1999. Shortage, priority and urban growth: towards a theory of urbanisation under central planning. *Urban Studies* 36(13):2217–2236. DOI: 10.1080/0042 09899239
Tähiste, A., and Mõniste, M. 2018. *Eesti 20. sajandi (1870–1991) sõjalise ehituspärandi kaardistamine ja analüüs: Vilaski raketibaas, Tinu külas Valga vallas Valga maakonnas* (Mapping and Analysis of Estonian Military Construction Heritage of the 20th Century (1870–1991): Vilaski Missile Base, Tinu Village, Valga Municipality, Valga County). Kärdla

Tammaru, T., and Kulu, H. 2003. The ethnic minorities of Estonia: changing size, location, and composition. *Eurasian Geography and Economics* 44(2):105−120. DOI: 10.2747/1538-7216.44.2.105

Tintĕra, J. 2019. Innovative housing policy tools for local governments in shrinking communities with a large share of privately owned apartments: a case study of Valga, Estonia. *Transylvanian Review of Administrative Sciences*, Special Issue, pp. 124−139. DOI: 10.24193/tras.SI2019.7

Tintĕra, J., Kotval, Z., Ruus, A., and Tohvri, E. 2018. Inadequacies of heritage protection regulations in an era of shrinking communities: a case study of Valga, Estonia. *European Planning Studies* 26(12):2448−2469. DOI: 10.1080/09654313.2018.1518409

Valga municipality (Estonia), 2018. *Valga vallas maastikupilti kahjustavate kasutusest väljalangenud ehitiste likvideerimine. Ehitusprojekt nr VV-01/2018* (Liquidation of out-of-use buildings damaging the landscape in Valga municipality. Construction project No. VV-01/2018).

Valga/Valka combined general plan, 1970. National Archives of Estonia. Available under code "ERA.T-14.6.14". Coloured version of the map available from: https://www.ra.ee/kaardid/index.php/et/map/view?id=359393 [accessed 24 August 2023]

Revision of Valga/Valka combined general plan, 1988. National Archives of Estonia. Available under code "VAMA.379.1s.163".

Valga general plan, 1954. National Archives of Estonia. Available publicly and digitally under code "EAM.3.1.467 leht 1". Available from: https://ais.ra.ee/et/description-unit/search?refcode=EAM.3.1.467 [accessed 09 May 2023]

Valka municipality (Latvia), 2023. *Valka shelter reconstruction project.*

Veldi, M., and Bell, S. 2019. A landscape of lies: Soviet maps in Estonia. In: Bell, S., Fisher, A., Maia, M.H., Pallini, C., and Capresi, V. (Eds.), *SHS Web of Conferences 63: Modernism, Modernisation and the Rural Landscape: Proceedings of the MODSCAPES_Conference2018 & Baltic Landscape Forum, Tartu, Estonia, June 11–13, 2018*. Published online (EDP Sciences), 11001. DOI: 10.1051/shsconf/20196308002

Westing, A. 1988. The military sector vis-a-vis the environment. *Journal of Peace Research* 25(3):257−264. DOI: 10.1177/002234338802500305

Woodward, R. 2005. From military geography to militarism's geographies: disciplinary engagements with the geographies of militarism and military activities. *Progress in Human Geography* 29(6):718−740. DOI: /10.1191/0309132505ph579oa

# 8 The role of architects in fighting the monotony of the Lithuanian mass housing estates

*Marija Drėmaitė*

**Introduction**

Recent scholarly studies have introduced a new approach on the design of the large Soviet mass housing estates, increasing interest in the unique architectural designs and regional diversity (Ritter et al. 2012; Meuser 2016). As Meuser and Zadorin (2015) demonstrate, the Soviet post-war mass housing was, despite the appearance of monotony, in fact substantively diverse. Michał Murawski (2018) has noted that the scholarly accounts of built socialism's shortcomings and disintegrations have contributed a great deal to the understanding of socialist modernity as a perverted version of modernity proper, failure-bound from the beginning. However, the exceptional nature of Baltic design within the Soviet mass housing context has been touched upon by several researchers, particularly in light of Baltic relations with, and orientation towards, the West and international modernism (Hess and Tammaru 2019; Kalm 2012:33–45; Drėmaitė 2017). David Crowley 2008; Crowley and Reid 2000; 2010; and Susan E. Reid 2014 emphasised the ways in which designers and consumers cultivated agentic creativity despite or in opposition to strictures imposed on them from above. Papers discussing specific Baltic aspects of mass housing have also shown the criticism of mass housing (Kurg, 2009), which led to alternative house design solutions. Findings of the research of the architects' role in designing large housing estates in Estonia suggest that regulations issued in Moscow played a less important role than previously assumed in town planning outcomes because international modernist city planning ideals, combined with local expertise, strongly influenced town planning practice (Metspalu and Hess 2018). Similar ideas were reflected in Lithuania (Maciuika 1999; Maciuika and Drėmaitė 2020). In this regard, the chapter will further explore the role of an architect and unique design in Lithuania in the field of mass housing.

The methodology of the research is based on the concept of the Baltic states as "the Soviet West". William Risch argues (2015) that different experiences of WWII and late Stalinism and contacts with the West ultimately led to this region (Baltic Republics and Western Ukraine) becoming Soviet,

DOI: 10.4324/9781003327592-11
This chapter has been made available under a CC-BY-NC-SA 4.0 license.

yet different from the rest of the Soviet Union. While "the Soviet West" was far from uniform, perceived differences between it and the rest of the Soviet Union justified claims at the end of the 1980s that the Soviet Union was an empire rather than a family of nations. The well-known Soviet-era cultural critic Yuri Gerchuk has observed (2000:82) that the Baltic republics (Lithuania, Latvia and Estonia) actively contributed to a transformation of the Soviet Union's aesthetic environment and to the formation of a new, modernist sensibility: "Annexed by the Soviet Union in the twilight of the Second World War, these republics were able to bounce back somewhat more rapidly than other regions during the era of the Khrushchev thaw; for this reason, cultural products from the Baltics inevitably came to symbolise the European culture". It can therefore be presumed that a smaller scale of the republics (Lithuania had almost three million residents, Latvia – up to two million, and Estonia – ca. one million); a developed housing stock from the pre-war period of the 1920s and 1930s; and later incorporation into the Soviet Union (the Soviet occupation of the three independent Baltic states by the Soviet armed forces occurred simultaneously in June 1940) resulted in a different planning and mass housing construction even under the all-union strict regulation.

Material for this research was selected from the USSR professional press covering the period of 1956–1990: the monthly journal *Arkhitektura SSSR* (*Architecture of the USSR*) and the Lithuanian professional journal *Statyba ir Architektūra* (*Construction and Architecture*). Original designs (including drawings, photographs and briefs) were examined at the Lithuanian State National Archives and the Vilnius Regional State Archives, as well as in the seven volumes of the "Collection of designs of the Lithuanian SSR towns, blocks and microrayons", published by the State Urban Design Planning Institute from 1967 to 1985 (in total, 290 designs).

Another methodological approach used in the research is based on the theory of expert cultures (Kohlrausch et al. 2010:9–30). The expert is not only seen as a trained professional but also as a mediator between the nation and the state. Expert status is also a cultural ascription largely dependent on social, economic and political environment. While one standpoint sees a static, top-down, highly controlled relationship between the totalitarian state and the professionals within it, the latest research reveals far more nuanced and complex reciprocal influences between the specialists and the state officials in charge. Indeed, Lithuanian and Estonian architects were rather closed professional groups (trained in local architectural schools with pre-war tradition – Tallinn (Estonian Art Academy), Kaunas (Polytechnic Institute) or Vilnius (Art Institute) with almost no administrative or leading specialists from Moscow, Leningrad or other Soviet republics). In the post-Stalinist period beginning in 1954, the all-union policy of "national specialists" in the national republics was introduced, and since then, all Lithuanian construction and architecture leadership became predominantly local, raised and trained in Lithuania. In 1959 they began to be assigned to leading positions in urban and regional offices, having successfully changed their "guardians" sent from Russia during the first post-war decade.

The state's increasing faith in its architects is corroborated by the fact that the architects did indeed enjoy greater freedom compared to representatives of the other creative professions (Maciuika and Drėmaitė 2020:70). Architects were regarded as experts (more from the technical than artistic standpoint), and as such they were granted greater decision-making authority, particularly in the field of city planning.

In this context, the recollections of the first generation of post-war modernists, the so-called founding fathers of the "Lithuanian modernist school", are important (Maciuika and Drėmaitė 2020). Born in the 1930s, raised in cultured family surroundings and finishing high school during the years of WWII (1940s) and the Soviet occupation, this generation began expressing itself in the 1950s by criticising Stalinist architecture by realising significant public buildings in urban centres and even in mass housing and by rising into ever more influential posts in architecture and academia. Architects (in published materials and in conversation) emphasised the Baltic and especially Lithuanian mass housing design as a special case within the entire Soviet Union. It can therefore be presumed that the Soviet cultural image of the "West" (Péteri 2010:1–13) and the group agency were determining factors in the self-understanding of Soviet-era Lithuanian architects and designers, becoming an underlying factor in the narrative of shaping different built environment in comparison to Soviet standardisation and even cultural resistance.

The main theoretical question of this chapter is how much impact local Lithuanian architects were able to make in a seemingly rigid system of Soviet housing production. Was it because of the peripheral nature of the Baltic republics, where regulations were less strict, or was it motivated by the self-perception as "the Soviet West" and professional aspirations of the architects as a professional group not satisfied by the Soviet standardisation? This chapter will therefore further explore the role of the architect and the individualised design approach in the field of mass housing.

**Standardisation of mass housing as an architectural problem**

The development of residential zones became a critical urban planning issue for the Soviet Union following the Communist Party's 1957 promise to provide every Soviet family with their own individual apartment (Decree No. 591). The housing construction industry had to focus on two issues in particular: standardisation and industrialisation of prefabricated housing types and the new residential district model for the housing blocks. Both undertakings were subject to strict regulation from the beginning: the adoption in 1954 of regulations known as the Construction Norms and Rules (known by their Russian acronym, SNiP – *Stroitel'nye Normy i Pravila*) served for years as a means to control residential housing planners. In the period from 1954 to 1991, the SNiP rules dealing with mass housing were thoroughly revised only four times: in 1957, 1962, 1971 and 1985 (Meuser and Zadorin 2015:21), resulting in slow development of residential architecture.

A Soviet version of the neighbourhood unit, the *microrayon* (microdistrict) was developed with the aim of grouping prefabricated blocks of flats. The composition abandoned the location of houses along the perimeter of a city block in favour of a more freestyle arrangement (called "open planning"), which followed three parameters: compass direction, topography and the economics of the assembly crane (Meuser and Zadorin 2015:153). New housing was to be grouped into large, functionally zoned *microrayons* with 9,000–12,000 inhabitants. The core unit of the *microrayon* was a group of blocks of flats serviced by kindergartens, schools and shopping centres. Several *microrayons*, in turn, would be joined together to create a residential district (*rayon*) with 40,000–50,000 residents, with its own central shopping and recreation centre, a medical services building and other similar public facilities. Green zones were introduced between buildings and roadways, while pedestrian walkways wound through interior courtyards.

The essence of this type of planning was a tiered system of public cultural and consumer services based on the estimated needs of 1,000 inhabitants and defined by frequency of use (Baranov 1967:168–242): daily use sites (kindergartens, schools, food shops), periodic use facilities (visited two to three times per week) and episodic use facilities (used two or three times monthly). Services accessed on a daily basis were located within the boundaries of a given *microrayon* and usually arranged no further than 400 m from a given home. All first-tier public buildings were also expected to follow standard designs and consist of prefabricated parts. Second-tier (or rayon/ district-level) facilities, such as cinemas, libraries, department stores and health care facilities, were meant to be used periodically and were thus located within 1 km of residential homes. The *microrayon* approach was extremely attractive for rapidly growing cities since planners could apply it continuously, linking one *microrayon* to another, achieving a limitless expansion of their socialist cities.

By 1961, the Third Congress of Soviet Architects was able to boast of huge quantitative progress (165 million square metres of residential floor space built in 1959–1960), but it also took note of significant shortcomings, including "a lack of creativity in use of standard designs", and "a one-sided perception of industrialisation in architecture" (Arkhitektura SSSR 1961:6). Even Nikita Khrushchev noted the "lack of aesthetics" in industrial construction in a report he presented to a plenary meeting of the Soviet Communist Party's Central Committee in November 1962. "Nowadays", the Soviet leader observed, "Soviet architects face many new problems, especially concerning large panel house construction. The technology of industrial construction demands simple forms and minimum variety. Even under such conditions, however, the question of expressiveness in architecture must not be ignored. Individual architectural and artistic undertones must reveal themselves without exceeding the limits of what is capable and reasonable" (Kosenkova 2013:65–6). Studies had been repeatedly conducted on the use of "artistic undertones", but the economy was the real reason why Soviet mass housing areas were full of elongated rectangular five-storey buildings with 60 to 80 units per structure (the Moscow Institute series I-464 were the most widespread industrial series in the Baltic cities) arrayed in extremely regular patterns.

To avoid monotony in thousands of new residential districts, diversity had to be introduced as a matter of urgency. In 1960, the architectural journal *Arkhitektura SSSR* introduced a new regular section titled "Residential districts and the scope of progress in the construction of *mikrorayons*". Between 1960 and 1962, institutes under the jurisdiction of *Gosstroi*, the All-Union Construction Committee, developed and published external finishing design recommendations and manuals for standard housing series (Arkhitektura SSSR 1960:9; 1960:10). Architects understood, however, that such measures were superficial and that more fundamental change was necessary. For example, the architect Albertas Cibas, an official with the Lithuanian *Gosstroi* (a republic's branch of the central institution), called for measures to attract the best and most experienced architects to work in standardised designs, providing them with a degree of creative liberty, particularly in the adaptation of standard designs for certain sites (Arkhitektura SSSR 1961:7).

**Introducing experimental design in Vilnius**

The tension between serialised and unique designs became a long-standing feature of Soviet architectural production. As Mart Kalm put it, "Standardised designs were already in extensive use during the Stalinist period but became an obsession during Khrushchev's Thaw, when economical building practices became the focus of attention. [...] The more the state demanded standardised designs, the more architects became irritated and felt oppressed by the restrictions" (Kalm 2012:39). In Lithuania, for example, such tasks were delegated to recent graduates who, in turn, hoped to escape their new duties as soon as possible.

However, the ambitions of a new generation of modernist architects could be seen in efforts to amend and improve standardised designs. Architect Vytautas Edmundas Čekanauskas recalled: "We referred to these buildings simply as bricks, for their slab shape and ungainly nature. We wanted to improve these buildings by changing those horrible Russian designs. An internal mini-competition was organised [in 1961 at the Vilnius Urban Construction Design Institute] to see what could be done with those buildings" (personal conversation with Čekanauskas, Vilnius, 11 December 2006). Indeed, proposals were already being made to design a series of residential buildings suited specifically for the Baltic republics, incorporating materials typically found in the region. Field visits to Finland organised for Soviet architects also inspired them to seek better solutions for mass housing architecture (Drėmaitė 2021).

Experimental design became an effective way of introducing improvements to the Soviet residential housing system. Architects and designers who could characterise their work as "experimental" (meaning that an experimental building would provide technical know-how for the rest of the building sector) could bolster their credentials as technical specialists and draw on greater resources and enjoy greater freedoms. The Vilnius Urban Construction Design Institute established a special office for this purpose in 1960. Between 1960 and 1965, numerous experimental apartment units and housing designs were produced, seeking alternatives that improved standardised designs. A group of young architects (Gediminas Valiuškis, Enrikas Tamoševičius and Algimantas and Vytautas Nasvytis brothers) drew up the experimental plans for apartment units in 1961. Algimantas' account illuminates some of the available strategies he employed in pursuit of his goals: "We looked particularly at developments in the West, because this has long been the predisposition in Lithuanian architecture. Our orientation was explicitly towards the West, and not the East. It was our purpose to soften the norms and requirements that were issued to us from Moscow. We always sought a way to adjust them to better fit our local conditions – or, wherever possible, to ignore them, to skirt them, or, in the end, to at least soften them" (Maciuika and Drėmaitė 2020:102–4).

Vytautas Nasvytis, Jaunius Makariūnas and Algirdas Jasinskas developed an improved version of the standard I-464 series house, with apartments that could be divided using light sliding partitions or room dividers that also served as closets, allowing for different configurations of each apartment. However, the price for one square metre increased by 5–6%, and the Vilnius factory producing the concrete elements refused to make changes. A chairman of the Board of Lithuanian Union of Architects complained: "This is a strange situation – on the one hand, architects are criticised for design flaws, yet on the other hand, their improvements are not accepted" (Cibas 1962:13).

In 1966, Vilnius hosted the third plenary meeting of the Soviet Architects' Union Executive Committee, during which Vytautas Balčiūnas, Senior Architect for the Vilnius Urban Construction Design Institute, voiced his criticism and called for allowing national republics to oversee the planning and construction of residential housing themselves: "We must review and repeal the planning and construction prohibitions which have been adopted en masse in recent years and which only serve to inhibit initiative and thwart progress. A proposal has been made to change the system of standardised planning and financing and to restore the previously enjoyed right to have a republic's construction committee plan and finance standardised projects being constructed in that republic. It is time to grant republics more self-sufficiency, which will also increase initiative and accountability" (Balčiūnas 1966). The proposal was not implemented, but the 1969 decree "On Measures to Improve the Quality of Residential Construction" (Decree No. 392) already aimed to produce greater architectural expressiveness, introduce unique cityscapes and imbue residential areas with a stronger sense of local identity.

### Shaping the individuality of the microrayon: Lazdynai and Žirmūnai as all-union models

Vilnius grew at a particularly fast rate. In 1945, the post-war Lithuanian capital had 110,000 inhabitants. By 1959, that number had more than doubled to 236,000 and in 1979 Vilnius was nearly at the half-million inhabitant mark. A new master plan for Vilnius (Master Plan Brief 1964), completed by architects Vaclovas Balčiūnas, Kazimieras Bučas, Vladislavas Mikučianis, Vilhelmas Sližys and Juozas Vaškevičius in 1967, foresaw the construction of ten new housing estates in massive neighbourhoods planned as separate city districts.

*Microrayon* D–18 was designed and built between 1962 and 1964 for 12,000 residents as a first part of the future Žirmūnai mass housing district in the northern periphery of Vilnius. The young urban planner Birutė Kasperavičienė (1926–1976) had previously collaborated on the design of a new industrial town named Elektrėnai (1960) and other *microrayons* in Vilnius (Figure 8.1). D–18 was to be an "experimental site", introducing the concept of diversity in skyline through the use of improved five-storey series I-464A panel houses (developed by architect Bronius Krūminis and structural engineer Vaclovas Zubrus), experimental nine-storey panel houses (designed in Lithuanian Urban Construction Design Institute by architect Enrikas Tamoševičius) and an open neighbourhood centre featuring public art. Kasperavičienė had also used the natural slope of the adjacent Neris River bank and adapted it into a park. The completed microrayon attracted an all-union interest.

In a continuing search for new ideas, the first Soviet-wide review of the country's architecture was organised in Moscow in 1967. From a field of 167 submitted designs, the first prize was awarded by a unanimous decision to the *Microrayon* D–18 of Žirmūnai thus "signalling a turning point in Soviet architecture" (Barkhin 1968). Žirmūnai, it was claimed, served as an example of urban housing perfectly matching the contemporary style of the new Soviet residential ideology calling for original architectural ensembles and profiles. Since it was the first mass-produced residential site to be awarded the prestigious architectural USSR State Prize, it was elevated to a new level of good practice. Reviewers singled out overall improvement in designs of standard five-storey houses: "The site's value stems from a successful implementation of mass housing" (USSR State Prizes 1968 April Session). It was explicitly stated that Žirmūnai served as proof "that industrial housing can be diverse: it can have its own character

*Figure 8.1* Architect Birutė Kasperavičienė at her drawing desk at the State Urban Construction Design Institute in Vilnius. Photo: A. Barysas, 1968.

Source: Lithuanian Central State Archives.

and it can avoid becoming a [nationwide] cliché" (USSR State Prizes 1968 October Session).

In 1962, two young architects, Vytautas Brėdikis (1930–2021) and Vytautas E. Čekanauskas (1930–2010) at the State Urban Construction Design Institute were commissioned to design Lazdynai, a large housing estate for 40,000 residents on the Western periphery of Vilnius. Both architects were already known for their modernist designs of public buildings and talked about their desire to improve standardised large housing estate image – in later interviews they mentioned considerable influence on their designs of Finnish (Tapiola), Swedish (Vällingby, Farsta), and modern French (Toulouse-Le Mirail) suburban projects (Maciuika and Drėmaitė 2020). The site for Lazdynai was naturally hilly and well forested – features that would be preserved in the final landscape design in contrast to usually levelled sited for large panel house construction. The project architects also suggested improvements to the series I-464 buildings and advocated for the placement of five- and nine-storey housing blocks perpendicularly across the sloping terrain to create a unique silhouette for the new community. For the first time in Lithuania they added large panel 12-storey towers as vertical landmarks of the site. The production of these new types of buildings was a challenge for the Vilnius Panel Construction Factory, but institutional nationalism (strong personal connections between architects and local Communist Party and municipal leaders) played a role when the need arose to defend the innovative designs to the Soviet Construction Committee (Figure 8.2).

Over time a kind of "institutional nationalism" took shape, strengthened by collegial ties with local Lithuanian government officials, which helped generate original solutions to material shortages and economic challenges. Local officials and state authorities in the memories of architects are mostly described as "favourable", well-disposed towards the architects as fellow Lithuanians, yet understanding nothing about architecture. In general, the architects' recommendations were locally respected because the architect was considered an authority.

Consideration of Lazdynai's nomination for the Lenin Prize in 1974 proceeded extremely smoothly at the Architectural Section and at the Plenary, because the uniqueness of the site was confirmed by *Gosstroi* and the Architectural Section members' visit to Lazdynai and a tour by helicopter (Lenin Prizes 1974 April Session). Thus, Lazdynai became the first mass housing urban design recognised with the most prestigious Soviet national prize (Drėmaitė 2019). The residential area showed a degree of Nordic influence with the semi-open courtyards and pedestrian avenues, the development of customised designs and adaptation of hilly terrain. Ideologically Lazdynai demonstrated the possibilities for a bright future in large panel mass housing construction, with only an added touch of "landscape design" and improvement of standard house series.

*Figure 8.2* Architects Genovaitė Balėnienė and Aida Lėckienė with colleagues working on the detailed plans of Lazdynai mass housing area. Photo: T. Žebrauskas, 1973.

Source: Lithuanian Central State Archives.

## Critique of *microrayon* and the pursuit of uniqueness over standardisation

The optimism of the 1960s had been replaced with the criticism voiced in the 1970s. Despite the success of Lazdynai, criticism of mass-constructed residential districts intensified both in Lithuania and throughout the Soviet Union. The problems of urban and architectural monotony of residential areas become a frequent topic of the professional press. Several main issues were named: First, it was a long-term and repetitive use of the same series I-464 with minor modifications from Žirmūnai district (in 1964) with 36,000 residents through Lazdynai (1967–1973) with 40,000 residents and further in Karoliniškės (planned for 45,000 residents, architects Genovaitė Balėnienė and Kazimieras Balėnas, 1971–1976), Viršuliškės (planned for 25,000 residents, architects Kasperavičienė and Jonas Zinkevičius, 1975–1980) and Baltupiai (planned for 20,000 residents by architect Nijolė Chlomauskienė, 1978). The second problem was a small selection of finishing materials and a lack of colour variety. In Karoliniškės and Viršuliškės, the colour schemes for the *microrayons* were designed; however, during the construction process, only dark red firewalls were done in Karoliniškės, whereas no colour was provided for Viršuliškės. In the case of Šeškinė, one more mass housing area built in 1977–1985 (architects Balėnas and Balėnienė) for 50,000 residents, instead of the planned clean white colour of the facades, grey was implemented because the factory ran out of white granite grains (Ruseckaitė 2010).

However, the most problematic thing was the lack of comprehensive implementation of urban projects – no residential area was fully built as envisaged in the approved projects. Economic considerations played a central role, with new regulations imposed that increased both the density and height of residential buildings in the *microrayon*. With a few exceptions, urban design for the remainder of this period was viewed as an endless row of tedious construction, made only worse by the low quality of work and partial project completion. Even Lazdynai came under criticism for falling short of better standards existing abroad, both in terms of aesthetics as well as technical execution – a low quality of sound insulation, panel construction, etc. (Gūzas 1971).

Criticism of the monotony of residential areas was followed by the proposals on how to avoid it. It was proposed to replace series I-464 with new types. Although experimental design in residential architecture flourished in the late 1960s and early 1970s, no actual construction was built. In 1970 Krūminis' group designed an experimental series for construction in Lithuania in 1971–1975, which served as the basis for the second-generation 120 V panel housing series (1973), distinguished by façade detailing, corner balconies and larger kitchens. These blocks with a shorter pitch were seen as a possibility to make apartment planning more convenient and to bring greater volumetric diversity to the *microrayon*. The need for the latter was highlighted again in Decree No. 392 "On Measures to Improve the Quality of Residential and Civil Construction", adopted in 1969 by the Soviet Council of Ministers and the Communist Party Central Committee, which aimed to achieve greater architectural expressiveness, introduce unique cityscapes and imbue residential areas with a stronger sense of local identity. Indeed, series 120, developed during the 1970s for different cities of Lithuania, was in construction up till 1990.

Original urban design ideas were proposed in Kalniečiai residential area in Kaunas, especially its 3rd *microrayon* (architect Alvydas Steponavičius, 1983). The 120 K series of five-, nine- and twelve-storey panel buildings, designed especially for Kaunas, were arrayed around pedestrian paths and courtyards, with a central public area accentuated by sixteen-storey monolithic concrete towers. In addition, each street featured different coloured building numeration plaques with unique graphic designs (Jankevičienė 1991:110–2). For the first time, the overall composition also incorporated existing old-style country homes with their surrounding garden plots. The biggest innovation in Kalniečiai, however, was the decision to forego the tiered system of consumer services, instead locating large shopping centres closer to principal streets.

Introducing concrete towers that were meant to be unique architectural landmarks of *microrayon* was seen as another solution. The first experimental sixteen-storey tower block was built using monolithic concrete in Lazdynai in

1980, designed by architect Česlovas Mazūras. In 1981–1982, architects Krūminis and Danas Ruseckas designed 13- and 16-storey monolithic concrete towers with rounded balconies for the Šeškinė residential district. In the end, however, poor construction quality, inferior materials and incomplete structures conveyed a sad image of squalor.

Urban sociology was gaining interest as yet another measure to improve mass housing areas. By the late 1970s, interest in urban sociology was on the rise, spurring research of the new residential districts and analysis of the quality of the living environment and its impact on human lifestyle, spiritual condition and health. A short time after residents moved into a new district, analysis of the neighbourhood's usage patterns began to paint a "sociological portrait" of the given area (Vanagas 1982). For example, in 1982 even 91.6% of Lazdynai's residents expressed satisfaction with their own district, emphasising the neighbourhood's suitability for pedestrians and a proper balance between architecture and the surrounding landscape (*ibid*). However, the results of such sociological studies had little impact on the construction of residential districts, where economic considerations always took precedence. Nevertheless, polling of residents in Vilnius' new neighbourhoods revealed one clear and strong preference for districts constructed within a more scenic natural environment (Vaškevičius 1974).

Indeed, environmental concerns began receiving more attention in the early 1980s. In 1980, Vilnius hosted a local conference for the planners of the new residential districts on landscape design and natural environment. The conference found that the intrusion by architects and builders into the natural environment during the construction of new residential neighbourhoods often harmed the existing ecological balance, causing irreparable damage to the environment. The observation was also made that new Vilnius construction sites merely used the natural environment, rarely doing anything to shape those surroundings (Jančiauskas 1981). Lithuanian planners began cooperating with Finnish architects over the question of how to preserve the natural environment in the design of new residential districts. An experimental planning project was developed in 1978, focusing on the Baltupiai district in Vilnius and Malminkartano in Helsinki – both low-rise construction areas with striking natural surroundings (a pronounced terrain, forests and a small river) and located on the urban periphery (Girčys and Katilius 1981).

**Regionalist approach to mass housing areas**

By the late 1970s and early 1980s, it was possible to see more numerous manifestations of regional identity and an ever more individualised approach to mass housing design, as with the series designed for the coastal city of Klaipėda in 1980 (by Krūminis, Sargelis, Zubrus and Jonas Stanislovaitis, an engineer with the Klaipėda Panel Building Factory).

The Baltic port city's volatile climate was also taken into account: terrace balconies were designed that could be transformed into enclosed glazed verandas. Another innovation in mass-produced apartment construction was the introduction of an 11 m$^2$ hall leading to the apartment balcony, heated attics, prefabricated roofs without rolled covering and more spacious kitchens (8.67 m$^2$).

Houses were decorated with red brick cladding, considered to be a style typical of the Klaipėda region, conceptually developed by architect Gytis Tiškus. While working on new residential districts in Klaipėda, Tiškus tried to maintain the unique architecture of each neighbourhood centre, seeking inspiration from local and regional characteristics. He changed and adapted standardised public buildings, conveying regional traits through colour and materials, using red brick or ceramic finishing. Despite these efforts, Klaipėda's originality was limited to its unique public buildings and red brick finishing – broader urban planning approaches, however, received their fair share of criticism.

Strong regionalist approach in architecture of residential neighbourhoods could be felt especially in smaller towns and resort settlements. It was motivated by the general Soviet design approach prevalent in resort areas that the built environment must please the eye, but more importantly, these areas were usually in natural resorts and protected areas. For example, mass housing area in a Baltic Sea resort town Nida on Kopu street was specially designed employing regional elements (architect Ramūnas Kraniauskas, 1980s). Nine 3- to 4-storey multi-apartment houses were constructed in yellow brick and finished with pitched roofs and decorative wooden elements (Drėmaitė et al. Neringa 2022:164–7). These large structures were harmoniously incorporated into a particularly fragile and protected natural environment. Residential area in another Baltic Sea resort town Palanga was specially designed in yellow brick, low-rise (2–4–5 storey) multi-apartment segments to avoid standardised five-storey prefabricated slabs (architects Juozas Šipalis, Edmundas Benetis, 1974–1980). Multi-unit two and four-storey apartment buildings in Birštonas, a small resort town along the Nemunas River in southern Lithuania, were specially adapted to suit the scale and surroundings of the natural environment.

**Conclusion**

Reviewing mass housing architecture in Lithuania, it is evident that architects sought to avoid Soviet standardised designs that were not valued as creative and prestigious within the professional environment. Despite standardisation and the very limited choice of materials and building types, there were attempts to improve the design of mass-produced architecture and neighbourhood planning. In the 1970s original district planning solutions were sought by using the modified series I-464 and composing them in a unique way in each district and decorating the buildings and the environment in a specific way to the district. Faced with urban monotony and its criticism, the spatial parameters of standardised designs began to be changed. In order to avoid urban and architectural monotony, in the 1980s, the identity and original character of residential areas began to be created with new series 120, specially designed high-rise towers, landscape design and regionalist approach to individualised house design. Such efforts were made easier by the existence of professional relationships developed between designers, local administration officials and the heads of construction material enterprises (especially the directors of housing construction factories).

Although architects in many Soviet republics began to shun mass construction projects and concede initiative to engineers, the design of mass housing in Lithuania was overseen by professional architects. The state's increasing faith in its architects is corroborated by the fact that the architects did indeed enjoy a greater freedom compared to representatives of the other creative professions. Architects were regarded as experts or specialists (more from the technical than artistic standpoint), and as such they were granted greater decision-making authority, particularly in the field of city planning. The increasing role of an architect as an expert in the field of mass housing illustrates the shift in late Soviet architecture, where decision-making in urban planning shifted from politicians to technocrats. This shift was validated because of the changing approach to an architect as a technical expert and growing expert culture in general.

Professional ambitions of architects and urban planners were reflected in design competitions and "experimental projects". However, the great majority of experimental designs were never realised or were implemented with considerable modifications because of the economic issues. It can be noted that the lack of prestige in mass housing urban planning lead to the fact that most of these areas were designed by female architects (e.g., Birutė Kasperavičienė designed 11 sites; Genovaitė Balėnienė – 11; and Nijolė Chlomauskienė – 15). This aspect in mass housing urban design can be researched further.

The numerous awards regularly given to Lithuanian urban planners in the late Soviet period can be viewed in two ways. Though a considerable role was played here by the good reputation earned by the designs of Žirmūnai D-18 and Lazdynai, Lithuanian approaches to *microrayon* design, in general, were notable within the general Soviet context for their architectural originality. First and foremost, these districts were constructed in suburbs well chosen for their natural characteristics, while the effort to give each new neighbourhood a sense of uniqueness drove improvements in industrialised housing construction and assembly as well as environmental clean-up projects. It could be said that these efforts became the defining characteristics of Lithuanian residential urban planning.

## References

Arkhitektura SSSR 1960:10. Varianty fasadov krupnopanel'nykh zhylykh domov serii 1–468 i 1–335 [Façade options for the large panel house series 1–468 and 1–335], 52–53.

Arkhitektura SSSR 1960:9. Novye reshenya fasadov krupnopanel'nykh zhylykh domov serii I–464 [New façade solutions for the large panel house series I–464], 22–24.

Arkhitektura SSSR 1961:6. Rezol'utsya III-go vsesoyuznogo s'ezda sovetskih arkhitektorov [Resolution of the 3rd All-Union Congress of Soviet Architects], 3–5.

Arkhitektura SSSR 1961:7. Tretyi vsesoyuznyi s'ezd sovetskih arkhitektorov [The Third All-Union Congress of Soviet Architects], 7.

Balčiūnas, V. 1966. Lietuvos architektų pasiūlymai [Proposals by Lithuanian architects]. *Statyba ir architektūra* 11:5.

Baranov, N. et al. (Eds.). 1967. *Osnovy sovetskogo gradostroitel'stva [Fundamentals of Soviet Urban Planning]*. Vol. II. Moscow: Stroiizdat.

Barkhin, M. 1968. Smotr dostizhenyi sovetskoi arkhitektury [Review of Soviet architectural achievements]. *Arkhitektura SSSR* 1:5–20.

Cibas, A. 1962. Lithuanian SSR Union of Architects Board Chairman's report 'Architects' tasks in the CPSU Programme' at the Board Plenum, February 2. Lithuanian Archives of Literature and Art, f. 87, ap. 1, b. 363, pp. 4–17.

Crowley, D. 2008. Paris or Moscow? Warsaw architects and the image of the modern city in the 1950s. *Kritika: Explorations in Russian and Eurasian History* 9(4):769–798.

Crowley, D., and Reid S.E. 2010. *Pleasures in Socialism: Leisure and Luxury in the Eastern Bloc*. Evanston: Northwestern University Press.

Crowley, D., and Reid S.E. Eds. 2000. *Style and Socialism. Modernity and Material Culture in Post-War Eastern Europe*. Oxford: Berg.

Decree No. 591. 1957. On the Development of Housing Construction in the USSR of the Soviet Communist Party's Central Committee and Council of Ministers 31 July 1957. Library of Legal Acts of the USSR. http://www.libussr.ru/doc_ussr/ussr_5213.htm

Drėmaitė M., Mankus, M., Migonytė-Petrulienė, V., and Safronovas, V. 2022. *Neringa. An Architectural Guide*. Vilnius: Lapas.

Drėmaitė, M. 2017. *Baltic Modernism. Architecture and Housing in Soviet Lithuania*. Berlin: DOM publishers.

Drėmaitė, M. 2019. Baltic *mikroraions* and *kolkhoz* settlements within the Soviet architectural award system. *The Journal of Architecture* 24(5):655–675. 10.1080/13602365.2019.1670717

Drėmaitė, M. 2021. Symbolic geographies, nordic inspirations, and Baltic identities: Finnish impact on the formation of the post war modernist architecture in Lithuania during the state socialist period. *Architectural Research in Finland* 4(1):21–35. https://doi.org/10.37457/arf.84565

Gerchuk, Y. 2000. The aesthetics of everyday life in the Khrushchev thaw in the USSR (1954–64). In: Reid, S.E., and Crowley, D. (Eds.), *Style and Socialism. Modernity and Material Culture in Post-War Eastern Europe*, p. 82. Oxford: Berg.

Girčys, G., and Katilius, S. 1981. Žmogus. Namas. Aplinka. Ar negriauname statydami? [The Man. The House. The Environment. Aren't we demolishing by constructing?]. *Statyba ir architektūra* 3:17–18.

Gūzas, E. 1971. Ant jausmo ir logikos svarstyklių [Between feelings and logic]. *Kultūros barai* 1:7–8.
Hess, D.B., and Tammaru, T. (Eds.). 2019. *Housing Estates in the Baltic Countries. The Legacy of Central Planning in Estonia, Latvia and Lithuania.* Cham: Springer. 10.1007/978-3-030-23392-1
Jančiauskas, H. 1981. Ateities miestai [Cities of the future]. *Statyba ir architektūra* 8:20–21.
Jankevičienė, A. (Ed.). 1991. *Kauno architektūra* [Architecture of Kaunas]. Vilnius.
Kalm, M. 2012. Baltic Modernisms. In: Ritter, K., Shapiro-Obermair, E., and Wachter, A. (Eds.). *Soviet Modernism 1955–1991. Unknown History.* Zürich: Park Books.
Kohlrausch, M., Steffen, K., and Wiederkehr, S. 2010. Expert Cultures in Central Eastern Europe. The Internationalization of Knowledge and the Transformation of Nation States since World War I – Introduction. *Expert cultures in Central Eastern Europe: The internationalization of knowledge and the transformation of nation states since World War I.* Osnabrueck: Fibre Verlag.
Kosenkova, Y. 2013. Predstavlenya o tselostnom organizme goroda v period izmenenya tvorcheskoi naprevlennosti sovetskoi arkhitektury. *Estetika ottepeli: novoe v arkhitekture, iskusstve, kul'ture.* Kazakova, O. Ed. Moscow.
Kurg, A. 2009. Architects of the Tallinn School and the critique of Soviet Modernism in Estonia. *The Journal of Architecture* 14(1):85–108. 10.1080/13602360802705171
Lenin Prizes for 1974 (April Session). RGALI, f. 2916, op. 2, d. 751, c. 28–29.
Maciuika, J.V. 1999. East block, west view: architecture and Lithuanian national identity. *Traditional Dwellings and Settlements Review* 11(1):23–35.
Maciuika, J.V., and Drėmaitė, M. (Eds.). 2020. *Lithuanian Architects Assess the Soviet Era: The 1992 Oral History Tapes.* Vilnius: Lapas.
Master Plan of Vilnius, the brief. 1964. Vilnius Regional State Archives, f. 1011, ap. 5, b. 155.
Metspalu, P., and Hess, D. 2018. Revisiting the role of architects in planning large-scale housing in the USSR: the birth of socialist residential districts in Tallinn, Estonia, 1957–1979. *Planning Perspectives* 33(3):335–361. 10.1080/02665433.2017.1348974
Meuser, P. 2016. *Seismic Modernism. Architecture and Housing in Soviet Tashkent.* Berlin: Dom publishers.
Meuser, P., and Zadorin, D. 2015. *Towards a Typology of Soviet Mass Housing. Prefabrication in the USSR 1955–1991.* Berlin: DOM publishers.
Murawski, M. 2018. Actually-existing success: economics, aesthetics, and the specificity of (still-)socialist urbanism. *Comparative Studies in Society and History* 60(4):907–937. 10.1017/S0010417518000336
Péteri, G. Ed. 2010. *Imagining the West in Eastern Europe and the Soviet Union.* Pittsburgh: University of Pittsburgh Press.
Reid, S.E. 2014. Makeshift modernity. DIY, craft and the virtuous homemaker in new Soviet housing of the 1960s. *International Journal for History, Culture and Modernity* 2(2):87–124. 10.5117/HCM2014.2.REID
Risch, W. 2015. A Soviet West: nationhood, regionalism, and empire in the annexed western borderlands, *Nationalities Papers* 43(1):63–81. 10.1080/00905992.2014.956072
Ritter, K., Shapiro-Obermair, E., and Wachter, A. (Eds.). 2012. *Soviet Modernism 1955–1991. Unknown History.* Zürich: Park Books.

Ruseckaitė, I. 2010. Sovietinių metų gyvenamieji rajonai Vilniuje: tipiškumo problema [Problem of standardisation in Soviet residential districts. The case of Vilnius]. *Journal of Architecture and Urbanism* 34(5):270–281. 10.3846/tpa.2010.26

USSR State Prizes 1968 (April Session). RGALI, f. 2916, op. 2, d. 396, p. 167.

USSR State Prizes 1968 (October Session). RGALI, f. 2916, op. 2, d. 397, p. 15.

Vanagas, J. 1982. Gyventojas apie savo butą (arba ką rodo sociologiniai tyrimai) [Resident [speaks] about his apartment or what do the sociological findings tell]. *Statyba ir architektūra* 5:10–12.

Vaškevičius, J. 1974. Architektūros ir landšafto harmonija [Harmony of architecture and landscape]. *Statyba ir architektūra* 11:4–7.

# Part III
# The non-politics of everyday life in spatial peripheries during socialism

# 9 Courtyards, parks and squares of power in Ukrainian cities

## Planning and reality of everyday life under socialism

*Kostyantyn Mezentsev, Nataliia Provotar, and Oleksiy Gnatiuk*

**Introduction**

Similar to Czepczyński (2008), it can be said that Ukrainian "socialist cities" were socialist not only because they were developed (constructed) during the socialist period and under the control of the socialist power, these cities were socialist because of the ideological contexts attached to almost every city planning approach, every project and every public space. Everyday use of public spaces was also endowed with socialist meanings.

Although everyday use of urban public spaces is considered as ordinary, day-by-day actions, encounters and interactions of residents, it is not perceived as a unity. It consists of many typical but not necessarily interconnected actions that are taken for granted by the city dwellers. Everyday practices differ in terms of their set, spatial configuration and mode of performance and are characterised by spatial diversity (Denysyk et al. 2020; Gnatiuk et al. 2021). Under socialism, the everyday was a space of contradictions being "simultaneously a site of alienation and liberation," and "its rhythms encompass both mundane cyclicality and the transformative potential of linearity" (Alekseyeva 2019:1). It was not something unequivocally established but changed with the mixing of residents resettled from different regions, from urban and rural areas and from different social strata who found themselves in the same public space at the same time (Mezentsev et al. 2019). By examining the everyday activities, events, experiences and also memories of residents in different types of public spaces in Ukrainian large ordinary cities Vinnytsia and Cherkasy, we can gain a deeper understanding of the planning aspects of their former, nowadays and even future development.

Often planned on the sites of the former market and cathedral squares, new "main public spaces" of Ukrainian large ordinary cities during socialism have become the spaces of power, the venue for parades and official celebrations, demonstrating the greatness of communist ideology. Such squares were not fenced but had invisible barriers separating power from the people. Meanwhile, the green public spaces became the loci of mass communication and even the self-organisation of residents. There was

DOI: 10.4324/9781003327592-13
This chapter has been made available under a CC-BY-NC-SA 4.0 license.

a variety of daily activities – summer cinemas, attractions, dance floors, playrooms and pavilions, kiosks of "cheap" ice cream and soda. Over time, invisible barriers that separated the main squares eased, and protest movements emerged there. At the same time, the everyday use of green public spaces was decreasing, and at the end of socialism, they often turned out into spaces of decline and danger.

Other kinds of public spaces are courtyards. If the urban planning documentation of the 1930s provided the principle of quarter building up the city (*Rules and norms* ..., 1930) with visually defined courtyard areas perceived "like a roomy park" (Staub 2005), the construction norms and rules of the 1950s introduced hierarchical city division into microdistricts, residential districts (consisting of several microdistricts united by a single community centre) and urban districts (in large cities), as well as free planning (*SNiP II-V.1* 1954). New microdistricts were built mainly on the urban periphery, where "space was immediately available but directly adjacent to established networks of services ... to which the new housing could be quickly and cheaply hooked up" (French 1995:75–6). Although the journal *"Architecture of the USSR"* ("Arkhitektura SSSR") idealised well-designed microdistricts as "modern satellites orbiting around older urban centres" (Harris 2013), the haste of construction (in accordance with the defined plans for new housing construction) caused the lagging far behind (often by years) the provision of the basic services considered as a fundamental feature of the scheme (French 1995). Microdistrict's courtyards were offering less everyday communication even compared to pre-WWII communal houses. Moreover, owing to industrialisation, former rural residents have settled in mass housing estates, causing intricate mix of modern urban infrastructure with some traditional rural everyday practices.

How was everyday life in urban space planned during socialism in Ukrainian cities? What was it really like? How has it evolved over time? We will attempt to answer these questions through the prism of Ukrainian large ordinary cities (*"звичайні великі міста"* – in Ukrainian), typical growing large centres of industrial-agrarian regions with no special privileges or exclusions in urban planning approaches and norms. This chapter is based on a study of three types of urban spaces during the socialist era – main squares, city parks and courtyards.

**Large ordinary cities – political periphery in the Soviet Union's urban network**

Ukrainian cities under socialism were part of the urban network of the USSR. Accordingly, the key principles of their planning and even the models of their public spaces used by the residents were determined in Moscow. The hierarchy of cities was key in Soviet structures, and "the Soviets was an urban empire" (Medvedkov 1990). The Soviet urban system operationally was highly hierarchical, mainly because of the chain-of-command nature of governmental functions (Adams 1977).

City planning in all Soviet republics was firmly based on administrative norms and instructions issued by supervising authority and directed by the communist party (Hess and Metspalu 2019). As a result, it was characterised by the almost non-existence of decentralised decision-making (with regards to the municipalities' autonomy to develop their cities) and strict vertical hierarchy and centralist principle of planning actors (Arzmi 2023). The General Scheme of Population Distribution on the Territory of the USSR, elaborated in the 1970s, established the multi-tier hierarchy of urban settlements with the core decision-making centre at Moscow (Vladimirov et al. 1986; Avdotiin et al. 1989; Kumo and Shadrina 2021). A separate tier was formed by the capitals of the union republics and intra-republican interregional centres of the Russian and Ukrainian Soviet Republics; these were up to 40 somewhat privileged second-order urban centres (Vladimirov et al. 1986). The rest of the cities were actually the periphery of the urban USSR's urban network, as they were opposed to the main focus of attention (Danson and de Souza 2012). The Soviet ideology aimed to assemble an egalitarian space with uniform cities evenly distributed across the USSR (Kumo and Shadrina 2021), and therefore, the same principles and approaches to the planning of public spaces were in force for these cities, differing only slightly depending on the category, number of residents, and taking into account special climatic conditions or location on the coast of large water bodies. Only for two cities – Moscow and Leningrad – some exceptions were allowed by the construction regulations and rules (*Rules and norms ...* 1930, *SN 41–58* 1959).

The term "large ordinary cities", used in this chapter, is not an official term from the planning documents but a specific type of Soviet city. We understand them as the large cities of the USSR that were not apparently privileged in terms of urban planning. In Soviet Ukraine, large ordinary cities were typical centres of industrial-agrarian regions with rapid population growth during the period of industrialisation essentially due to the inflow of rural residents. Such cities had large areas of low-rise detached residential development, and since the 1950s new standard microdistricts were built for workers of industrial enterprises. In contrast to the republic's capital (Kyiv), interregional centres (Kharkiv, Odesa, Lviv), closed cities (Dnipro, Zhovti Vody) or mining and metallurgical cities, which had their own formalised or informal rules and traditions for planning and arranging public spaces, large ordinary cities did not have such freedom and acted as an arena for the passive implementation of planning projects handed down from above.

Large ordinary cities of Ukraine cannot be considered peripheral in the full sense of this term. Most of them were centres of administrative regions, and important industrial nodes, and were characterised by significant population growth during the period of socialism. However, the concept of peripherality is multifaceted. Paasi (1995) distinguishes four aspects of peripherality – political, economic, cultural and ideological, and later Luukkonen (2010) – five dimensions of peripherality – economic, political, social, physical and cultural. Peripherality in some aspect does not mean that the city would also be peripheral in some other aspects (Luukkonen 2010; Carter 2015). Thus, Ukrainian large ordinary cities were peripheral in a political sense with a weak ability to influence the governance and the decision bearing on its interests, they had weak institutional structures and lack of local embeddedness in the urban network (in terms of composition and structure of network relations within and beyond the city, including networking with the other cities of similar rank in the state hierarchy) and weak civic society (Luukkonen 2010). Interpreting peripherality as a "product" of power relations (Nagy et al. 2015), such cities should be considered as the objects of policy-making rather than active participants in shaping it (Arter 2001). To some extent, they can be considered also as peripheral in a cultural sense with imposed cultural standardisation, domination of a particular (so-called Soviet) culture and loss of identity (Paasi 1995; Luukkonen 2010).

Concerning the understanding of the Soviet cities' network, we agree with Domański and Lung (2009) that the (political and partly cultural) peripherality of Ukrainian large ordinary cities can only be understood in the context of its relationships to the core (the centre of the decision-making in urban planning) and other peripheries. So, these cities cannot be characterised as spatial periphery on the Soviet Union margins but rather by network-shaped peripherality ("in-betweenness") (Herrschel 2012), being located behind cities that in the Soviet urban hierarchy could have a greater influence on the public spaces planning. This hierarchical political peripherality remains visible in today's Ukrainian network of cities (Mezentsev et al. 2015), although these cities have risen one step higher.

**Data and methods**

This research is based on the analysis of the interviews on everyday use and perception of the selected public spaces in the 1960–1980s with residents and local experts conducted in two Ukrainian large ordinary cities, Vinnytsia and Cherkasy, the local print media from the mid-1950s to the end of the 1980s, photographs from the respondents' family archives and urban planning documentation and cities' master plans elaborated in the 1930–80s. These different time periods are considered as a basis for a deeper understanding of public space planning and everyday use in large ordinary cities after 1957 until the collapse of the Soviet Union, i.e., from the so-called third phase in the Soviet urban development (according to French (1995)), which was marked by the beginning of mass residential construction and a change in principles and approaches to planning and ends with the crisis of urban development and planning in the late 1980s.

The first step included an analysis of urban planning documents relevant in the 1930s and 1980s, primarily building codes and regulations, as well as master plans of Vinnytsia and Cherkasy. The master plans of the early 1930s, the second half of the 1940s and the 1950s and 1960s were analysed based on indirect sources, in particular (Dmytrenko 2016; Vecherskyi and Zlyvkova 2011), while master plans of the 1980s – based on the official resolutions of the Council of Ministers of Ukraine (Resolution... ..., 1984; Resolution... ..., 1987).

The second step involved a selective analysis of local print media, Vinnytska Pravda and Cherkaska Pravda newspapers, as well as local history publications, which include analytic materials, quotes from the local press related to urban planning, arrangement and use of public spaces (Karoieva et al. 1998; Horodskykh et al. 2012; Fedoryshen 2015; Yukhno 2013, 2016, 2019), as well as relevant photographs from this period.

In January–February 2022, 15 semi-structured, in-depth interviews including four thematic blocks were conducted. A total of 12 interviews were conducted with residents of Vinnytsia (6) and Cherkasy (6) aged 50 to 82. The first three thematic blocks concerned frequency of visits to the main square of the city, the central city park and the courtyard area, as well as details of their use, arrangement and perception. The interviews were conducted retrospectively – to recall memories and feelings of the interviewees with regard to the selected public spaces in the 1960s, 1970s and 1980s. They were encouraged to express their personal life narratives from the childhood, adolescence and youth. In particular, they were asked to "tell your story", to "recall a memorable incident", to specify "what were your feelings" and to find "epithets to describe them". The specific terminology of the Soviet period was used by the interviewers to revive memories of the informants. Some questions were asked about whether the interviewees took photographs in different types of public spaces – when, against which background and under what circumstances. The fourth block concerned the assessment of everyday life in general and the evolution of everyday use of public spaces by interviewees in the 1960s, 1970s and 1980s.

Based on the identified features of planning and everyday use of public spaces in planning documents, local media and interviews with residents, three additional interviews were conducted with local experts on the development of the cities of Vinnytsia (1) and Cherkasy (2), who helped to interpret certain facts, events and trends related to these cities and their public spaces.

During and after the interviews, some interviewees provided the opportunity to view photos from family archives taken in public spaces, which became an important additional source of information for this chapter.

As the case public spaces were selected two main (central) squares – Lenin Square (now Maidan Nezalezhnosti) in Vinnytsia and also Lenin Square (now Soborna Square) in Cherkasy; two central urban (municipal) parks – Maxim Gorky Vinnytsia Central City Park (now Leontovych Central City

Park) in Vinnytsia and May 1 Park in Cherkasy (now Sobornyi Park); and courtyards of multi-apartment buildings in microdistricts built during the Soviet era, mainly on the outskirts of the cities.

**The square of power and ... for power?**

According to planning documents, the main square in a large ordinary city was intended for the placement of administrative and public institutions, as well as for holding demonstrations, parades and holidays ("mass public festivities"). Its functions were to be social, cultural and political, and the traditional central-city (economic) functions were to be downgraded; it should be a nucleus of urban social and political life, and its role as a central business district was to be reduced and dispersed to secondary centres throughout the city (Bater 1980; Hausladen 1987). Traffic on the main square was expected to serve mainly the buildings and structures located on it. It was recommended to organise a detour of the intensive flows of urban traffic (*SN 41–58* 1959, *SNiP II-K.2–62* 1967, SNiP II-K.3–62, 1963), and some later to arrange parking lots for cars (*SNiP II-K.2–62* 1967).

In terms of everyday use of the main squares in the 1960–80s, it is important to emphasise several points. First. The main squares were *places for demonstrations, parades and official celebrations.* Under the absence of land rent they were planned as large open spaces which creates "best possible mise-en-scène" for commemorative parades or any obligatory gathering (Czepczyński 2008, French 1995) like Lenin squares in both Vinnytsia and Cherkasy. In fact, they played the role of a "symbolic city centre" (Hausladen 1987), places of public manifestations supporting the regime.

Second. During the celebrations, such immense squares remained *exclusive spaces for power holders* and simultaneously "only a container for the anonymous crowd" (Czepczyński 2008). Talking about main squares, most interviewees recall holidays (as May 1, Victory Day, October Revolution Day) and related arrangement and events like buildings and street lamps decorated with flags and banners, passing the square along the tribune by the representatives of various institutions and organisations who carrying flags, posters and banners, greeting the authorities and being greeted by them (Figure 9.1b,c). Moreover, even the placement of the tribune on the square was specifically regulated in the urban planning documentation: "to the right of the movement of the demonstrators' columns" (*SN 41–58* 1959). However, for those who did not participate in the demonstration, the entrance to the square was restricted, and the squares were guarded by a significant number of militiamen.

Third. In spite of exclusiveness and visible and invisible separation in times of official celebrations, the main squares were *the places for ordinary people's short-time obligatory but funny meetings,* sometimes starting point for the later activities outside them. As interviewees recall, the mood of the public usually was elevated (Figure 9.1a):

*Figure 9.1* Main square in the large ordinary city; a – May Day demonstration in 1963, b – on holiday in 1970, c – on weekday in 1967 (Lenin Square, Cherkasy), d – wedding photo sessions with laying flowers to the Lenin monument in 1978 (Lenin Square, Vinnytsia).

Source: Photos from the family archives of LC01 (a,b,c) and LV09 (d) interviewees.

*On holidays, we had a euphoric mood: either in the snow and rain in November or in the sun in May, any weather. We always walked with joy, positive emotions (LV09)*

*We did not really want to come and carry banners. However, when we came, there was an upswing... ... We didn't walked depressed and shouted loudly (LC02)*

After the demonstration, the cordon was removed, so people could walk around the square and communicate with the acquaintances they met. Some corporate teams had a tradition of taking photos after the demonstration (Figure 9.1b): *"When we worked at the plant in 1982–1983, we always took group photos on the square" (LV13).*

At the same time, many eyewitnesses recall that people did not stay long on the square, searching for more comfortable and intimate places: *"everyone wanted to get rid of the heavy posters and banners as soon as possible and disperse along the side streets to the parks and courtyards, or hiding in the side corners to have a festive drink" (LV09); "after the manifestation, we agreed with friends where to go for a picnic and quickly scattered away" (LV10).* During the celebrations, alcohol stores and restaurants were usually closed in and around the main square (in order to ensure safety and order in the city), so immediately after the demonstrations and parades most of participants hurried to leave the main square and spend their leisure time, in the forest parks on the outskirts of the cities or in the suburban area, participating in the so-called *"mayovkas"*. It was both a tradition and to some extent a protest – to go out into nature and get rid of "ideologisation". As Czepczyński (2008) noted, although large public squares were designed as agora-like spaces they were not really used for democratic practices but only to enhance the communist power, becoming "anti-agoras of socialist cities". Thus, the participants of the festive demonstrations on the main city square had two "masks". One was ideologically correct with the desire not just to join the "important event", but also to support it actively by shouting predetermined slogans. It was accompanied by feelings of pride, joy and patriotism. And the other one with the desire to leave the main square as soon as possible and go outside choosing different activities for the "celebration". It was confirmed by the local expert who emphasised on the existence of "double standards:" *"First, compulsory [demonstrative] participation in official demonstration, and then free [from officialdom] gathering in nature" (EC07).* Moreover, most personal photos were taken on the "mayovkas".

Fourth. In the 1960–80s, the festive events on the main squares took place only a few times a year, so the rest of the time they served *as empty transit spaces.* In Vinnytsia and Cherkasy, the main square was rather a transit space for locals and visitors to the city centre. There were many shops around on the main street, as well as a hotel and main post office or a cinema, and many people were walking along the street *on affairs* (specific Soviet era expression meaning going shopping, to the pharmacy, to the post office, for administrative services, etc., but not to work), to public transportation stops or taking a walk. Therefore, there were always a certain number of people in the central squares, but they were crossing it simply as a part of the main street (Figure 9.1c). Some interviewees expressed their attitude in the following way: *"We had no walks there [to the Lenin Square]. We didn't care about that place" (LV10),* "There were not many people. Deserted space" *(LC05)* or *"Typical central square. No special emotions" (LV14).*

Such emptiness of large main squares located next to the administrative and communist party buildings was supposed to show the grandeur and to a certain extent separation, unreachability of power for ordinary residents who preferred to bypass these squares without urgent need. Most of our interviewees answered that they had never taken pictures on the central square. Some photos found in the interviewees' family archives confirmed that on ordinary days they were empty.

Monuments of the communist leaders, mostly Lenin, were raised on the main squares (both case squares in Vinnytsia and Cherkasy had Lenin monuments). Arrangements of the area around these monuments usually included flower benches, but there were no benches for sitting and communication, and the concrete slabs were scorched by the sun in clear weather during the warm season. *"This 'pan' is cold in winter and hot in summer. Not good to have a walk" (LC06)*. In view of this, the squares were ill-suited for leisure and recreation. Moreover, there was constant control over the behaviour of visitors – an additional pushing factor that turned formally public spaces into exclusionary spaces of power and a subject of political control (Hou 2010). Since the central squares were unattractive and generally uncomfortable as public spaces. As our local expert summarises:

*What is Lenin Square – an open space? Nothing except the monument to Lenin. Few people thought of taking pictures against such a background (EC07)*

Fifth. In the 1960–80s, some new practices with ideological connotation were cultivated at the main squares as *places for memorable events*. Thus, Lenin Squares were a canonical place for wedding photo sessions, with a tradition of laying flowers to the Lenin monument (Figure 9.1d). The local expert from Cherkasy suggested that it could be considered as *"a new Soviet ritual" (EC07)*. In some period, teenagers were taken to the Komsomol near the Lenin monument.

**Central city park: regulated everyday life and satisfaction of residents' demand**

Urban greening was one of the basic principles of socialist urban planning that was implemented in Ukrainian cities. Special attention was paid to the main parks of cities, which were called "city parks" and were designed as "parks for culture and rest". To this day no city "is so small that it does not have its Park of Culture and Rest" (French 1995:47).

In the early 1960s, numerous projects for the creation of green public spaces were developed. In the local newspaper "Cherkaska Pravda" the issue was discussed in an interview with the architect:

*Cherkasy will change unrecognisably. It will be a wonderful garden city on the shore of the azure sea. The embankment will turn into a park zone ("A word to M. I. Korablin, a chief city architect", 15 October 1961)*

This project was not fully implemented, and the idea of a garden city remained only an attractive metaphor since the master plan actually fixed the transformation of Cherkasy from a "resort city" to a "city of the great chemical industry". As Conterio (2022) notes, although urban planners argued that Soviet cities were becoming garden cities, they just were a city filled with gardens.

City parks were important foci of the residents' everyday life. However, the set of activities in the parks was clearly regulated. It was envisaged to divide the city park into zones by use and purpose: entertainment, cultural and educational events, physical activity, recreation for children, "quiet rest" for adults, as well as farm buildings (*SN 41–58* 1959, *SNiP II-K.2–62* 1967).

In the 1960s, Vinnytsia Central City Park and Cherkasy May 1 Park were the *main vacation spots*. Sunday was the only one day off, and people, having finished their household chores, went to the park in the afternoon. *"People visited the park to have a rest, mainly with children, at least once a week" (LV11)*. It was like a ceremonial exit for the whole family with the aim of *"seeing others and showing themselves" (LV09)*. Therefore, to go out, people usually put on the most beautiful clothing, *"not anything but the best dress" (LV10)*, and looked at other people, *"like at a fashion show" (LV09)*.

In most cases, our interviewees emphasised that parks offered a wide range of activities – attractions, game pavilions, exhibitions, etc., confirming that urban parks were in fact considerably more elaborate and would normally comprise, in addition to green spaces and flowerbeds, places of entertainment (French 1995). *"There was something for young people to do... ... There were a lot of people" (LC01)*. The most repeated mentions concern visiting dance floors in parks. Live music was played on the site, including a large symphony orchestra. *"There was a dance floor. There was such a wooden board and fence where the orchestra sat" (LC01)*.

Children could visit all kinds of amusement rides (apparently, the top attraction was an observation wheel) and play on a swing. There was a checker and chess club, an exhibition of paintings, courses on cutting and sewing, a bookstore (it was problematic to buy books in the city at that time) and library. *"Chess tournaments. Library. To take a book to read sitting on the bench" (EC07)*. At the stadium, the visitors could watch sport competitions. During the holidays, there were organised exhibitions of bouquets and handicrafts: *"the exhibition of flowers remains in mind for the rest of my life" (LC01)*.

On weekdays, the park was also full of people. During daytime, it was visited by pensioners and mothers with children from the surrounding quarters, and in the evening mostly by young people.

Everyday activities in the parks took place throughout the year with certain specifics in different seasons. In summer, a children's camp operated in the park. The Summer Theatre in Vinnytsia Central City Park regularly received troops from Leningrad and Moscow theatres and top pop music performers. In winter, a New Year tree was set up in the park, and a fair town was arranged nearby. The stadium turned into a skating rink in winter. *"It was very romantic; people of all ages were skating in the light of lanterns, with music playing" (LV09)*. The hilly part of the park served as a place for mass sledging and skiing. As a witness of those times remembers, *"we went there the holiday evening, squealed and squeaked with joy, happy ..." (LV12)*.

*Courtyards, parks and squares of power in Ukrainian cities* 163

*Figure 9.2* Central city park in the large ordinary city a – May 1 Park in Cherkasy in 1970s, b – Gorky Central City Park in Vinnytsia in 1960s.

Source: Photos from the archives of LC01 (a) and LV09 (b) interviewees.

The park has been a popular destination for personal and family photographs. They were most often taken in walkways and alleys, near decorative sculptures, in front of the summer theatre and the fountains, sitting on benches and at the main entrance arches (Figure 9.2):

*"I remember sitting with friends on the parapet of the fountain … when son was growing up, we also gathered in the park near the fountain; we keep a plenty of photos" (LV12)*

Our local experts confirmed this, pointing that they have many of such photos from the park in their archives, like *"Fountain – geese and swans. I have more than 20 photos of these geese. Standard … " (EC07)*.

When planning city parks, special attention was paid to catering. Restaurants, cafes and ice cream kiosks were to be located in the most visited places in all areas of the park. Such commercialisation elements within the green public space were free from "ideological" burden and played a role of an incentive for the visitors. In the summertime, eating ice cream was an intrinsic part of leisure time. *"The highlights were the ice cream and carousel. We went there to eat ice cream … " (LC04)*. Also, there were many sale points of soda with different syrups.

In contrast to the main squares, citizens felt freer and safer in city parks. Law enforcement officers ensured supervision and order maintenance in the parks, but visitors did not feel alienated and excluded. The municipal trusts of green construction (miskzelenbud) were responsible for arranging and maintaining the city parks; however, industrial enterprises also took "mastership" over certain green spaces. The secondary schools, located in the city centre, were responsible for cleaning several parts of the park on the public cleansing day (*subbotnik*) as well. Citizens were also involved in the arrangement of city parks on a voluntary basis, particularly children and members of the Komsomol, and "garden weeks" and "tree planting weeks" were part of the plan for the cultivation and maintenance of urban green space (Conterio 2022). *"I still have a photo – I am three years old. Parents on subbotnik. And I plant trees with them" (LC05)*, and *"Being schoolchildren, we plant trees in the park" (LC06)*.

**Large housing estate's courtyard: less order, more freedom?**

With the transition to microdistrict, courtyards were no longer considered as a separate structural element of planning, but indirectly their planning was discussed and specified. In particular, when planning microdistricts, it was recommended to "allocate a residential area for the placement of residential buildings with landscaped courtyards for public rest and children's games... ..., areas for economic purposes, parking spaces, intra-block passageways, and building entrances" (*SNiP II-V.1* 1954). In the broader context of the microdistrict planning, requirements were set for the location of playgrounds for children, the rest of the adult population, physical activities, dog walking, areas for economic purposes, and other special purpose areas (for placing garbage collectors, cleaning furniture and clothes, etc.), as well as landscaping the territory.

In microdistricts on the city's periphery, there were more opportunities for the local community to arrange courtyards according to their own ideas. To some extent, yard areas have become centres of informal socialisation and satisfaction of individual housing needs. Such processes can be called "socialist peripheral urbanisation" which differs from classical peripheral urbanisation. While the latter is characterised by the lack of centralised planning and "slow temporality" and incompleteness, when "homes and neighbourhoods grow little-by-little, in long-term processes of incompletion and continuous improvement led by their own residents" (Caldeira 2017:5), socialist peripheral urbanisation was centrally planned and rapid, providing much better living conditions and amenities. However, as a peripheral, socialist urbanisation on the city edges was distinguished by a creative and transformative nature (Caldeira 2017), when residents brought their own vision to the arrangement of the area.

Getting from the street to the courtyards created a feeling of semi-privacy, "intimacy in the urban milieu" (Alekseyeva 2019), and even disorganisation, lack of effective planning requirements and control. The courtyards' actual appearance was sometimes going far beyond the initial idea of architects and planners. Even if urban planners included clearly demarcated functional areas, residents' everyday use of them played an equal, if not more important, role in determining their actual function (Harris 2013). Such a situation can be considered as a certain strategy of the authorities: limiting the spaces of the main squares for citizens' everyday use, leaving them as spaces for the aggrandisement of power and fear of power, they gave a certain unofficial freedom in the arrangement of the courtyards and spontaneous agreement of the residents about its planning. Soviet architects even expressed an opinion about "positivisation of space" (Ikonnikov 1988) which included practices such as planting trees and creating gardens with the explicit demarcation of the personal territory, fixing own areas from external "intrusion" of children, neighbours, etc. (Alekseyeva 2019).

The less defined the urban spaces (as courtyards in the Soviet planning regulation) are, the more residents have been encouraged to develop their own systems (Staub 2005). That is why the unofficial yard arrangement has become more extensive. Moreover, the courtyard was positioned as "a space of escape from the state" (Dixon 2013), as a "personalised space of ownership" (Alekseyeva 2019). On the other side, the quality of such public space left much to be desired for (Durmanov and Dubbeling 2004). Consequently, the dwellers modified the courtyards according to their views on comfortable everyday space, appropriated public spaces for their own purposes (Harris 2013) – enclosed individual areas with improvised fences, planted flowers and trees, made cellars under balconies for the preservation of agricultural products, laid footpaths to the nearest transport stops and service facilities, etc.

Moreover, urban planning documentation even regulated certain elements of privacy in courtyards. In particular, it was envisaged to place sheds for storing fuel in the yards, as well as sheds and cellars for storing agricultural products; somewhat later it was allowed to placing garages for individual cars (*SNiP II-V.1* 1954, *SNiP II-K.2–62* 1967, *SNiP II-60–75\*\** 1985). The courtyards often became locations for arbitrarily installed tin garages, sheds used for keeping rubbish, and building materials. People that came to the city from the rural area tried to snatch a piece of land near their windows to plant fruit trees, vegetable gardens and flowerbeds and sometimes kept chickens and rabbits, as well as dovecote houses. *"Young people from the village with their communication tradition ... They brought elements of their everyday life with them" (LC05)*. Our local expert explained that *"reactively settled in the newly created regional centre ..., from villages ..., people transferred the Ukrainian tradition of arranging the yard to their new living area" (EC07)*. Such innovations caused a mixed reaction among residents. Someone expressed approval or at least understanding: *"Urban gardens were a fine thing, because everything was in bloom – the more greenery, the better. The land should not be vast ... " (LV13)*. Other interviewees condemned these things: *"Vegetable gardens spoil the aesthetic appearance of the microdistrict. It is no longer a city, but it is not a village either" (LV09)*.

The socialist microdistrict was built to supply egalitarian and equal housing for everybody, intending to accommodate residents from all strata of society, to foment a tight-knit residential collective, and facilitate classless neighbourhoods. Standardised high-rise housing estates became the most important spatial manifestation of the ideology to create a collective and just society (Leetmaa et al. 2015). However, in practice, it simultaneously enhanced anonymity, lack of personal identification with home, and lack of neighbourhood relations (Czepczyński 2008; Alekseyeva 2019). As our interviewees noted, closer relations with neighbours and the formation of "courtyard collectives" were typical for the construction of the 1930s, even if they lived in barrack-type houses with primitive living conditions. Over time, a more communal lifestyle in housing estates shifted to an individualistic one (Hess and Tammaru 2019).

Although the courtyards were considered as focal points for recreation and congregation after work (Dixon 2013), as *spaces for communication and joint activities of the neighbours*, and some of our interviewees report quite intense common activities and festivities in their courtyards, in most cases, celebrations were usually held in apartments and together with friends and relatives, not neighbours; the same for common photos. Compare:

*"Relatives are far away, but these [people] are nearby ... in the house for meat processing plant workers... ... They are together both at work and here" (LC03)* and *"We did not participate. We had no time" (LC01)*.

In many courtyards there was a tradition to hold community clean-ups, a kind of *"subbotnik"*: *"on the appointed date, the residents went out to clean the courtyard from rubbish, to plant bushes, trees and flowerbeds, and to arrange playgrounds for children" (LV09)*. Another kind of common activity was when men knocked up tables and benches somewhere in the courtyard, there played dominoes or cards. But in some instances *"such tables were often used by drinkers making noise at night and thus causing inconvenience to residents" (LV09)*.

It is also worth emphasising that courtyard areas were usually planned as more convenient for children and the elderly, oppressing other age groups, which comprise the majority of the population (Kogan 1967). This resulted in *social divisions* of the local community based on age. In particular, the "monopolisation" of the courtyard space by children (and parents with children) was observed when the playgrounds were located in the central part of the courtyard or by older people when, for example, benches near entrances were "owned" mainly by the elderly female residents (Alekseyeva 2019). They could *"sit there for hours, embarrassing the residents entering or going out the building, performing a kind of a border control" (LV11)*. They collected, exchanged and disseminated all the information about the residents and events in the courtyard and beyond. Also, the courtyards were the realms of preschool and primary school-age children accompanied by their parents or grandparents, or playing under the parental supervision from the apartment windows. Older children could also play "cops and robbers" in the courtyard, making shelters and tree houses, but more often they spent long hours outside the courtyard at an open school stadium playing football or other sports games.

Another social division emerged between residents by their social status. At the beginning, microdistricts were settled by residents of roughly the same status, who often worked at one or several enterprises. For example, the first residents of Vyshenka, the largest residential estate of Vinnytsia, built up in the 1960s, were employed mostly in machinebuilding enterprises (Gnatiuk and Kryvets 2018). However, the polarisation of residents intensified over time. The tool intended to create an equal society, in reality did not encourage people to interact with neighbours with whom they did not share similar interests and values, and physical integration served to increase psychological segregation (Janušauskaitė 2019).

Another division arose regarding the placement of individual cars and the construction of makeshift garages in courtyards between those who had automobiles and those who did not. To their neighbours, car owners, their automobiles, and single-car garages dirtied new housing estates and generally got in the way of people's everyday lives (Harris 2013:218).

In this way, the desire of urban planners to create egalitarian neighbourhoods collided with the production of everyday social divisions in the courtyards.

**Everyday public spaces in Vinnytsia and Cherkasy in the 1960s–80s – between standardisation, hierarchisation and modernisation**

As Crawford (2018) notes, socialist urbanisation took place through *standardisation*. Soviet architects and planners considered standardisation as an effective tool to embrace rationality, new technology, "to assist in the construction of environments appropriate to the new socialist way of life" (Crawford 2018:72), and to ensure the conformity of individual citizens to a socialist "ideal" (Staub 2005).

Standardisation of planning and everyday use of public spaces in Ukrainian large ordinary cities was manifested in various aspects. First, urban planning involved standard approaches to planning public spaces. The result was nearly identical large open main squares and even their architectural ensembles of landmark buildings. For example, there were similar monuments of Lenin on the main squares of Vinnytsia and Cherkasy. Such standardisation has to some extent contributed to the erasure of local identity. Accordingly, the use of these public spaces was similar in different cities.

Second, urban planning laid down standard approaches to landscaping and creating green spaces. City parks with similar names, functional zoning, sets of attractions and even sculptures were built according to standard templates. Practically every city had a central park for culture and rest, many of which bore the name of Gorky or were dedicated to certain anniversaries or Soviet holidays. Staying in such parks created a sense of placelessness since such a park could be found in most Ukrainian large ordinary cities. Some impressions and memories of our interviewees about leisure time in the parks of Vinnytsia and Cherkasy were almost the same.

Third, urban planning since the mid-1950s has established standard approaches to planning residential blocks/microdistricts. As a result, standardised building projects were the norm (Alekseyeva 2019), and "urban spaces became a uniform carpet of residential neighbourhoods joined end on end" (Staub 2005:340). They were arranged in accordance with standard norms for the entire country with regard to the provision with public spaces, social infrastructure, roads, etc. (see Underhill 1990). Only in the courtyards of multi-apartment buildings, with general approaches to standard planning, individual elements of local flavour were allowed.

Thus, standardisation was a means to produce a recognisable Soviet identity (Staub 2005), however with certain local features carefully hidden behind (standardised as possible) facades.

The ordinary Soviet city was a *hierarchically organised* administrative structure (Staub 2005) with an officially recognised "socialist hierarchy of places" (Czepczyński 2008) which constructed contrasted official spaces of power, semi-official green spaces, and almost unofficial local spaces of "dosed self-expression".

Thus, the main squares of Ukrainian large ordinary cities under socialism were planned only for everyday access to work (mainly for the urban elite) and for certain public holidays for selected residents, merged into an impersonal crowd of government supporters. Only the government determined the requirements for arranging such public spaces. Interviews with locals in Vinnytsia and Cherkasy confirmed that ordinary citizens were actually separated from the "quiet and deserted" everyday life of the main squares. There were usually no benches, ice cream or beer kiosks, etc., and a long stay there could only generate the interest of the law enforcement officers. Residents used the main square as a transit point on the way to public transport, administrative and public buildings, and work. On holidays, the use of the square took place according to a predetermined scenario, and after the end of festivities, residents hurried to leave the square for further informal communication. Therefore, in relation to the main squares, we can talk about *the "powerful"* (monopolised by the state) *production of space aimed to account for the interests (priorities) of the government*.

City parks were available for everyday use by all residents; however, their planning was carried out without the involvement of citizens, although paying attention to their interests. The interviews show that the daily activities in the both Vinnytsia and Cherkasy central parks generally met the needs and interests of the local dwellers. A certain manifestation of individuality was allowed here, but under the control of the security authorities. Thus, in relation to parks, we can talk about *the "powerful" production of space but taking into account the interests (demands) of citizens.*

The courtyards of multi-apartment residential buildings had free access for all, and citizens were widely involved in their arrangement. As a result of rapid industrialisation and "socialist peripheral urbanisation," rural practices were preserved/brought/adapted to urban conditions by former rural residents who moved en masse to new microdistricts in cities. Thus, in this case, we can talk about *the mixed production of space* (state plus community/citizens) *and the planned spontaneous "positivisation" of it.*

In the 1960–80s, the strategic vision for the planning and everyday use of public spaces in Ukrainian large ordinary cities generally remained unchanged. However, *social changes* brought their own corrections, in particular regarding the diversification of everyday use of public spaces, the growing role of individualisation, motorisation, the spread of the consumerist model of behaviour and corruption. Contrary to the original idea of standardisation and equity of microdistricts, the mass housing community evolved into a heterogeneous body with new social divisions (Harris 2013). By 1980s, Soviet cities entered a new stage in their development, in which the role of individual choice was increasingly bringing new patterns of public space use, sometimes essentially different than what was planned (French 1995). Enhanced living standards, the growth in private car ownership, the availability of services at home (e.g., television), centralised gas supply significantly influenced the arrangement of both green spaces and courtyard areas. "On a more everyday level, citizen initiatives and informal activities have created other new uses and forms of public space" (Hou 2010:9).

In the 1980s, city parks were losing some functions and, accordingly, visitors. As our interviewees noted, open dance floors and summer theatres in parks often could not withstand competition from indoor venues and activities and gradually declined. Visits to the parks on weekends and in the summertime became less frequent as many people received summer cottages (*dachas*) from the state and started to spend more free time there. Moreover, the newcomers contributed to the fast growth of Vinnytsia and Cherkasy were mainly former villagers who were used to visiting their native localities on weekends and holidays and thus visited the park only rarely, on a special occasion: *"Village, garden, household. On weekends, the city was dying out... ... Everyone went to the village... ... In the 80s, everyone got summer cottages and gardens, and people rushed there" (LC05).* Thus, although city parks showed some flexibility in responding to changing user needs (Low et al. 2005), in the late 1980s, they were not ready to overcome new challenges and gradually deteriorated.

Over time, microdistricts continued to play a central role in the residential organisation of large ordinary cities, and by 1980, around half of the Soviet Union's urban population lived in microdistricts (Alekseyeva 2019:60). However, everyday life in the courtyards of large housing estates has undergone changes. The most significant were the changes associated with the weakening of "neighbourly communication," the curtailment of joint leisure time, including increasing the scale and openness of courtyard areas and the number of residents in multi-story buildings. As Dixon (2013) mentioned, the high-rise buildings stand grouped around a "courtyard", but this space tended to be too large to be recognisably enclosed, and a sense of belonging might not arise in this larger yard. While earlier living with the shared use of some premises encouraged communication, the following increase in the standard and comfort of living have reduced this need. Moreover, this was facilitated by the spread of a more individualised lifestyle, increased incomes and strengthened segregation based on property and status, as well as the emergence of alternative places for communication based on interests. *"When we got an apartment, we felt like civilised people. Yes, we kept friendship with neighbours. However, we already felt closeness. This is mine" (LC02).* In the 1970s and to larger extent in the 1980s, the spread of television in every household caused expansion of the information channels, and communication with neighbours was no more the main source.

Another tangible change is related to the weakening of residents' self-organisation regarding the arrangement and maintenance of courtyards. Some of them continued to be involved in community clean-ups, but others believed that the improvement of the courtyard was the responsibility of the city (local) authorities. However, they, especially in the second half of the 1980s, became insufficiently capable of arranging and maintaining the area, providing only minimal needs. As a result, many yards in microdistricts became neglected and cluttered.

The main city squares underwent the least changes in terms of planning approaches. Some changes were related to their use when, in the late 1980s, they hosted oppositional anti-Soviet political rallies. The protesters gathered there no longer by order from above but of their own will and at their own risk. So, contrary to the intentions of the planners, spacious *anti-agoras* became real *agoras*, where people manifested their disappointment against ruling regimes (Czepczyński 2008:71). However, the concept of their planning and everyday use was practically not revised, and the problem of striking a reasonable balance between occasional public functions and everyday use (Bater 1980) has not been solved. Moreover, in a number of Ukrainian large ordinary cities, this remains noticeable even today. In this regard, a local expert from Cherkasy emphasises that there was no attempt to rethink the former Lenin Square: *"[t]his is traffic for public transport and parking for cars. The architects have neither the courage nor the ability. There is no unifying idea" (EC07).*

The interviews show that words used by our informants to describe their everyday life in public spaces have predominantly positive connotations (like stability, happiness, joy, pride), associate with certain activities and leisure places (dances, walks, park and river), as well as the stage of the life cycle (youth, childhood). However, the key characteristics of the everyday life have changed over time in line with societal changes and stage of their life course. The interviewees characterise the 1960s as *"true, interesting" (LC01)* when they *"lived with faith in the future, did not delve into internal troubles" (LC02)*, the 1970s *"as it should be" (LC01)*, *"full of emotions" (LC04)* when they *"always have waiting for something" (LC03)* and *"many things were looked at with different eyes" (LC02)*, but still *"everything could be planned for a long time" (LC06)*, with a *hope for the future" (LC04)*, and the 1980s as associated with *"work, work, work, no time to rest, in nature on weekends" (LC06)*, *"less leisure, more village and garden" (LC05)*, but with *"more opportunities" (LC05)* when *"everything goes by so fast" (LC03)*.

To sum up, the intention of the Soviet authorities to establish certain standardised patterns of planning organisation and the use of public spaces in large ordinary cities met with a "hierarchical" reaction of residents who accepted the main squares as spaces of power and officiality, while considering the city parks as spaces that could be adapted according to their demand and interests, and the courtyards of large housing estates as "oases for allowed freedom". However, over time, open public spaces lost some of their advantages, and still retain a tangible imprint of the socialist planning in everyday use.

**Conclusion**

Summarising the analysis of the planning and everyday use of public spaces in Ukrainian large ordinary cities of Vinnytsia and Cherkasy in the 1960–80s, the following key points can be assumed:

1 Everyday use of the public spaces in large, ordinary cities by residents and their involvement in planning and arrangement were clearly hierarchical. The production of public spaces was, at first glance, clearly regulated in accordance with the top-down urban planning system. However, less visible public spaces were planned and used with regard to the vision of the residents.
2 Planning and everyday use of public spaces in the large ordinary cities were conditioned by standardisation. It caused the emergence of similarly planned and named central squares, main city parks and even courtyards. However, in the latter, the residents had the most freedom to influence their arrangement and deviate from the standards.
3 Planning and everyday use of public spaces in large ordinary cities were changing differently in accordance with the change in planning approaches and the resident's demands caused by growing incomes and technological progress. The inevitable modernisation of society from 1960s to 1980s gradually changed both statewide planning decisions and local initiatives.

In view of this, the lessons to the urban planners by the Ukrainian large ordinary cities with relation to public spaces are the following: (a) a high level of standardisation in planning decisions, which led to the assimilation of such cities and their public spaces, was a successful short-term strategy to satisfy the demands of people but dropped out of race in long-term period; (b) the extreme standardisation and overregulation of everyday behaviour in certain places (e.g., central squares) may be compensated by allowing people more freedom and self-expression in other places (e.g., courtyards) – a kind of informal *social contract* that was "tested" under socialism but can be adopted in the other socio-political systems as well; (c) highly standardised urban facades may hide a more diverse space behind reflecting individual and city-level specificity.

**References**

Adams, Russel B. 1977. The Soviet metropolitan hierarchy: regionalization and comparison with the United States. *Soviet Geography* 18(5):313–328. 10.1080/00385417.1977.10640179

Alekseyeva, A. 2019. *Everyday Soviet Utopias. Planning, Design and the Aesthetics of Developed Socialism*. New York: Routledge.

Arter, D. 2001. Regionalization in the European peripheries: the cases of Northern Norway and Finnish Lapland. *Regional and Federal Studies* 11(2):94–114. 10.1080/714004693

Arzmi, A. 2023. Planning GDR and Czechoslovakia. The Scale question under state socialism. In: Guerra, M.W., Abarkan, A., Castrillo Romón, M.A., and Pekár, M. (Eds.), *European Planning History in the 20th Century. A Continent of Urban Planning*, pp. 153–162. Routledge. 10.4324/9781003271666-16

Avdotiin, L.N., Lezhava, I.G., and Smoliar, I.M. 1989. *Urban design*. Moscow: Stroiizdat, (In Russian) [Авдотьин Л. Н., Лежава И. Г., Смоляр И. М. Градостроительное проектирование].

Bater, J.H. 1980. *The Soviet City: Ideal and Reality*. Beverly Hills: Sage.

Caldeira, T. 2017. Peripheral urbanization: autoconstruction, transversal logics, and politics in cities of the global south. *Environment and Planning D: Society and Space* 35(1):3–20. 10.1177/0263775816658479

Carter, H. 2015. Peripheralization through Planning: the case of a golf resort proposal in Northern Ireland. In: Lang, T., Henn, S., Sgibnev, W., and Ehrlich, K. (Eds.), *Understanding Geographies of Polarization and Peripheralization Perspectives from Central and Eastern Europe and Beyond*, pp. 98–111. New York: Palgrave Macmillan.

Conterio, J. 2022. Controlling land, controlling people: urban greening and the territorial turn in theories of urban planning in the Soviet Union, 1931–1932. *Journal of Urban History* 48(3):479–503. 10.1177/00961442211063171

Crawford, C.E. (2018). From tractors to territory: socialist urbanization through standardization. *Journal of Urban History* 44(1):54–77. 10.1177/0096144217710233

Czepczyński, M. 2008. *Cultural Landscapes of Post-Socialist Cities. Representation of Powers and Needs*. Burlington: Ashgate.

Danson, M., and de Souza, P. 2012. Periphery and marginality: definitions, theories, methods and practice. In: Danson, M. and de Souza, P. (Eds.), *Regional Development in Northern Europe: Peripherality, Marginality and Border Issues*, pp. 1–15. Abingdon: Routledge.

Denysyk, G., Mezentsev, K., Antipova, E., and Kiziun, A. 2020. An everyday geography: spatial diversity of the everyday life. *Visnyk of V. N. Karazin Kharkiv National University, Series "Geology. Geography. Ecology"* 52:130–138. 10.26565/2410-7360-2020-52-10

Dixon, M. 2013. Transformations of the spatial hegemony of the courtyard in post-Soviet St. Petersburg. *Urban Geography* 34(3):353–375. 10.1080/02723638.2013.778663

Dmytrenko, V. 2016. History of planning and development of the city of Cherkasy. In: Dymczyk, R., Kryvosheia, I., and Morawiets, N. (Eds.), *Architectural and Cultural Heritage of Historical Cities of Central and Eastern European Countries*, pp. 15–21. Uman-Poznan-Chenstohowa. (In Ukrainian).

Domański, B., and Lung, Y. 2009. Editorial: the changing face of the European periphery in the automotive industry, *European Urban and Regional Studies* 16(1):5–10. 10.1177/0969776408098928

Durmanov, V., and Dubbeling, D. 2004. Ukraine. Inheritance of centralised planning. In: Turkington, R., van Kempen, R., and Wassenberg, F. (Eds), *High-Rise Housing in Europe. Current Trends and Future Prospects*, pp. 203–214. Delft: DUP Science.

Fedoryshen, O. 2015. *Vinnytsia. The History of the Olden Days (Historical Chronoscope)*. Kyiv. (In Ukrainian)

French, R.A. 1995. *Plans, Pragmatism and People. The Legacy of Soviet Planning for Today's Cities*. London: UCL Press.

Gnatiuk, O., and Kryvets, O. 2018. Post-Soviet Residential Neighbourhoods in Two Second-order Ukrainian Cities: Factors and Models of Spatial Transformation. *Geographica Pannonica* 22(2):104–120. 10.5937/22-17037

Gnatiuk, O., Mezentsev, K., and Provotar, N. 2021. From agricultural station to luxury village? Changing and ambiguous everyday practices in the suburb of Vinnytsia, Ukraine. *Moravian Geographical Reports* 29(3):202–216. 10.2478/mgr-2021-0015

Harris, S.E. 2013. *Communism on Tomorrow Street. Mass Housing and Everyday Life after Stalin*. Baltimore: The Johns Hopkins University Press.

Hausladen, G. 1987. Planning the development of the socialist city: the case of Dubna New Town. *Geoforum* 18(1):103–115. 10.1016/0016-7185(87)90024-8

Herrschel, T. 2012. Regionalisation and marginalisation. Bridging old and new divisions in regional governance. In: Danson, M. and de Souza, P. (Eds.), *Regional Development in Northern Europe: Peripherality, Marginality and Border Issues*, pp. 30–48. Abingdon: Routledge.

Hess, D.B., and Metspalu, P. 2019. Architectural transcendence in Soviet-era housing: evidence from socialist residential districts in Tallinn, Estonia. In: Hess, D.B., and Tammaru, T. (Eds.), *Housing Estates in the Baltic Countries. The Legacy of Central Planning in Estonia, Latvia and Lithuania*, pp. 139–160. Springer. 10.1007/978-3-030-23392-1

Hess, D.B., and Tammaru, T. 2019. Modernist housing estates in the Baltic countries: formation, current challenges and future prospects. In: Hess, D.B., and Tammaru, T. (Eds.), *Housing Estates in the Baltic Countries. The Legacy of Central Planning in Estonia, Latvia and Lithuania*, pp. 3–27. Springer. 10.1007/978-3-030-23392-1

Horodskykh, O., Denysova, L., and Voloshyna, T. 2012. Historical research project "Creators of the architecture of Vinnytsia. In: Zahorodnia, L. (Ed.), *Architectural Vinnytsia: Time, Space, Personalities. Almanac*, pp. 108–128. Vinnytsia: PRADA ART. (In Ukrainian)

Hou, J. 2010. (Not) your everyday public space. In: Hou, J. (Ed.), *Insurgent Public Space. Guerrilla Urbanism and the Remaking of Contemporary Cities*, pp. 1–17. New York: Routledge.

Ikonnikov, A.V. 1988. Design in an urban environment or design of an urban environment?, *Trudy VNIITE. Series Technicheskaia estetika. Dizain i gorod* 57. (In Russian)

Janušauskaitė, V. 2019. Living in a large housing estate: insider perspectives from Lithuania. In: Hess, D.B., and Tammaru, T. (Eds.), *Housing Estates in the Baltic Countries. The Legacy of Central Planning in Estonia, Latvia and Lithuania*, pp. 181–202. Springer. 10.1007/978-3-030-23392-1

Karoieva, L.R., Lysa, L.S., and Filin, O.O. 1998. *The Silver Moment (Postcards and Photos from the Museum Collection)*. Vinnytsia: Vinobldrukarnia. (In Ukrainian)

Kogan, L.B. 1967. Urbanization – communication – microdistrict. *Arkhitektura SSSR* 4:39–44. (In Russian)

Kumo, K., and Shadrina, E. 2021. On the evolution of hierarchical urban systems in Soviet Russia, 1897–1989. *Sustainability* 13(20):11389. 10.3390/su132011389

Leetmaa, K., Tammaru, T., and Hess, D.B. 2015. Preferences towards neighbour ethnicity and affluence: evidence from an inherited dual ethnic context in post-Soviet Tartu, Estonia. *Annals of the Association of American Geographers* 105(1):162–182. http://www.jstor.org/stable/24537954

Low, S., Taplin, D., and Scheld, S. 2005. *Rethinking Urban Parks: Public Space and Cultural Diversity*. Austin: University of Texas Press.

Luukkonen, J. 2010. Territorial cohesion policy in the light of peripherality'. *Town Planning Review* 81(4):445–466. http://www.jstor.org/stable/40890973

Medvedkov, O. 1990. *Soviet Urbanization*. London and New York: Routledge.

Mezentsev, K., Pidgrushnyi, G., and Mezentseva, N. 2015. Challenges of the post-Soviet development of Ukraine: economic transformations, demographic changes and socio-spatial polarization. In: Lang, T., Henn, S., Sgibnev, W., and Ehrlich, K. (Eds.), *Understanding Geographies of Polarization and Peripheralization Perspectives from Central and Eastern Europe and Beyond*, pp. 252–269. New York: Palgrave Macmillan.

Mezentsev, K., Provotar, N., Gnatiuk, O., Melnychuk, A., and Denysenko, O. 2019. Ambiguous suburban spaces: trends and peculiarities of everyday practices change. *Ekonomichna ta Sotsialna Geografiya* 82:4–19. 10.17721/2413-7154/2019.82.4-19

Nagy, E., Timár, J., Nagy, G., and Velkey, G. 2015. The everyday practices of the reproduction of peripherality and marginality in Hungary. In: Lang, T., Henn, S., Sgibnev, W., and Ehrlich, K. (Eds.), *Understanding Geographies of Polarization and Peripheralization Perspectives from Central and Eastern Europe and Beyond*, pp. 135–155. New York: Palgrave Macmillan.

Paasi, A.A. 1995. The social construction of peripherality: the case of Finland and the Finnish-Russian border area. In: Eskelinen, H., and Snickars, F. (Eds.), *Competitive European Peripheries*, pp. 235–258. Berlin: Springer-VS.

Staub, A. 2005. St. Petersburg's double life: the planners' vs. the people's city. *Journal of Urban History* 31(3):334–353. 10.1177/0096144204272418

Underhill, J.A. 1990. Soviet new towns, planning and national urban policy: shaping the face of Soviet cities. *The Town Planning Review* 61(3):263–285. https://www.jstor.org/stable/40112920

Vecherskyi, V., and Zlyvkova, O. 2011. Historical and town planning research of Vinnytsia. In: Vecherskyi, V. (Ed.), *Historical and Urban Planning Studies: Vasylkiv, Vinnytsia, Horlivka, Izmail*, pp. 133–194. Kyiv. (In Ukrainian)
Vladimirov, V., Naimark, N., and Subbotin, G. et al. 1986. *District Planning*. Moscow: Stroiizdat. (In Russian)
Yukhno, B. 2013. *Cherkasy Mystories. Travels in TIme from Sosnivka to Kryvalivka*. Cherksy: Brama-Ukraina. (In Ukrainian)
Yukhno, B. 2016. *Retro Grad*. Cherksy: Brama-Ukraina. (In Ukrainian)
Yukhno, B. 2019. *Cherkasy. Puzzle*. Cherksy: Brama-Ukraina. (In Ukrainian)

*Construction norms and rules, urban planning documents*

Resolution of the Council of Ministers of the Ukrainian SSR "On the Master plan for the development of the city of Cherkassy" dated March 29, 1984 N144. Retrieved from https://ips.ligazakon.net/document/KP840144 (In Ukrainian)
Resolution of the Council of Ministers of the Ukrainian SSR "On the Master plan for the development of the city of Vinnytsia" dated July 25, 1987 N259. Retrieved from https://ips.ligazakon.net/document/KP870259?an=160 (In Ukrainian)
Rules and norms for the settlements development, design and construction of buildings and structures. Moscow: Gosudarstvennoie tekhnicheskoie izdanie, 1930. (In Russian)
SN 41-58. Rules and norms for city planning and development. Moscow: Gosudarstvennoie izdatelstvo literatury po stritelstvu, arkhitekture i stroitelnym materialam, 1959. (In Russian)
SNiP II-60-75**. Construction norms and rules. Part II. Design standards. Chapter 60: Planning and development of cities, towns and rural settlements. Moscow: Tsentralnyi institut tipovogo proektirovaniia, 1985. (In Russian)
SNiP II-K.2-62. Construction norms and rules. Part II, Section K. Chapter 2: Settlements planning and development. Design standards. Moscow: Izdatelstvo literatury po stritelstvu, 1967. (In Russian)
SNiP II-K.3-62. Construction norms and rules. Part II, Section K. Chapter 3: Streets, roads and squares of populated areas. Design standards. Moscow: Gosudarstvennoie izdatelstvo literatury po stritelstvu, arkhitekture the same i stroitelnym materialam, 1963. (In Russian)
SNiP II-V.1. Construction norms and rules. Part II. Building design standards. Moscow: Gosudarstvennoie izdatelstvo literatury po stritelstvu the same i arkhitekture, 1954. (In Russian)

# 10 Planning urban peripheries for leisure

## The plan for Greater Tallinn, 1960–1962

*Epp Lankots*

### Introduction

In 1976, to illustrate an article about the geographies of leisure published in the Estonian nature journal *Eesti Loodus* (Lausmaa 1976:292), Edgar Valter drew a caricature depicting a relaxed couple on a roadside picnic, the contours of a housing estate on the edge of the city still in sight (Figure 10.1). The article by Ene Lausmaa, an Estonian urban geographer, described new urbanised uses of nature relating to advances in technology and mobility during the Soviet post-war decades. Similarly, in the 1980s, the landscape around Moscow's outskirts was described as "the mosaic scenery composed of fields, meadows, and forests dotted by villages, small and mid-sized towns, industrial facilities, and dacha communities" (Brade, Makhrova and Nefedova 2014:97). These depictions visualise the spatial forms that were characteristic of land use around Soviet urban agglomerations and as they were stipulated in the general plans. They also designate the shift from city-scale to regional thinking that took place around the mid-20th century when the peri-urban area, or *prigorod* in Russian, emerged as a concept in Soviet urban planning. The idea of the peri-urban zone, in principle, was based on CIAM's (*Congrès internationaux d'architecture moderne*) functional city with clear zoning of different functions such as living, working, recreation, transport, industry and agriculture. The new plans aimed at "providing the Soviet people the best conditions for work and leisure time" (Romanov 1963:19), and so the peri-urban zones emerged as the destination for relaxation in nature while at the same time they were directly linked to the housing construction on the edge city. This quest for leisure was frequently called the most important "sign of present times" in urbanised societies and in the 1960s recreation was declared to be one of the core ideas in planning the peripheries (Tippel 1967:42; Romanov 1963:19; Lunc 1968:13).

Technological development and general economic growth in the 1960s had made possible the shortening of the working week from six to five days, the raising of living standards with paid holiday and the increasing availability of private car ownership. Propelled by the techno-scientific

DOI: 10.4324/9781003327592-14
This chapter has been made available under a CC-BY-NC-SA 4.0 license.

*Planning urban peripheries for leisure* 177

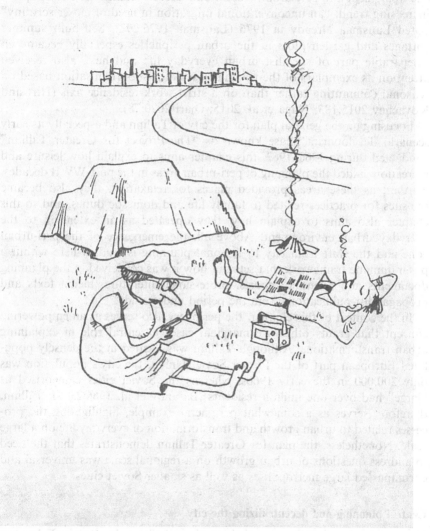

*Figure 10.1* A caricature by Edgard Valter in the nature journal *Eesti Loodus*.
*Source:* Lausmaa 1976.

revolution, leisure studies became one of the new areas of research into everyday life, with a particular focus on the distribution of work and free time and planning leisure infrastructure for different categories of recreational activity (short- and long-term leisure, organised and free leisure, sedentary and mobile leisure, etc.). These advancements also changed the dynamics between urban and rural areas. Commuting between the city apartment and peri-urban recreational areas became a new and rapidly increasing trend: "an unconventional migration in need of closer scrutiny" noted Lausmaa already in 1976 (Lausmaa 1976:292). Self-built summer cottages and garden plots in the urban peripheries especially became an inseparable part of socialist urban everyday life and have also received attention as exemplars of the peculiarly socialist suburbanisation based on seasonal commuting rather than on a strict work-residence axis (Hirt and Kovachev 2015:183; Nuga et al. 2015; Ojari 2020:78).

Focusing on the general plan for the city of Tallinn and especially its early scenario development phase known as "The Project for Greater Tallinn" conducted during 1960–1962, this chapter aims to explain how leisure and recreation guided the planning of peri-urban areas in the post-WWII decades. Further, as these areas provided places for relaxation, they also became the sites for practices related to family life and domestic duties, and so this chapter also aims to explain how they operated as an extension to the everyday urban environment. Above all, the emergence of the peri-urban zone and the shift from city to regional planning is studied here within a programmatic framework that sets out how it was conceived in the planning documents and scientific reports, professional literature, media texts and propagandist publications from the period under study.

In the context of this chapter, the periphery also serves as an empowering concept that grants otherwise marginal cases a central role in explaining urban transformations. Although Tallinn was located in the densely populated European part of the former Soviet Union, the city's population was only 300,000 in the early 1960s, whereas the Soviet cities categorised as "large" had over one million residents (Baranov et al. 1966:23–4). Tallinn, therefore, serves as a somewhat peripheral example, highlighting the processes related to urban growth and transformation of everyday life on a large scale. Nonetheless, the plan for Greater Tallinn demonstrates that the need to address questions of urban growth on a regional scale was universal and encompassed large metropolises as well as smaller Soviet cities.

**Central planning and decentralizing the city**

Urban planning (*planirovka* and *gradostroitel'stvo* in Russian) became increasingly rationalised in the Soviet Union during the late 1950s and was strictly subjected to the national-economic planning (*planirovanye*) of the state (Shaw 1983:393–84; Taylor and Kukina 2019:192). Detailed and complex rules were worked out in Moscow to regulate the planning activities in each of the republics (Nuga, et al. 2015:38). Based on thorough research and prognostics on economic and demographic development for longer and shorter periods, as well as implications in terms of housing, services and land-use zoning, the general plans for the Soviet cities prescribed the city's development for 25–30 years and, in theory, at least, were subject to a five-yearly review (Shaw 1983). This also affected urban planning as a professional practice, which had previously been dominated by the architect-planner, as a range of new actors like economists, demographers and especially experts in technical disciplines like energy and transport turned planning into a largely technical exercise (Nuga et al. 2015:38). The planning process consisted of various stages. The first stage was the compilation of technical-economical terms, which meant devising the development scenario for the city and its economic zone of influence. After the general plan was conceived, a series of more detailed planning projects were developed during its lifetime. Different geographical scales and focus themes (rural and urban areas, detailed and thematic plans) determined also the distribution of work between the different state design institutes. In Estonia, for example, the biggest design institute "Eesti Projekt" was responsible for the general planning of cities, including also the planning of recreational areas and resorts (Kerde 1983:2).

While the prospective view and the systematic consideration of different functions in the peripheral zone became an established practice in urban planning during the Khrushchev era, governments had previously attempted to reorganise the peripheries in the 19th century when these areas became leisure destinations for the more affluent and privileged classes, and also in the early Soviet decades (Skorobogatyj 1936:236). However, it was not until 1935 that the General Plan for the Reconstruction of the City of Moscow (architects Vladimir N. Semenov and Sergey E. Chernishev) – a *tour-de-force* of Stalinist urban planning that envisaged a metropolis of the first communist state in the world (Cohen 1995:246–8) – attempted to prevent the indefinite expansion of the city by creating a 10-kilometre-wide green belt or a "forest-park zone" around the city to serve as a reservoir of fresh air and a place of recreation for workers (Moscow 1935:12–3). In 1935, some 500,000 Muscovites would relax in the rural areas of Moscow oblast, and by 1967, when the peri-urban area already stretched some 50–60 kilometres from the city boundary, this figure had multiplied five-fold (Shaw 1979:132–5). Moscow set the standard for all subsequent general plans, with recreational functions gaining increasing importance in the planning of the urban peripheries. For example, in Leningrad, 30% of the city's population travelled outside the built-up area of the city for leisure purposes and a 60-kilometre-wide peri-urban zone was stipulated in the general plan to "satisfy the multiplicity of needs of the modern city imposing its influence far beyond city borders" (Kamenskij 1972:78).

By comparison, these vast green areas on the urban periphery were rare in Western metropolitan areas. The Soviet Union was considered one of the very few industrialised nations in the world whose planners could still seriously contemplate the drawing of recreational zones around their cities with separate zones for daily, short-term and long-term recreation (Shaw 1979:132). Although the 1960s had brought a general rise in living standards and mobilised city dwellers, the reason for retaining such vast green areas is often thought to be due to the expense and relative inaccessibility of the private car, thus making the dependence on public transport inescapable for the masses, and this enabled Soviet planners to control access and prevent the over-exploitation of green areas (Shaw 1979:132). The Soviets, however, emphasised that the sparse settlements and extensive natural areas and opportunities for outdoor leisure were a deliberate, progressive and distinctly Soviet phenomenon as opposed to the large metropolitan agglomerations and uncontrolled sprawl of Western countries (Baranov et al. 1966:23–4).

**The Greater Tallinn plan and the quest for peri-urban leisure**

When Tallinn received its first socialist general plan (architects Anton Soans, Harald Arman and Otto Keppe) shortly after WWII and the annexation of Estonia by the Soviet Union, it focused primarily on the reconstruction of the central area in a socialist-realist grand manner that was loaded with symbolism of the new social order. Beyond the centre, the plan shaped the mostly isolated, smaller residential areas by re-planning the streets to create avenues where blocks of housing alternated with smaller urban parks and green areas (Estonian Museum of Architecture 1950). More complex scenarios for regional development were practically missing in the plan, and the peri-urban zone demarcated existing industrial and holidaying settlements that had already taken shape before the war.

The change in the political course of the Soviet state under Khrushchev and the goal of increasing the material well-being of Soviet citizens by solving the housing question led also to the reconsideration of the city's development. The title Greater Tallinn was used only in the preliminary scenario-drawing phase of the second general plan of Tallinn, which lasted from 1960 to 1962 and was conducted by architects Harald Arman, Dmitri Bruns, Otto Keppe and Voldemar Tippel, and by economist H. Heinvere. The planning process itself took longer, from 1965 to 1968, and the final plan was adopted only in 1971. Directions for the future development of the urban area were set by the Institute of Economic Sciences of the Academy of Sciences, and prognoses for the growth of the national economy to 1980 were produced by the State Planning Committee: the "Gosplan", including prognoses on economy, industry, transportation and housing in Tallinn and its metropolitan area, population growth and the resultant expansion of city territory (Estonian Museum of Architecture 1961:3). It was thus the first urban plan in Estonia to manifest the new techno-scientific orientation in urban planning.

While the plan for Greater Tallinn was pivotal in its pursuit of resolving urban growth by reserving new and extensive areas for locating new industrially produced housing estates, its second important feature was the planning of new land uses also in the peri-urban zone. As the morphology of the city of Tallinn is determined by its location in the narrow area between Tallinn Bay in the Baltic Sea and Lake Ülemiste, the exploitation of the coastline has been central to the formation of the urban area. At the beginning of the industrial era, numerous factories and manufacturing enterprises appeared around the old town and along the shores of Tallinn and Kopli bays, with workers' settlements concentrated around the historical centre. During the interwar independence period, the city had expanded mainly inland together with Nõmme, previously a summer resort that grew into a satellite town in the 1930s and was united with Tallinn in 1940. There were also a few smaller settlements established in seaside areas within 10–30 kilometres of the city – the summer houses and weekend cabins in Rannamõisa and Kloogaranna to the West and Merivälja garden city to the East. These areas were also the sites for the first Soviet-era summer houses and gardening plots that were allocated to the residents of Tallinn at the end of the 1950s.

The plan's architects envisaged developing the city as a compact semicircle opening towards the sea (Bruns 1993:151). The idea of developing two new wings for the city, the new mass-housing estates Lasnamäe and Õismäe, gave balance to the city's structure by highlighting the historical core and also secured the entire city's connection to the sea (Bruns 1993:154). Also, two satellite towns, Keila and Aruküla, were planned in the peripheral zone – these were historical settlements redeveloped by relocating several industrial enterprises there from the city (Estonian Museum of Architecture 1961:10). Besides other functions such as agriculture, industry and transport, the plan also considered the organisation of leisure time and provided a spatial framework for this as part of its core task, thereby asserting that Tallinn's influence extends far beyond the borders of the city and that its economic influence on nearby areas primarily lies in the recreational uses of natural areas (Sirp ja Vasar 1961:4; Palm 1963:68). Thus, for the first time, the peri-urban zone would be brought under the jurisdiction of the city of Tallinn.

A large proportion of the studies on the peri-urban land uses and particularly on recreation in the Greater Tallinn project relied on a young architect, Asta Palm, who simultaneous to her work on the project was also working on a doctoral dissertation on the principles of architectural planning in the Tallinn recreational area (Figure 10.2) (Palm 1964). Palm was a pioneer in leisure studies in Estonia, and for three consecutive decades, she was responsible for the majority of the research conducted prior to those major planning exercises that featured also leisure and recreation. As construction activity increased on the edge of the city and traffic intensified, Palm emphasised the need to coordinate the land uses in the peri-urban zone in order to avoid problems such as potential polluting of the recreational areas by nearby industries. Following her work, the Tallinn peripheral zone was structured according to long-term or short-term holidays. Similar to the first

182  *Epp Lankots*

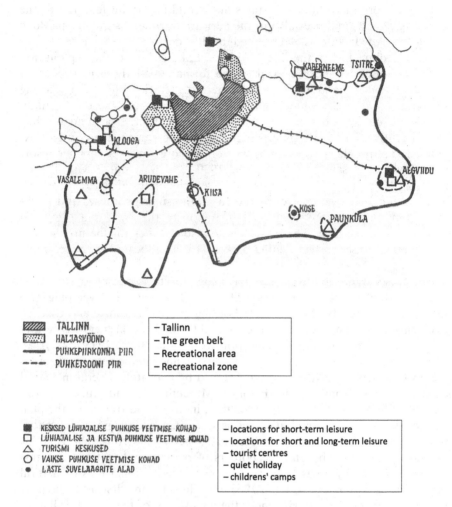

*Figure 10.2* The scheme by Asta Palm depicting the Tallinn recreational area in 1963.
*Source:* Palm 1963.

general plan for Moscow from 1935, the inner ring, a 2–5 kilometres wide forest park zone, was planned for the short-term leisure of 30% of the city population and to provide a reservoir of clean air for the city, and Palm emphasised the need to increase forested areas by two to three times. Construction activity in the forest park zone was to be restricted, with the exception of sports facilities. Long-term leisure was to be organised within a radius of 70 kilometres, and the Northern Estonian coastline was planned to be in the most intensive use for approximately 30% of the population, or 100,000 holidaymakers. This was the zone where a variety of different holiday institutions were to be developed by the state: young pioneer camps, holiday complexes for state enterprises, campgrounds, tourist complexes, motels, garden houses and summer cottages including areas with a quieter regime separated from the more intensely used collective areas (Palm, 1963:69). The plan also proposed improved connections to the centre and prioritised the establishment of railway lines (Estonian Museum of Architecture 1961:21–2).

Besides the planning of the coastline, the use of forests as a reservoir of fresh air had particular importance in the development and use of the peri-urban area. Historically, these areas had primarily been utilised to serve urban life, mainly as a resource of wood for heating and for construction material, but the industrial era attached new meanings to the forests (Malev 1974:45). Leisure was one of these meaningful frameworks that ascribed an independent value to forests that differed from its traditional utilitarian use. In Estonia, this shift has a prehistory in the interwar period when a considerable effort was made by the state to promote nature tourism, and this was sustained by the Romantic idea of the city as being essentially polluted and unhealthy (Jonuks and Remmel 2020:466), but the shift towards appreciating the cultural value of the forests gained full momentum only during the 60s. It appeared not only in the professional terminology ("the forest park zone") that was used in the plan for Greater Tallinn but also in the formation of bureaucratic apparatus like the Forest Management of Tallinn Green Belt, which was established in 1963 and was responsible for planning and overseeing the functioning of the recreational areas (Malev 1974:7). Another example of the new culturised uses of nature is the evaluative discourse that emphasised the need to determine the criteria for the aesthetic function of forests and other types of landscapes appreciated primarily for their "painterly" and "artistic" qualities (Baranov et al. 1969:106; Malev 1974:66; Bugaev, Piskunov and Rakov 2021:306). Also, articles discussing recreational uses of landscapes as an enrichment to the overall understanding and complex meanings of nature and landscape planning as nature protection (Eilart 1964; Kumari 1964; Zobel 1979) were published. Together with the ideas of the "nature complex" as one component in the peculiar socio-economic system called the "leisure system" that was inspired by systems theory and prevailed in the leisure studies of that time (Lausmaa 1983:30–2), these rational operations enable the planning of peri-urban areas around Tallinn and beyond to be considered as a peculiar green-technological activity designed to sustain the good quality of urban everyday life (Bugaev, Piskunov and Rakov 2021).

The plan for Greater Tallinn led to more detailed and substantial considerations of the different aspects and uses of the peripheral area also in the general plans for other Estonian cities. For example, the Tartu general plan from 1970 involved not only detailed assessments of economic, industrial and natural perspectives for the year 2000 and the planning of recreational functions by different types of holiday institutions, but it also considered sightseeing, tourist attractions and national monuments, as well as protected natural objects and issues of environmental protection as constitutive parts in the planning of recreational areas (Estonian National Archives 1974). By the 1970s, the planning of peri-urban zones had become a normative practice subjected to SNIP (*stroitelnye normy i pravila*), the all-union building and planning codes for the cities (Estonian National Archives 1974), which meant that other towns of various sizes also received their zoning plans for peripheral areas (Estonian National Archives 1974a).

## Summer house settlements and everyday life in Tallinn's urban periphery

While everyday life in a new flat in a mass-produced building certainly added a degree of comfort and a change of domestic environment as opposed to life in the previous overcrowded communal apartments, the compression of people and activities in the flats that sometimes led to uncomfortable intensity in social interaction remained an issue. As the parks and other green areas in the city satisfied the needs for outdoor leisure only to a small degree (Luik 1983:10), the peri-urban recreational zone indeed prescribed a richer variety of possibilities for the urban residents to spend their leisure time, including both active and more passive ways of holidaying and inviting them to engage with nature in various ways.

However, the reality would not always meet the ideal scheme as insufficient capital investment and poor coordination of different administrative levels often resulted in rather ad hoc development of leisure facilities and holidaying spots. The state tried to alleviate the shortage of organised holidays outside one's home also by encouraging personal contributions from the citizens in the form of personal, self-built summer houses. Accordingly, the summer house settlements, which were largely intended for productive holidaying such as gardening, grew into a peculiar mass phenomenon in the Soviet Union and occupied extensive areas designated for recreation within the peri-urban zone. The areas initially reserved for collective leisure in general plans were turned into peculiar urbanised landscapes that combined holidaying and subsistence farming, leisurely consumption and utilitarian production at the same time (Zaviska 2003:788).

In Estonia, the first wave of summer houses appeared in Tallinn peri-urban area in the late 1950s, before the plan for Greater Tallinn was drafted. To grant the summer houses a correct socialist basis, it was declared that society was not yet ready for radical collectivization and therefore parts of family-centred models of habitation needed to be preserved temporarily, especially as the material base for spending most of the free time collectively was still insufficient (Estonian National Archives 1975:15). Also, Asta Palm had discussed personal summer houses as a typology for organising mass holidays while acknowledging it as being a temporary phenomenon that would perish in the future and be replaced by what she called "the summer towns". According to Palm, these would be agglomerations, established both by the state and by institutions and companies for their employees, consisting of 1,000–5,000 small cottages for family stays with different cultural and service buildings such as a canteen, library, cinema, cafes and sports facilities for collective use (Palm, 1963:67). In some settlements, the possibility of practising gardening in the form of collective gardens was also foreseen (Palm 1963:67).

The state-built "summer towns", however, were never implemented, and contrary to the envisioned plans, the building of personal summer houses intensified due to a new decree that allowed the establishment of settlements on a cooperative basis. The land for summer houses and plots for gardening – land insufficiently fertile for intensive large-scale agriculture – was allocated by state enterprises and institutions, which distributed the plots to those employees who wished to form a cooperative and contribute their savings to building a summer house. Members of the cooperative had to order the plan for the settlement, wherein the principles of housing design were also decided: the cottages were each built according to either standard or one-off designs (Ojari 2020:84). To prevent year-round usage and the growth in ownership of a second fully functional family dwelling in addition to the flat in the city, the summer houses were not allowed to be insulated and were in use only from May to October (Ojari 2020:87). The employer often helped with arranging infrastructure works such as building roads within the settlement or installing water supply pipes. This mixed type of ownership and management pattern made the gardening and summer house areas a peculiar type of ambiguous, semi-public, socialist space where cooperative ownership and institutional framework were combined with private holidaying and family food supply.

Another feature that helped to reconcile the materialist aspirations of private life with the socialist way of living was the sense of collectivity that the summer house settlements upheld. The community life in the cooperative has been characterised as very active, ranging from collective work done in the common areas of the settlement, exchanging building and gardening know-how as well as seeds and plants, to social gatherings like collective midsummer celebrations (Ojari and Lankots 2019; Caldwell 2011). There were also settlements where the members of the cooperative commissioned and built a small community building where the meetings, as well as game evenings or other social gatherings like birthdays, were celebrated, like the clubhouse in the cooperative "Tagaoja Vigvam" in Vääna-Jõesuu. The small community and the shared practices and values that were also legally grounded on the cooperative ownership of the settlement helped the residents to build up long-term relationships with the area and with each other and to form social networks quite similar to stable middle-class areas in Western Europe. Thus, the new suburban leisurely settlements not only helped to extend the everyday environment of the residents of the mass housing estates outside the city limits but also helped to create a sense of community that the large *mikrorayon* had failed to achieve in the city. The summer house also functioned as a peculiar kind of gravity point for family life, achieving a degree of democratisation within the household that the well-equipped modern flat in the city had abolished, leaving a larger share of domestic duties to women (Attwood 2010:166). In the summer house, all the family members were engaged in various activities around the house and garden, from construction to managing the house and maintaining the garden to simply enjoying the holiday.

The actual usage of the houses and the frequent, sometimes daily commuting to the plot during the season that was vital for family food provision suggest that the summer houses operated as functional extensions to the small-size flat in the city, compensating for the shortcomings in planning of the housing areas and everyday environment. As acknowledged by a few planners and architects, the spatial separation of these two functions – residential and leisure – had led to an uncontrolled spread of summer house settlements in the Tallinn peri-urban zone, and that became an obstacle to finding available land for prospective areas for organising mass leisure close to the city (Tallinn City Government archive 1977). The initial idea of the "summer towns" intended for cultural leisure, as discussed by Palm, had transformed into small-scale-agricultural production units that formed large agglomerations and sometimes looked more like monotonous urban settlements and created a strange "anti-nature effect" (Hanson 1970:4). By the 1980s, with the advent of *perestroika* and criticism of the extensive residential construction that had produced vast alienating environments, the economic arguments that questioned the combined model of a flat in the city with a summer house in the periphery grew louder. It was argued that a residential model that required two properties – one for living and the other one for leisure and subsistence farming – is more expensive than residing in a single private house (Volkov 1983:58). According to the critics, the apparent efficiency of this double system was also sustained artificially by interest rates for building loans being considerably higher for building a private house than for a summer house (Kraak 1987:3). By 1990, when over 52,000 families (about 10–13% of the population of Estonia) were spending their short- and long-term holidays in a summer house (Estonian National Archives 1990), discussions about holidaying in the summer house had shifted from being about ways of spending leisure time to being about the models of living that would form the ground for urban lifestyles to emerge in the post-socialist suburbia of the 1990s – lifestyles where questions of privacy and family life became prevalent.

**Conclusion**

Due to the industrially driven, centrally planned economy and the state housing policy, the development of peripheral urban areas in the Soviet Union took a different form from the post-war sprawling cities of Western market-led economies. These peripheral areas were shaped by diverse patterns of land-use and a variety of different landscapes, but agriculture, industry and related residential functions have tended to dominate discussions about regional development in the histories of socialist urban planning. However, the new housing programme initiated by Khrushchev and the general plans conceived for Soviet cities, such as the plan for Greater Tallinn, reveal how land uses in peri-urban areas were rebalanced in the 1960s. The need to organise the free time of the residents of the new mass-housing estates on the outskirts of the city led to increased consideration of urban peripheries for leisure. Thus, recreation and the production of nature for that purpose became paramount practices that determined also a considerable share of the migration between the socialist city and its periphery (Logan 2021:79).

The interconnectedness of the new housing estates and the peri-urban leisure-scapes was related to the idea of a domestic environment as a network of different services and functions, which was the dominant rationale also behind the plan for Greater Tallinn (Tippel 1963:41). The housing estates and their smaller units, the *microrayons*, were planned according to a multi-stage domestic service system that divided the different functions like healthcare, education, kindergartens, social services, shops, etc., between the city, larger residential areas and local neighbourhoods. Similarly, green spaces and recreational landscapes were also part of this rational framework and were considered an equally functional part of everyday life, forming a similar web-like structure that extended out to the larger peri-urban territory (Tippel 1963:42; Palm 1963:67; Tobilevich 1968). There were, however, a few endeavours in socialist countries to integrate housing and recreational functions within a unitary living environment; such as, for example, the unbuilt model city of Etarea in former Czechoslovakia (1967, architect Gorazd Čelechovský) (Krivý 2019; Logan 2021:87), and the scientific research city Akademgorodok in Russian Siberia (founded in 1957 by Soviet academic Mikhail Lavrentyev) (Bugaev, Piskunov and Rakov 2021). Yet in most cases, like Tallinn, leisure spaces were predominantly developed within the peri-urban zone, and this was further stimulated by advancements in transport. In reality, this led to the proliferation of summer house settlements in urban peripheries instead of a fully functional leisure system for collective use. In this way, aspirations for a fulfilling and pleasant domestic life were realised, compensating for the shortcomings of the everyday environment of the city. Although ideological arguments based on social values and productive holidaying were often applied to justify the massive spread of the summer houses, there was indeed a tension between the collective ideology and rational planning on the one side and family-centred uses of leisure spaces on the other. This tension was central to the socialism of the 1960s and onwards, which sought to reconcile the central tenets of socialism with the changing relationship between public discourse and private aspirations.

## Acknowledgement

This work was supported by the Estonian Research Council grant No. PSG530.

## References

Attwood, L. 2010. *Gender and Housing in Soviet Russia: Private Life in a Public Space*. Manchester: Manchester University Press.

Baranov, N. et al. (Eds.). 1966. *Osnovy sovetskogo gradostroitel'stva = Principles of Town Planning in the Soviet Union:* TOM 1. Moskva: Central'nyj nauchno-issledovatel'skij i proektnyj institut po gradostroitel'stvu, Strojizdat.

Baranov, N. et al. (Eds.). 1969. *Osnovy sovetskogo gradostroitel'stva = Principles of Town Planning in the Soviet Union:* TOM 4. Moskva: Central'nyj nauchno-issledovatel'skij i proektnyj institut po gradostroitel'stvu, Strojizdat.

Brade, I., Makhrova, A., and Nefedova, T. 2014. Suburbanization of Moscow's urban region. In: Stanilov, K., and Sýkora, L. (Eds.), *Confronting Suburbanization. Urban Decentralization in Postsocialist Central and Eastern Europe*, pp. 97–132. Wiley: Chichester.

Bruns, D. 1993. *Tallinn. Linnaehituslik kujunemine*. Tallinn: Valgus.

Bugaev, R., Piskunov, M., and Rakov, T. 2021. Footpaths of the late-soviet environmental turn: the "forest city" of Novosibirsk's Akademgorodok as a sociotechnical imaginary. *The Soviet and Post-Soviet Review* 48:289–313. DOI: 10.3 0965/18763324-bja10043

Caldwell, M. 2011. *Dacha Idylls: Living Organically in Russia's Countryside*. Berkeley: University of California Press.

Cohen, J-L. 1995. When Stalin meets Haussmann. The Moscow Plan of 1935. In: Ades, D., Benton, T., and Whyte, I. B. (Eds.), *Art and Power: Europe under the Dictators, 1930–45*, pp. 246–248. London: Hayward Gallery.

Eilart, J. 1964. Puhkemaastikud, nende planeerimine ja kujundamine *Eesti Loodus*, no. 2:90–98.

Estonian Museum of Architecture. 1950. *Tallinna generaalplaan (graafilised materjalid)*. Arhitektid Otto Keppe, Harald Arman, Anton Soans. 1950. EAM.3.1.507.

Estonian Museum of Architecture. 1961. *Suur-Tallinna põhiteesid (tehno-ökonoomilised alused)*. EAM.18.5.33

Estonian National Archives. 1974. *Tartu generaalplaani korrektuur. I köide. Linnaümbruse tsoon, haljasvöönd ja linnalähitsoon*. ERA.T-14.4-6.33787.

Estonian National Archives. 1974a. *Haljasvööndite ja parkmetsade piiride plaanid*, Eesti Projekt, Tallinn. ERA.T-14.4-6.34021

Estonian National Archives. 1975. *ENSV puhketsoonide generaalplaan*, Eesti Maaehitusprojekt, Tallinn. ERA.T-14.4-6.615

Estonian National Archives. 1977. *Tallinna generaalplaani praktilise ellurakendmise analüüs ja kontroll*, ERA.T-14.4-6.8967

Estonian National Archives. 1990. *Eesti puhkepiirkondade generaalplaani täiendamine*, Eesti Maaehitusprojekt, Tallinn. ERA.T-18.1-1.50

Hanson, R. 1970. Kas teie aed on kaunilt kujundatud? *Rahva Hääl*, 11 January:4.

Hirt, S., and Kovachev, A. 2015. Suburbia in three acts: the east European story. In: Hamel, P., and Keil, R. (Eds.), *Suburban Governance: A Global View*, pp. 177–197. Toronto: University of Toronto Press.

Jonuks, T., and Remmel, A. 2020. Metsarahva kujunemine. Retrospektiivne vaade müüdiloomele. *Keel ja Kirjandus* no. 6:459–482. DOI: 10.54013/kk751a1
Kamenskij, V. A. 1972. *Leningrad: general'nyj plan razvitija goroda.* Leningrad: Lenizdat.
Kerde, H. 1983. Rajooniplaneerimisest detailplaneerimiseni, *Ehitus ja Arhitektuur*, no. 3:2.
Kraak, V. 1987. Vaatame ka medali teist külge. *Edasi*, 22 March:3.
Krivý, M. 2019. Automation or meaning? Socialism, humanism and cybernetics in Etarea. *Architectural Histories* 7(1):3. DOI: 10.5334/ah.314
Kumari, E. 1964. Looduskaitse probleemid ootavad lahendamist, *Eesti Loodus*, no. 1:25–31.
Lausmaa, E. 1976. Puhkusest üldse ja puhkuse geograafiast, *Eesti Loodus*, no. 5: 292–294.
Lausmaa, E. 1983. Looduskaitse lühiajalise puhkuse veetmise aladel, *Looduskaitse ja puhkus*. Tallinn: Valgus, pp. 29–34.
Logan, S. 2021. *In the Suburbs of History: Modernist Visions of the Urban Periphery.* Toronto: University of Toronto Press.
Luik, H. 1983. Puhkuse ja vaba aja veetmine kui sotsiaal-majanduslik problem. In *Looduskaitse ja puhkus*. Tallinn: Valgus, pp. 17–23.
Lunc, L. B. 1968. Razvitie gorodov i osushhestvlenie meroprijatij po ih ozeleneniju, predusmotrennyh general'nymi planami, *Rahvusvahelise nõupidamise materjalid teemal "Linnade haljastus kui keskkond elanikkonna tööks, eluks ja puhkuseks I"*, pp. 4–29, Estonian National Archives ERA.R-1951.1.210.
Malev, M. Ed. 1974. *Eesti NSV puhkealad.* Tallinn: Valgus.
Moscow. 1935. *Moscow General Plan for the Reconstruction of the City.* Moscow: Union of the Soviet Architects. Available at: http://tehne.com/library/generalnyy-plan-rekonstrukcii-goroda-moskvy-general-plan-reconstruction-city-moscow-moscow-1935 (Accessed 4 November 2022).
Nuga, M., Metspalu, P., Org, A., and Leetmaa, K. 2015. Planning post-summurbia: from spontaneous pragmatism to collaborative planning?. *Moravian Geographical Reports* 23(4):36–46. 10.1515/mgr-2015-0023
Ojari, T. and Lankots, E. 2019. Interview with U. Kõresaar. 6 August, Kabli.
Ojari, T. 2020. Under the watchword of holidays. Development of summer cottage architecture during the Soviet period. In: Lankots, E., and Ojari, T. (Eds.), *Leisure Spaces. Holidays and Architecture in 20th Century Estonia*, pp. 74–119. Tallinn: Estonian Museum of Architecture.
Palm, A. 1963. Tallinna ümbruse puhkekohtade planeerimise printsiipe. In *Linnaehituse küsimusi Eesti NSV-s*. Tallinn: Eesti NSV Ministrite Nõukogu Riiklik Ehituse ja Arhitektuuri Komitee, pp. 66–75.
Palm, A. 1964. *Suurte linnade puhkepiirkondade arhitektuurplaneerimise printsiibid: (Tallinna näitel).* Tallinn: Tallinna Polütehniline Instituut.
Romanov, S. 1963. Eesti NSV linnade generaalplaanide olukorrast. In *Linnaehituse küsimusi Eesti NSV-s*. Tallinn: Eesti NSV Ministrite Nõukogu Riiklik Ehituse ja Arhitektuuri Komitee, pp. 19–28.
Shaw, D. J. B. 1979. Recreation and the Soviet city. In: French, R. A., and Hamilton, F. E. I. (Eds.), *The Socialist City: Spatial Structure and Urban Policy*, pp. 119–143. Chichester: Wiley.

Shaw, D. J. B. 1983. The Soviet urban general plan and recent advances in Soviet urban planning. *Urban Studies* 20:393–403.
Sirp ja Vasar. 1961. 'Suur-Tallinn', *Sirp ja Vasar*, 10 February:4.
Skorobogatyj, A. F. 1936. Lesoparki u nas i za rubezhom. In: Korzheva, M. P. et al. (Eds.), *Problemy sadovo-parkovoj arhitektury: sbornik statej*. Moskva: Sojuz sovetskih arhitektorov. Sekcija planirovanija gorodov, Izdatel'stvo Vsesojuznoj Akademii arhitektury.
Taylor, M., and Kukina, I. 2019. Planning history in and of Russia and the Soviet Union. In: Hein, C. (Ed.), *The Routledge Handbook of Planning History*, pp. 192–207. London and New York: Routledge.
Tippel, V. 1963. Suur-Tallinna probleemidest. In *Linnaehituse küsimusi Eesti NSV-s*. Tallinn: Eesti NSV Ministrite Nõukogu Riiklik Ehituse ja Arhitektuuri Komitee, pp. 39–46.
Tippel, V 1967. Puhkuse organiseerimisest, *Ehitus ja Arhitektuur*, no. 3:41–44.
Tobilevich, B. P. 1968. *Sistema organizacija otdyha v gorodah*. Rahvusvahelise nõupidamise materjalid teemal "Linnade haljastus kui keskkond elanikkonna tööks, eluks ja puhkuseks I", pp. 39–52, Estonian National Archives ERA.R-1951.1.210.
Volkov, L. 1983. Maa-asustuse arendamisest seoses toitlustusprogrammiga. *Ehitus ja Arhitektuur*, no. 3:52–60.
Zaviska, J. 2003. Contesting capitalism at the post-Soviet Dacha: the meaning of food cultivation for urban Russians, *Slavic Review* 62(4):786–810. DOI: 10.2307/3185655
Zobel, M. 1979. Loodusmaastik: milleks, kellele?. *Eesti Loodus*, no. 5:634–642.

# 11 Gldani
## From ambitious experimental project to half-realised Soviet mass-housing district in Tbilisi, Georgia

*David Gogishvili*

### Growth of Tbilisi and the birth of mass-housing districts

A large part of the contemporary built environment of Tbilisi, the capital of the Republic of Georgia with around 1.2 million residents (GeoStat 2020), came together during the 70 years of the Soviet urban planning and architectural practice. During this period, Tbilisi grew from a small or medium-sized city of about 240,000 people, at the beginning of the Soviet occupation in 1921, to a large metropolis of over 1.2 million people in 1991, after the dissolution of the Soviet Union (GeoStat 2016:3; Jaoshvili 1989:109). In the 1970s, the population of Tbilisi exceeded one million, increasing its importance on the Soviet scale and providing additional funds for development (Jaoshvili 1989). An industrialisation process fueled by the evacuation of Soviet factories from Eastern Europe, which continued slowly after WWII, contributed significantly to the growth of Tbilisi. The majority of large factories were built along the railway line on the left embankment of Tbilisi, as well as on some other lands on the urban fringe not too far from the new Soviet housing districts built to house the increasing workforce required for the growing capital. Thus, Tbilisi population and territorial growth are strictly tied to the industrialisation. Around the early 1970s, the number of people employed in various sectors of industry and construction reached 42% of the total employed population of Tbilisi. The number gradually declined in late years due to the general advances in industrial production (Jaoshvili 1989:121).

Due to this growth, Tbilisi expanded rapidly into the northeast and east, building new residential quarters in these areas (Jaoshvili 1989). As a result of the First (for the period of 1934–1954 authored by Kurdiani, Malazomov and Gogava), Second (1954–1970) and Third (1970–2000) Soviet Master Plans (or General Plans as they were called) of Tbilisi, developed and conceived by Georgian architects and planners, vast areas in the city were built up. Beginning with the First Master Plan of Tbilisi in 1934, housing districts were planned on a larger scale, and street widths were significantly extended. Residential districts occupied approximately five or six hectares and housed a population of up to 4,000 people,

DOI: 10.4324/9781003327592-15
This chapter has been made available under a CC-BY-NC-SA 4.0 license.

primarily living in apartments of four or five floors. As the population of Tbilisi began to grow further and density of the built areas had to follow, the Second Master Plan, approved in 1953, further extended the size of the residential districts and the new housing units. The Third Master Plan, created by architects Chkhenkeli, Jibladze, Japaridze, Shavdia, Lortkipanidze and Bolkvadze focused primarily on building the large housing districts consisting of microrayons and was approved in 1969. From this period onwards, industrial growth of Tbilisi slowed down and mass housing turned into the main driver of its territorial growth (Salukvadze and Golubchikov 2016). The district of Gldani is one of the key projects realised within the Third Master Plan in the capital of Georgia. While a variety of residential structures stand out in the more recent urban fabric of Tbilisi,[1] most residents live in apartment buildings erected during Georgia's forceful presence in the Soviet Union. These are mostly multi-storey prefabricated estates in the mass-housing districts built from the late 1950s onwards and located in mid-city territories, early suburbs and peripheral locations of Tbilisi.

To meet the increasing housing demands, the Georgian Soviet Socialist Republic (Georgian SSR) constructed Gldani, a mass-housing district on the northern edge of Tbilisi (Figure 11.1). Tbilisi City Council commissioned Gldani project to TbilQalaqProject, an institution involved in the

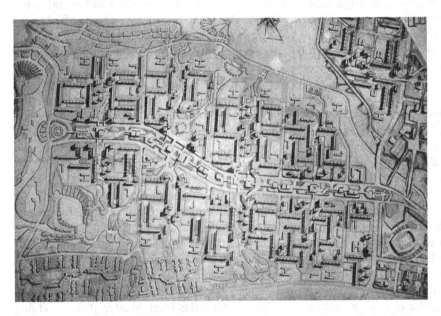

*Figure 11.1* The plan view of Gldani mass-housing district. The vertical axis running between the microrayons was never completed as well as some other features on the outskirts of the district (bottom right corner). Author: National Archive of Georgia, 1968.

urban planning practices in Tbilisi. The future author of Gldani project, Temur Bochorishvili,[2] a 27-year-old architect at that time, worked at Tbilisi Zonal Scientific Research and Project Institute (Зональный Научно Исследовательский и Проектный Институт) that was involved in developing typical and experimental residential and public buildings. Taking advantage of this opportunity, Bochorishvili submitted his winning project to the commissioning agency. According to the author, his work caused widespread excitement, and as it was approved, he joined the TbilQalaqProject that was assigned the task of implementing the project.

Construction of this residential district began in 1969 and lasted for approximately ten years. There are eight microrayons in Gldani, plus "microrayon A" with approximately half the size of the usual microrayon standard. The district was planned to house up to 147,000 people. Without a doubt, if we consider the number of residential units and the intended or actual number of residents, Gldani is the largest housing district compared to others built in Tbilisi, also designed by various Georgian planners and architects and developed since the early 1960s (Afterthesoviets 2009b). According to the municipal election data from 2014, Gldani had over 132,000 voters (residents of age 18 years old and more), but according to some unofficial sources, the actual population is even higher. The purpose of Gldani's construction was to both improve the living conditions of Tbilisians and provide an urban home for rural immigrants who had come to work in the newly established or expanding factories in the area (Kvirkvelia 1985). In the early 1970s, the first residents moved into the buildings in Gldani. However, some of its components were either significantly delayed or never completed which made living in the district hard and forced residents to find alternative solutions to the issues experienced as a result.

Like every other urban area in Georgia, from the late Soviet years, Gldani has experienced a rapid and marked process of social and physical transformation. Some of these changes were caused by the unfinished infrastructure in the district, while others were brought by the new economic system. The effects of the transition between a state-planned and a market-based economy were particularly evident. During the first decade of transition, this process was accompanied by weak institutions, poor governance, and murky corruption practices (Van Assche et al., 2012). This followed the general pattern of transition experienced by many cities from the postsocialist Global East (Hirt 2013; Sýkora and Bouzarovski 2012).

This chapter is based on observations and research conducted as part of individual and collective projects beginning in 2009. Throughout 2009, 2011 and 2017, data were collected primarily through meetings with planners and architects (working during and after the Soviet era) and through interviews conducted on-site with residents of Gldani mass-housing district. Interviews with the author of Gldani project, Temur Bochorishvili, who passed away in 2014, were conducted in 2009 and 2011. Additional data were collected during fieldwork in Gldani in the same years. The research is framed by a critical analysis of significant aspects of Soviet housing policy, including prefabricated mass-housing estates and their transformation from the late Soviet year and the post-Soviet period. Throughout this chapter, I will describe the most noticeable features of this transformation on the physical landscape of Gldani and tell the story of its conception.

## The Soviet housing issue and its evolving stages

From the early years of the Soviet Union, the Communist Party officials recognised that housing problems were a concern for the general population (Andrusz et al. 1996; Hirt 2012). Industrialisation and urbanisation exacerbated the housing shortage during the post-WWII era in the Soviet Union. For a long time, housing was considered a public good in the USSR, but it was given a lower priority than other essential services such as steelmaking (Gentile and Sjöberg 2010; McCutcheon 1989). To address the housing question, building standards and construction practices were developed from the late 1950s, and mass-housing programmes were introduced to accommodate the ever-growing urban population and escalating housing needs (Harris 2013). Initially, this led to a growing number of low-quality, standardised apartment blocks known as Khrushchevkas. The buildings were constructed in the early 1960s during Nikita Khrushchev's tenure as head of the Communist Party. Even though these apartment blocks improved the living conditions for millions, the buildings were also known for their draughts, water leaks, poor acoustic insulation between flats and poor thermal insulation due to concrete walls and metal window frames. In addition, the amount of living space allocated per resident was also low (Hess and Metspalu 2019; Hirt 2012).

The development of more elaborate prefabricated multi-apartment dwellings began in the late 1960s and 1970s, following standardised plans for high-density, multi-storey buildings. Ever more prefabricated and mass-produced residential units were developed by the industrial building sector. Eventually, construction began incorporating full prefabrication: finished panels for "room-sized boxes" (McCutcheon 1989:44). All over the Soviet Union, concrete plants manufactured the elements for apartment blocks. With the increasing demands of housing construction, the production capacity of these plants increased, including the ones located in Tbilisi and other Georgian cities. The housing programmes, however, were not sufficient to meet the needs of the growing urban population nor to meet the demands of higher standards of living. While these measures provided shelter and improved living standards for many citizens, they did not solve the problem of limited residential space or comfort. The number of housing units provided was not adequate (Bouzarovski et al. 2011). The housing shortage in the USSR in the 1980s led to some of the transformations experienced by the housing district discussed in this chapter.

The industrialisation of residential construction also led to a standardisation and simplification of housing units (Morton 1980). The institutionalised uniformity of most of the Soviet residential estates and buildings was underpinned by the ideological motive that all Soviet citizens were equal, at least on a discursive level. Therefore, their housing should be homogeneous as well. Soviet guidelines defined the construction and planning processes and left little room for creativity for architects and planners (Afterthesoviets 2009b). Building projects were based on state-provided building catalogues and approved in Moscow. The architects' role was subordinated in the ear of mass-produced housing developments in the Soviet Union to the construction firms that produced reinforced concrete panels and assembled them on site in identical residential buildings (Harris 2013:31). In certain projects, including the one discussed in this chapter, some individual features were achieved, although they remained limited. In this historical context, the mass-housing district of Gldani and its microrayons were built on the outskirts of Soviet Tbilisi.

**The birth of the largest mass-housing district in Tbilisi**

In August 1970, the Council of Ministers of the Georgian SSR approved the Third Master Plan of Tbilisi, which included the mass-housing district of Gldani (Neidze 1989). Gldani is one of eleven housing districts built as microrayons in Tbilisi to house the growing urban population of the city starting from the late 1950s and going on until the very end of the Soviet era in 1991. The Master Plan was designed to facilitate the development of Tbilisi until 2000 and prevent it from sprawling and merging with nearby towns, a scenario that seemed likely at the time (Kvirkvelia 1985).[3] By constructing large, dense microrayons of prefabricated housing, the new master plan aimed to direct territorial growth towards the north and northeast of the capital.

*Gldani mass-housing district*

Temur Bochorishvili, an architect and planner working at Tbilisi Zonal Scientific Research and Project Institute, designed and planned Gldani. TbilQalaqProject was commissioned by the Tbilisi City Council to develop the general building plan for the Gldani district. Bochorishvili soon applied with his proposal, which was sent to Moscow and received high praise from the competition commission headed by the Chief Architect of Tbilisi at the time, Ivane Chkhenkeli. Despite the strict rules and standards, Bochorishvili managed to include some original and experimental features in the project. One of them is the balconies the architect was allowed to design by the government to "counter the discontent of the local population stemming from the uniformity of the buildings" (Bochorishvili 2009). The vertical axis running between the microrayons in the author's plan was the second feature that was particularly favoured by the architect (Figure 11.2).

*Figure 11.2* The aerial photo of Gldani shot in 1981. The undeveloped territories, initially devoted to the vertical axis since the 1990s was filled up by various small-scale developments as the privatisation of urban land started further separating disconnected microrayons on both lines of the axis. Photo: National Archive of Georgia, 1981.

The axis was crossed by the unique bridge modelled after Florence's Ponte Vecchio which was "designed to offer shops, restaurants, barber shops, gyms, clothes workshops, and a variety of other functions that would bring residents of two districts together" (Bochorishvili 2011). Besides these features, Gldani is a typical Soviet urban planning project which was constructed in accordance with strict standards approved by Communist Party leadership (Afterthesoviets 2009b).

In Bochorishvili's view, the success of his project was due to its ambitious and experimental nature, as he was quite young and not as conservative as his senior colleagues (Bochorishvili 2011). Gldani was constructed on land used by the inhabitants of a nearby village with the same name (Kharadze 1997). Soon one smaller district was developed just north of the district that was also authored by Bochorishvili and was called Gldanula, meaning small "Gldani" in Georgian. Gldanula is significantly smaller compared to Gldani and concentrates only four 16-floor residential buildings and six 9-floor buildings, as well as some additional structures for service and trade.

Gldani, situated on the outskirts of Tbilisi, is part of the district of Gldani-Nadzaladevi. It is located near the railway line that connects industrial zones with the central railway station (Neidze 1989).[4] After the first part of the district was built between 1969 and 1971, the first residents moved in (Kverenchkhiladze 1989). However, due to the grand scale of the project, its construction continued for over a decade. While all residential units were completed in the 1970s, the public halls and other recreational areas remained unfinished.[5] Bochorishvili and those involved in the Gldani project were required to adhere to a construction catalogue regarding the design of the buildings and the rules for each microrayon (Afterthesoviets 2009a). The structural elements of the building, such as the facade panels and the exterior landscaping, were erected using prefabricated structures produced in Georgia (Kvirkvelia 1985). All windows, doors and other smaller parts were produced industrially and used throughout the district. The design of green and recreational spaces, streets and transportation systems was left to the discretion and imagination of the working group (Afterthesoviets 2009a).

**Gldani physical features**

Gldani mass-housing district covers over 4,200 square metres of which up to 1,300 square metres are dedicated to housing of over 147,000 people (Kvirkvelia 1985:169). Microrayons and micro-districts were Soviet planning units consisting of apartment buildings (with 9 to 16 floors) housing 5,000 to 12,000 people (Gurgenidze 2016). Soviet legislation also defined a residential norm of nine square metres per person, which also defined the standards for the development of Gldani (Bouzarovski et al. 2011:2700). Overall, the district had 13,231,000 square metres devoted to housing out of the 4,200,000 square metres plot (Bochorishvili 2011).

Each microrayon was supposed to provide necessary amenities like kindergartens, schools, health care and grocery stores (Afterthesoviets 2009a). This was also the ambition of the author of the district, Bochorishvili wanted to show his professional abilities and planned "to develop a modern and a self-sufficient city plan where almost all the requirements for employment and living would be concentrated" (Bochorishvili 2009). According to the master plan, two metro stations were supposed to be built in Gldani to ensure good connectivity of the district with other parts of Tbilisi; however, only one station was completed and launched in 1989, while the district welcomed its residents in the early 1970s. Because of this delay in the development of the transportation infrastructure, the district was difficult to access.

Gldani's microrayons are arranged along a vertical axis that extends almost 2.5 kilometres. Various social and public services would be provided along this axis, mostly for the residents of the area (Figure 11.2). With a set of horizontal pedestrian bridges that run alongside all the microrayons, the axis was connected to all the microrayons (Figure 11.1). Another parallel road – every 500 metres – connects these streets with the rest of the microrayons. It created a multilevel transit system that separated pedestrian traffic from public transportation and other traffic via bridges that connected housing areas to the central axis. One of the bridges realised, modelled and inspired after the bridge in Florence which was supposed to act as a multifunctional pedestrian bridge connecting the two parts of the district. A large public park was also envisioned on the fringes of what became Gldani in the original plan (Afterthesoviets 2009a). Due to a lack of funds and the chaos that followed the dissolution of the Soviet Union, this part of the project was never materialised (Gurgenidze 2016). Despite these shortcomings, Gldani was an innovative project for Georgia and Tbilisi – it was the first linear and multilevel mass-housing district with horizontal public and commercial spaces connecting its microrayons (Bochorishvili 2011).

There was also a plan to construct parking spaces along with the vertical line, but this feature was not deemed important by the officials that oversaw the construction and was dropped, as the project author recalls (Bochorishvili 2011). This approach was different compared to the vision that accompanied the planning and construction of new residential districts later in the USSR, where the number of parking spaces increased compared to the previous approach (Siegelbaum 2008). This was not the case when Gldani was realised. Thus, the district provided only a limited amount of parking for residents and some temporary parking for visitors. A lack of parking spaces was also caused by the incomplete vertical axis, which resulted in an insufficient number of parking places. This led to the DIY urban transformations initiated by residents in the late 1980s and intensified further after the collapse of the Soviet Union. Gldani's garage count was also calculated based on the overall (limited) role that the automobile played in urban life in Tbilisi during the Soviet period. According to Siegelbaum (2008), this absence of parking was intended to encourage the use of other modes of public transportation and reflected the scarcity of private automobiles in the city. As an example, in 1975 there were 125 cars per 1,000 inhabitants in the United Kingdom, but in Georgia this number reached only 35 and a decade later, in 1985, 71 (Siegelbaum 2008:9; Tuvikene 2010:515).

**Gldani social features**

Populated by the residents from Tbilisi and other rural parts of Georgia, Gldani was built to accommodate residents relocating from various geographies and types of settlements. Residents of historical neighbourhoods in Tbilisi relocated to the newly built district as their houses had become dilapidated or had been damaged by floods a few years ago (Jaoshvili 1989:131). Despite relocating to the district far from Tbilisi's central area, their move was considered an improvement in living standards compared to the old and overcrowded central living quarters of Tbilisi where they lived before (Bochorishvili 2009). A large part of the residents of Gldani moved from Georgia's rural areas to work in the factories expanding along the railway line in the Soviet era. This is the

*Gldani* 199

reason behind the high concentration of ethnically Georgian population in this district despite the relative ethnic heterogeneity of Tbilisi until the dissolution of the USSR (Jaoshvili 1989).

**Apartment building extensions**

The socialist housing construction marathon was unable to meet the housing needs of all and thus provided the basis for extension policies and practices starting from the late 1980s that particularly accelerated following the collapse of socialism. A significant political decision made by the late Soviet government shaped the current urban morphology of Gldani and other cities in Georgia. During *Perestroika*, state policies were "humanized" by acknowledging the societal problem of inadequate living conditions (Bouzarovski et al. 2011:2694). A project titled *"Zhilishche 2000"* (Housing/ Habitat 2000) was launched in 1988 to soften the rigid housing rules. This programme aimed to address the persistent housing problem and ultimately provide a home for every Soviet family by increasing the available residential space in situ. In accordance with this initiative, Georgia's socialist government permitted the extension of state-owned residential apartments in compliance with a number of regulations, including planning, construction and technical controls, as well as size and volume regulations for the extension of apartments (Salukvadze and Sichinava 2019).

From 1988 to 1991 Georgian cities saw the widespread erection of metallic frames for apartment building extensions for thousands of five- to nine-storey block buildings. In the beginning, this work was carried out by state companies. Following the collapse of the Soviet Union, state-owned construction companies were disbanded, and "do-it-yourself" practices became widespread. Over the following decades, this process continued at a varying pace until it was fully banned in the second half of the 2000s (Gogishvili 2021). Thus, in Gldani and other parts of Tbilisi, residents have been able to manage and extend their living spaces. Numerous apartment building extensions were done by residents using a variety of materials and in a variety of forms, often violating safety standards (Bouzarovski et al. 2011). It was possible for residents to encroach on public spaces by disregarding former construction regulations. Although these developments provided additional living spaces and occasionally improved living conditions, they also limited the amount of public space available within neighbourhoods.

Extensions to apartment buildings took a variety of forms and sizes: from enclosing balconies without enlarging the living space significantly to constructing extensions on the ground floor or making use of stairwells. The result was the occupation of previously public spaces and their conversion into residential areas. Moreover, various façade-attached extensions took place either by using extension from balcony, cantilever or frames.[6]

## Multiple transitions of Tbilisi: from the 1980s onwards

Similar to many post-Soviet cities, Tbilisi stepped on the postsocialist transition treadmill (Salukvadze and Golubchikov 2016). Urban built environments were dramatically affected by the shift from a centrally planned to a market economy, and this was particularly visible at the urban level (Gogishvili 2021; Stanilov 2007; Sýkora and Bouzarovski 2012). In Gldani, we can observe the results of more than two decades of transformations that started with the decision of the Soviet government to improve the living conditions of the Soviet citizens (Bouzarovski et al. 2011), but importantly changes that were defined by the Soviet legacy and the conditions of the time. These changes have manifested in various aspects of the built environment, be it residential areas, public infrastructure such as the vertical axis designed by Bochorishvili or massive proliferation of garages (Figure 11.2). In Gldani as in other parts of Tbilisi, residents demonstrated social resilience and tailored homes and outdoor spaces according to their existing and newly formed needs rather than conforming to existing structures (Bouzarovski et al. 2011; Gurgenidze 2016). Multiple, often conflicting actors have initiated these changes, which have often resulted in a deterioration of the built environment and living conditions.

The privatisation of housing has been one of the defining urban processes of transition and has had a lasting impact on Gldani. It began in the early 1990s and was followed by the privatisation of urban land and non-residential buildings. In the early stages, becoming an apartment owner was possible only through relatively tightly controlled state procedures and costs. This was soon replaced by an almost automated process of apartment privatisation that lacked a coherent strategy (Salukvadze and Golubchikov 2016) but was mostly a populist move of the government struggling in different spheres. The privatisation of the housing stock reached almost 95% by 2004 (Vardosanidze 2010). Control of the apartment blocks was chaotically transferred from the state to newly formed groups of homeowners and private developers. This soon led to the rapid deterioration of residential buildings and their related infrastructure (such as courtyards, gardens and access routes). Since 2007, local governments have reclaimed some housing management responsibilities and have also established homeowner associations that have assumed responsibility for building maintenance and management (Gogishvili 2021). Later, these associations were involved in the privatisation of adjacent plots of public land. This has led to significant changes in both the built environment and the daily lives of the residents of Gldani.

A lack of government support and control over urban land distribution in the early 1990s led to the appropriation of vacant spaces between buildings and factories, as well as green spaces between residential areas and collectively owned spaces such as courtyards. Often these spaces were converted for commercial purposes, but cars were also parked there. Starting from the late

1980s to the 1990s and especially the 2000s, the inner courtyards of Gldani, as well as other cities, have become increasingly crowded with cars and garages (Gogishvili 2021). Many were constructed using whatever materials were available and without any permits or approvals from the local authorities. In the 1990s, the motorisation rate was rising but remained still low, so some saw this as an opportunity to occupy a portion of land in front of their home regardless of whether they owned a vehicle. While the country was in the midst of a deep socio-economic crisis, owning or controlling an additional land parcel was highly valuable. This led to even greater densification and overloading of residential areas, converting large and open green spaces into disconnected parking lots, causing traffic problems, disturbing the peace and quiet in urban areas and giving private individuals access to valuable public spaces (Salukvadze and Golubchikov 2016).

**Commercial activities**

The introduction of new commercial initiatives, mostly initiated by individuals or small businesses, was another significant change that occurred in Gldani from the late 1980s onwards. Most of these activities, which had previously been almost entirely alien to the district, ended up being concentrated in several key areas of each microrayon. First, each microrayon had a centrally located street that served as a commercial district. Microrayons located near major transport hubs or in the centre of the district are particularly affected by this phenomenon. In these areas, most commercial goods and services were closely related to the needs, desires and conventions of the surrounding community. Loaves of bread and computer lessons were exchanged and sold. Seasonal fruits and vegetables and ice cream were available from informal kiosks. In Gldani, the central strip is lined with kiosks and market stalls, as well as shops built into adjacent apartment blocks. Most of the formal and informal shopping areas can be found in the central nodes of the district. A vertical axis that was originally assigned to be constructed along all microrayons was supposed to be the primary area where the commercial and other public functions were to be concentrated. But as this plan was only partly realised and the strict state control disappeared from 1991, these functions started to spread in various locations described above. The territory that stayed vacant due to the failure of the plans related to the vertical axis as been filled in various parts as well (see Figure 11.2). This has been mentioned with a regret by the architect, who experienced the loss of function and transformation of the commercial axis and the bridge inspired by the example from Florence into a self-managed and unregulated commercial centre (Bochorishvili 2011).

The area around the Gldani metro station was also envisioned as a commercial, leisure and transport hub. Eventually, this vision became a reality and remains so today. At present, the area is also home to small local businesses, including currency exchange kiosks, shopping centres, street vendors and cafes. The concentration of commercial activity decreases as one moves away from the metro station. Initially, provisions and other services were sold from unlicensed garages and small buildings. Recent changes in the microrayons include the arrival of shopping malls and chain stores, which outcompete and displace kiosks and corner shops. Often, large grocery chains have taken over the physical space of corner shops as well as their local customers.

**Gldani today**

Gldani is the largest residential area in Tbilisi and is part of an even larger district, Gldani-Nadzaladevi, located on the northern edge of the Georgian capital. The changes discussed above, combined with current realities, create a challenging environment for the residents of Gldani. It is important to note that some of these challenges are the result of unfinished work on the Gldani project, while others are the result of the transition from a state-planned economy to a free market economy during the first two decades of Georgia's independence. As in other parts of Tbilisi, Gldani's Soviet-era housing is slowly falling into disrepair. In addition to the age of the structures, the main source of the problem is the Soviet government's decision in the 1980s to improve housing standards by allowing the expansion of private living space. While residents were primarily responsible for these changes, the process was largely controlled by the state, and with the dissolution of the USSR, state control mechanisms disappeared (Bouzarovski et al. 2011).

Mobility issues are often cited by residents as another major concern. Connectivity of the district with the rest of the city and within the district needs to be improved. Part of the problem can be attributed to the limited capacity and coverage of the Tbilisi metro system and unrealised plans for the construction of the second metro station in the district. The number of private cars in Gldani and throughout Tbilisi is increasing, leading to congestion, pollution and loss of public parking spaces as organised parking spaces are scarce, leading drivers to convert recreational areas for parking.

The collapse of the Soviet Union caused many industries in Tbilisi and other parts of Georgia to shrink or cease operations entirely. This had a significant impact on the lives of those who had moved from rural to urban areas in search of work. This problem was experienced by many households in Gldani. As a result of this collapse, many of the residents were unemployed or underemployed. The proliferation of informal economic activities and the attempt to reclaim public land for economic use are examples of this. Despite the lack of clear data on unemployment or household income at the district level, it is likely that Gldani has one of the lowest household incomes in Tbilisi.

## Conclusion

Thousands of Soviet citizens were provided with flats in large prefabricated housing estates such as Gldani as a partial response to the acute housing shortage in the USSR. However, the improvements they brought were often marginal and failed to solve the housing problem at the end. Overall, Gldani failed to meet the growing needs of the rapidly expanding urban population and the growing demand for higher standards of living. Living space in such housing units was strictly limited and planned according to the standard minimum of nine square metres per Soviet citizen. Tbilisi's Third Master Plan in 1970, which significantly improved living standards for the majority of the population, contributed to more than half of the current housing stock. Despite the improvement in living standards, housing remained inadequate.

Gldani mass housing district, like many other Soviet-era projects, was not fully implemented. This led to problems later on. From the late 1980s, Gldani's built environment underwent radical changes, largely driven by the concerns of its residents and the failures of the original project. This process was out of control of the weak local and central governments. While many initiatives, such as garages and apartment extensions, have significantly altered the cityscape in an uncontrolled and unplanned manner but also partly addressed the problems left from the previous era, these initiatives have also damaged the built environment and often resulted in an unequal distribution of space among residents. Despite its involvement in maintenance and renovation issues, the municipality does not have a clear vision for the future direction of the district. It is imperative that future interventions address the district's problems.

## Notes

1 These currently cover over one-third of Tbilisi built-up area, which is around 50 square kilometres.
2 Bochorishvili later designed an extension of the Gldani district called Gldanula which is regarded as a separate neighrbourhood and one of the microrayons of Temqa which is another mass-housing area developed to house the increasing population of Tbilisi. He is also the architect of many other individual buildings in Tbilisi.
3 It was part of the Soviet failed project that would turn peripheral "Tbilisi", a reservoir and an artificial lake located northeast from Gldani, into the heart of the city. This vision was based on the fact that Tbilisi population would increase further and reach two million by the end of the 20th century (Jaoshvili 1989).
4 This proximity to the railway line was one of the main reasons for locating the mass-housing district here.
5 This is particularly visible from the central axis which is suddenly disrupted somewhere after the third and fifth microrayons.
6 More detailed categorisation of the apartment building extensions created through my participation is provided on the following link: https://www.researchgate.net/publication/313853402_Micro-rayon_Living_-_Everyday_Life_Strategies_and_DIY_Practices_in_the_Post-soviet_Micro-rayon

## References

Afterthesoviets 2009a. Dreams vs. Catalog. In: *Social Housing after the Soviets*. Available at: https://afterthesoviets.wordpress.com/2009/06/29/dreams-vs-catalog/ (accessed 31 August 2022).

Afterthesoviets 2009b. Talking to experts on Gldani. In: *Social Housing after the Soviets*. Available at: https://afterthesoviets.wordpress.com/2009/06/28/experts-on-gldani/ (accessed 30 August 2022).

Andrusz, E.G., Harloe, M., and Szelenyi, I. (Eds.). 1996. *Cities After Socialism: Urban and Regional Change and Conflict in Post-Socialist Societies*. Studies in Urban and Social change. Oxford and Cambridge: Blackwell Publishers.

Bochorishvili, T. 2009. Personal communication.

Bochorishvili, T. 2011. Personal communication.

Bouzarovski, S., Salukvadze, J., and Gentile, M. 2011. A socially resilient urban transition? The contested landscapes of apartment building extensions in two post-communist cities. *Urban Studies* 48(13):2689–2714. 10.1177/0042098010385158

Gentile, M., and Sjöberg, Ö. 2010. Soviet housing: who built what and when? The case of Daugavpils, Latvia. *Journal of Historical Geography* 36(4):453–465. 10.1016/j.jhg.2010.01.001

GeoStat 2016. *Mosakhleobis 2014 Tslis Sakoveltao Aghtseris Dziritadi Shedegebi. Zogadi Informatsia*. Tbilisi: National Statistics Office of Georgia.

GeoStat 2020. *Statistical Yearbook of Georgia: 2020*. Tbilisi: National Statistics Office of Georgia. Available at: https://www.geostat.ge/en/single-archive/3351

Gogishvili, D. 2021. Competing for space in Tbilisi: transforming residential courtyards to parking in an increasingly car-dependent city. *Eurasian Geography and Economics* 0(0):1–27. 10.1080/15387216.2021.1993292

Gurgenidze, T. 2016. Archive of Transition. Available at: https://archiveoftransition.org (accessed 31 August 2022).

Gurgenidze, T. 2019. Standartizebuli Tskhovreba. Available at: http://danarti.org/ka/article/standartizebuli-cxovreba—tinatin-gurgenidze/69 (accessed 13 February 2023).

Harris, S.E. 2013. *Communism on Tomorrow Street: Mass Housing and Everyday Life after Stalin*. Illustrated edition. Washington, DC: Baltimore: Woodrow Wilson Center Press / Johns Hopkins University Press.

Hess, D.B., and Metspalu, P. 2019. Architectural transcendence in Soviet-era housing: evidence from socialist residential districts in Tallinn, Estonia. In: Hess D.B., and Tammaru, T. (Eds.), *Housing Estates in the Baltic Countries: The Legacy of Central Planning in Estonia, Latvia and Lithuania*. pp. 139–160. The Urban Book Series. Cham: Springer International Publishing. DOI: 10.1007/978-3-030-23392-1_7

Hirt, S. 2012. *Iron Curtains: Gates, Suburbs, and Privatization of Space in the Post-Socialist City*. Hoboken, NJ: Wiley & Sons.

Hirt, S. 2013. Whatever happened to the (post)socialist city? *Cities* 32:S29–S38. 10.1016/j.cities.2013.04.010

Jaoshvili, V. 1989. Mosakhleoba. In: Jaoshvili, V. (Ed.), *Tbilisi: Ekonomikur-Geograpiuli Gamokvleva*, pp. 102–133. Tbilisi: Sakartvelos SSR Mecnierebata Akademia: Vakhushti Bagrationis Sakhelobis Geografiis Instituti.

Kharadze, K. 1997. *Gldani – Istoriul-Geografiuli Narkvevi*. Tbilisi.

Kverenchkhiladze, R. 1989. Tbilisi: Ekonomikur-Geograpiuli Gamokvleva. In: Jaoshvili, V. (Ed.), *Tbilisi: Ekonomikur-Geograpiuli Gamokvleva*, pp. 251–264. Tbilisi: Sakartvelos SSR Mecnierebata Akademia: Vakhushti Bagrationis Sakhelobis Geografiis Instituti.

Kvirkvelia, T. 1985. *Arkhitektura Tbilisi*. Moskva: Stroizdat.

McCutcheon, R. 1989. The role of industrialised building in Soviet Union housing policies. *Habitat International* 13(4):43–61. DOI:10.1016/0197-3975(89)90037-4

Morton, H.W. 1980. Who gets what, when and how? Housing in the Soviet Union. *Soviet Studies* 32(2):235–259.

Neidze, V. 1989. Qalaqis Dagegmarebiti Taviseburebani da Mikrogeographia. In: Jaoshvili, V. (Ed.), *Tbilisi: Ekonomikur-Geograpiuli Gamokvleva*, pp. 327–364. Tbilisi: Sakartvelos SSR Mecnierebata Akademia: Vakhushti Bagrationis Sakhelobis Geografiis Instituti.

Salukvadze, J., and Golubchikov, O. 2016. City as a geopolitics: Tbilisi, Georgia — A globalizing metropolis in a turbulent region. *Cities* 52:39–54. 10.1016/j.cities. 2015.11.013

Salukvadze, J., and Sichinava. D. 2019. Changing times, persistent inequalities? Patterns of housing infrastructure development in the South Caucasus. In: Tuvikene, T., Sgibnev, W., and Neugebauer, C.S. (Eds.), *Post-Socialist Urban Infrastructures*. 1st ed. Abingdon, New York: Routledge. DOI: 10.4324/9781351190350

Siegelbaum, L.H. 2008. *Cars for Comrades: The Life of the Soviet Automobile*. New York: Cornell University.

Stanilov, K. (Ed.). 2007. *The Post-Socialist City: Urban Form and Space Transformations in Central and Eastern Europe after Socialism*. GeoJournal Library 92. Dordrecht: Springer.

Sýkora, L., and Bouzarovski, S. 2012. Multiple transformations: conceptualising the post-communist urban transition. *Urban Studies* 49(1):43–60. 10.1177/00420980103 97402

Tuvikene, T. 2010. From Soviet to post-Soviet with transformation of the fragmented urban landscape: the case of garage areas in Estonia. *Landscape Research* 35(5): 509–528. 10.1080/01426397.2010.504914

Van Assche, K., Salukvadze, J., and Duineveld, M. 2012. Speed, vitality and innovation in the reinvention of Georgian planning aspects of integration and role formation. *European Planning Studies* 20(6):999–1015. 10.1080/09654313.2012.673568

Vardosanidze, L. 2010. Qalaquri Sabinao Fondis Aswliani Peripetiebi: RUsetis Imperiidan Damoukidebel Saqartvelomde. In: *Tbilisi Tsvlilebebis Khanashi (Urbanuli Sivrtsisa Da Qalaqdagegmarebis Sotsialur-Kulturuli Ganzomilebani*, pp. 134–147. Tbilisi: Tbilisi State University.

# Part IV
# Ecology and environment in the socialist periphery

# 12 New ecological planning and spatial assessment of production sites in socialist industrial Yekaterinburg (formerly Sverdlovsk) in the 1960s–80s

*Nadezda Gobova*

**Planning of Yekaterinburg in the context of general Soviet planning paradigms**

Yekaterinburg was founded on the eastern side of the Ural Mountains in Russia in 1723 and was established on the model of an enclosed, comprehensively planned and state-regulated city-factory. Throughout the history, the city and the factory had shifting economic balances, but at the end of the 19th century the manufacturing function significantly shrunk in size giving a way to other dominating economies such as retail and service economy. Between the foundation of the city and the Soviet Revolution in 1917 at least nine city masterplans of Yekaterinburg had been produced with the purpose to project its future development, on the one hand, and to control its natural sprawl and informal growth, on the other.

The Soviet history of Yekaterinburg was predetermined during the Congress of the VKP (b)[1] held in 1925 in Moscow, where the city was renamed in Sverdlovsk and was declared as one of the important centres selected for the realisation of the state programme of rapid industrialisation of the country. This decision elevated the administrative role of the city, turning it into the economic, industrial and cultural capital of the Urals, and leading to the development of many new large-scale constructions on its territory.

The Soviet architectural and urban planning development of Yekaterinburg broadly corresponded with the general architectural and planning agenda of the whole Soviet Union and was changeable depending on the shifts in the political, economic and social development of the country. The first major milestone in the Soviet Union and Yekaterinburg's urban planning history was defined by the ideology of a Soviet Socialist Revolution in 1917 and can be characterised by extensive theoretical searches for distinctive spatial and functional structure of the ideal Soviet socialist industrial city. These explorations are generally known in the Soviet planning history as a debate between *urbanists* and *disurbanists*[2] which was represented by two groups of architects, economists and planners. The first group advocated for construction of compact self-sufficient urban settlements in proximity to production sites, and the second group supported disurbanisation process where residential settlements were distributed and located remotely from the industries.

This Soviet planning discourse echoed theoretical concepts and practical propositions in other countries at the beginning of the 20th century, as increase in production and construction of industrial enterprises within the cities across the world changed their economies, ecologies and exacerbated social problems. Thus, the argument of Soviet *disurbanists* echoed the ideology of decentralisation of English Garden Cities movement, which was widely popularised across many European and non-European countries at the time.[3] Another *disurbanists* proposal for the liner function-flow planning model of industrial city by Soviet economist Nikolay Milyutin and architect Moisey Ginsburg was influenced by the conveyor-belt production system developed by Henry Ford, which was extended to the work of the whole structure of the city (Milyutin 1930:7).

The second milestone in Soviet urban planning development is related to the post-WWII period of Soviet reconstruction, known as Stalinist classicism. It characterised by the creation of grand city plans, beatification of streets and decoration of facades in neo-classical style. During this period Yekaterinburg's entry plazas to industrial sites and enterprises' gate buildings received neo-classical ornamentation and forms and were aligned by axes with major streets to designate the focal points in the city. The aim for visual embellishment of Soviet cities deviated from the pursuits in many other post-WWII countries in Europe, where the planners and authorities facing citizens' devastating living conditions embraced the principles of rationalisation and economic construction, reflected the ideas of social justice and equality and aimed to achieve improvement via building for all urban groups (Highmore 2010:87).

The last milestone in Soviet city planning starts from the end of the 1950s and marks, on the one hand, the development of rational and scientific approach in production and management of urban-industrial environment and, on another hand, the creation of closed and isolated infrastructure as a response to ideology and pressures of the Cold War (Gobova 2020:293–304).

Yekaterinburg's development officially followed and complied with central political, economic and planning agendas descended by the central Moscow's authorities and reflected in city's official masterplans and planning briefs. However, practical realisation of these formal strategies in the context of rapidly industrialised and opportunistically formed urban setting was rather problematic and went against official directives. The specifics of these processes are revealed and discussed further in this chapter.

**Planning and construction in Yekaterinburg in the 1930s–50s**

The first Soviet masterplan called the Greater Sverdlovsk Masterplan was created in 1930 as the emblematic plan for the transformation of the old imperial city into a new Soviet industrial centre. The general scheme of the Masterplan envisaged the expansion of the existing Sverdlovsk via the creation of a "large grouped city, comprising a number of settlements, organised at their industrial bases, and connected with each other by common production purposes and a centralised management system of communal and socio-cultural services".[4] The Greater Sverdlovsk Masterplan was based on an industrial construction plan and mainly envisaged the effective and economic development of the largest machine-building plants, copper-processing plants, enterprises for the production of building materials, the food industry and other types of productions.

Thus, the old Verkh-Isetskaya metallurgical factory, the Nizhne-Isetsk metallurgical factory and the Uktuz factory, along with their associated settlements, were supposed to form the basis for the deployment of new Soviet enterprises. Additionally, a number of new enterprises and workers' settlements (*sotsgorods*) were proposed for construction on previously undeveloped land to the north of the old city of Yekaterinburg, including Sotsgorod Uralmash and Sotsgorod El'mash.

The Greater Sverdlovsk Masterplan was largely a theoretical and planning manifestation, which reflected the early Soviet debate on the future of a socialist city and specifically on the type of spatial and functional relationships which should emerge between industrial enterprise and residential settlement. The main argument concerned the positioning of an industrial enterprises either in the proximity to a city or at a distance from it. The first allowed to minimise the construction and maintenance costs of transportation networks connecting a settlement and a factory as well as to save workers' time on a daily commute. It also reduced the scale of ecological influence from this transportation. The distancing allowed isolating unsafe and polluting industries from the cities but required creation of prolonged and complex transportation infrastructure.

In the reconstruction of tsarist Yekaterinburg into Soviet Greater Sverdlovsk, planners adopted a hybrid scheme which proposed the decentralisation of economic resources and industrial infrastructure through the spatial distribution of a number of industrial enterprises while assuming situation of workers' towns – *sotsgorods* – in the proximity to relevant factories. The Masterplan projected the industrial capacities and growth of the city's population, noting that "unlike the spontaneous growth of capitalist cities, the growth of Sverdlovsk will be strictly regulated by the amount of labour required for industrial enterprises, municipal economy, regional institutions and organisations. Once these limits are reached, the population surplus, unless new production facilities are deployed, will be resettled in other industrial cities that require a labour force".[5]

Responding to the course of the state-planned and regulated economy Soviet architects and planners sympathetically embraced the vision of the future city as a precise clockwork mechanism where all social and economic processes should be predicted, pre-planned and regulated. The Soviet aim for rationalisation, labour management and spatial control echoed and maximised earlier formulated principles of American industrial scientific management reflected in economic concepts of Taylorism and Fordism, which proposed application of the studies of human efficiency and rational elaboration of manufacture processes for the increase of productivity and economic growth.[6] While these theories encouraged informed link between technological methods of production and spatial organisation of manufacturing processes, they could be hardly realised in early Soviet practice, where planning proposals for socialist industrial cities often coincided with the lack of knowledge in industrial technologies and specifics of production demands. This gap between the planned ideal urban-industrial scenario and the ground conditions and technological needs of industrial enterprises led to perplexity and inability to face the reality of highly complex and difficult tasks once the actual realisation and construction has begun.

The contradictions between the ideal and reality inevitably disrupted the implementation of the Greater Sverdlovsk Masterplan, while a decade later the outbreak of WWII in which the Ural region became a back-centre of economic and military support for the Soviet army, significantly complicated this process. During the war, the scale of industrial production in Sverdlovsk increased seven-fold, while the population of the city doubled. The existing industry of Sverdlovsk and the whole industrial transportation system of the region were fully restructured in order to be able to serve military production and other logistics in the war period. In parallel with the restructuring of the existing enterprises, Sverdlovsk accommodated additionally on its territory approximately 200 factories relocated from various western regions, including Ukraine, Belorussia, the Baltic regions, Moscow and Leningrad (Alferov 1980:68).

Some of the relocated enterprises were urgently deployed on the sites reserved by the Greater Sverdlovsk Masterplan for industrial needs or on other unoccupied peripheral territories of the city, while others were randomly located in the inner areas of the city in the existing repurposed facilities. Thus, the "Bolshevik", the factory from Kiev, was located 12 kilometres southeast of Sverdlovsk and largely contributed to the establishment of the new centre for chemical production in the region. Smaller relocated factories randomly occupied existing buildings in the centre of the city. For instance, the "Uralcable" factory was located in the former workers' club of the Verkh-Isetsky metallurgical factory, the Tools factory was located in the Regional Library building, the Penicillin factory was in the Gosstrakh (State Insurance Agency) building and the Radio factory was in the House of Industries (Tokmeninova 2013:20). Thus, during WWII Sverdlovsk accommodated a large number of industries on territories that were not originally reserved for production purposes.

During the period of the 1930s and 1940s, multiple influential state actors – industrial and military agencies – emerged in the city. They maintained largely autonomous positions, holding their own budgets and resources for the infrastructural development and creating economic, social and infrastructural feudal-like enclaves in the city. Funded and coordinated by various branches of the central state authorities, they pursued their own independent territorial interests and maintained advanced positions in local disputes over the city's land.

Despite the centrally planned and strictly predicted development of the city, which was documented in approved city's masterplans, the actual building process in Yekaterinburg was often deviated from the planning purposes as most of the state enterprises pursued their own spatial and functional goals in land distribution and construction and used their advanced economic positions to achieve them. Such "high priority enterprises"[7] would not consult with the city planning authorities about their construction needs and will rather seek a direct approval from a higher-level state central authorities. This process increasingly complicated the coordination and planning of the urban development in Sverdlovsk as a whole.

Practically, the city authorities, local architects and planners had limited powers and restricted control over the growth of the production sites in the city. Despite being responsible for implementation of centrally approved masterplans, they could hardly control their realisation. The negotiations between the industrial interests (or state economic interests) and public interests (or city's interests) were hardly possible, as economic priority was always directed towards industrial production.

The general specifics of uneven development of Soviet highly industrialised cities were described in the post-Soviet study by Michael Gentile and Örjan Sjöberg. They named such cities as "multi-brunch economic bases", where "the same factory – settlement pattern may be reproduced within the city limits, causing formation of 'towns within the town' and where economies of priority distribution of resources allowed some enterprises and related settlements to be more independent and advanced in gaining access to special financial resources and land" (Gentile and Sjöberg 2006:708).

The example of Yekaterinburg is largely illustrative of this process, and it is also a good representative case for the authors' theoretical model of Soviet "intra-urban landscapes of priority", which they argue were formed by economic priorities and their influence on allocation of land and housing distribution (Gentile and Sjöberg 2006:716). While this chapter describes similar processes in Yekaterinburg, it also discusses them as one of the underlying reasons for formation of complex, unsystematic and environmentally problematic urban-industrial infrastructure. However, this chapter focuses further on the planning specifics of this process and shows how conflicting spatial and functional situations in the city's development led to escalation of ecological problems and how local authorities and planners addressed them in the later Soviet period.

After the end of WWII, many industries remained permanently in what had been intended to be their temporary facilities, creating randomly distributed production areas within the fabric of Sverdlovsk and challenging the realisation of the functional zoning of the city as initially planned. The result of such ad hoc spatial and functional growth led to an increase in autonomous construction and the clustering of urban and industrial infrastructure in Sverdlovsk. In the following decades, the situation that formed during WWII period became highly problematic in terms of the ecological function of the city and its internal logistics.

During WWII, the territory of Sverdlovsk significantly expanded primarily because of substantial growth of industrial territories. The spatial change, however, was not conforming with the planning objectives set out in the 1930s. To address this contradictory situation, a work on the development of the new Masterplan of Sverdlovsk was started in 1947, following the release of a governmental decree for the approval of General Masterplans to reconstruct the most important Soviet cities (Kosenkova 2005:373). The latest version of the Greater Sverdlovsk Masterplan developed by Lengiprogor in 1936 was now taken for redevelopment by a local team of Sverdlovsk's Gorproekt under the lead of architects Petr Oransky and V.A. Arkhangelsky and economist N.T. Strashko. The major aim was to reflect and consolidate the substantial spatial change which had occurred in the city in the war period and to propose a strategy for the city's future development (Tokmeninova 2013:20).

The masterplan proposed a compact structure of the city attempting to connect the independent infrastructure belonging to various military and industrial state agencies into functionally and logistically coherent urban form. However, the architects and planners whose task was to create a new masterplan had little authority to question whether certain developments located randomly in the city during the war should remain in the same location, be reduced in size or relocated. Such questions were not under the jurisdiction of the city and planning authorities at the time. During the post-war reconstruction period, it was already predictable that any further growth of industries would cause an interlocking urban situation with a number of residential zones, enclosed and surrounded by the industrial production sites and railways.

**New ecological assessment of industrial and residential areas in the 1960s–80s**

Until the middle of the 1950s, most of the industrial enterprises in the Ural cities were separated from the residential areas by a relatively small ecologically protected zone, which usually did not exceed one kilometre in width. This parameter was defined by the SNiP[8] and did not create much difficulty for locating the residential areas relatively close to industrial sites. In the 1960s, following the increase of the production capacities of metallurgical and other industries in the Urals as well as reassessment of their ecological footprint, the Soviet Ministry of Health introduced into planning regulations a new parameter which defined the maximum allowed concentration of harmful substances in the atmosphere of residential districts (known as PDK[9]). The calculation of this parameter depended on the industrial capacities and the type of fuel and primary materials used in the production. The introduction of PDK into the planning regulations required a full revision of the existing ecologically protected zones between the industries and residential areas, and many of the residential districts had to be moved away to much greater distances from the industrial sites.

The initial solution which the local planners suggested at the time in response to the new regulations was to locate new residential districts at least 10–15 kilometres away from the industrial enterprises (Lakhtin 1977:45). Such a requirement transformed the previous vision of an industrial city as a complete and calculated urban form with strictly outlined boundaries, predetermined population and calculated functions into a city whose planning had to rationally respond to ecological situations and required decentralisation of the infrastructure and the creation of new flexible and adaptable planning models.

However, such a new intention for decentralisation of the industrial cities inevitably conflicted with another important determining parameter in the state planning system, namely, economic viability. While the remote location of the new residential districts could improve living conditions of future residents, it would require significant investment in the construction of transportation infrastructure to connect these districts with the industries as places of work, and the existing ecological problems within the already built environment would not be resolved.

While the new health and safety standards issues concerned all Ural industrial cities, requiring the relocation of their population, almost no radical actions were taken in the subsequent construction, apart from in a few cases when smaller satellite settlements were developed at a distance from the industries (for instance, Kopeisk near Chelyabinsk). In many industrial cities the new residential districts built during the 1960s and 80s were located on the peripheries but still relatively close to industrial giants.

Unable to move the population of industrial cities far enough away from harmful enterprises, the planners started to search for other ways to improve the ecological condition in the cities. This period was distinguished in local city planning by the development of adoptive planning methods, which could help to mitigate severe ecological conditions within specific planning areas and were more easily put into practice.

The Fragment of the Plan of Mikrorayon Shartash (Figure 12.1) shows the interweaving condition of industrial and residential infrastructure. The significance of this plan is that while its main intention was to propose new residential and public construction in the district, it also outlines the boundaries of existing industrial infrastructure and designates specific functions of industrial sites, production facilities, connecting industrial railways and roads.

The raise of environmental concerns related to industrial pollution and industrial waste was characteristic of the overall development of many post-

216 *Nadezda Gobova*

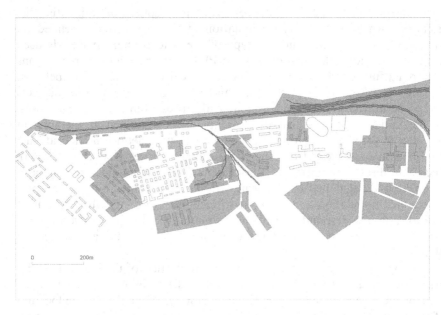

*Figure 12.1* Fragment of the Plan of Mikrorayon Shartash showing the interweaving condition of industrial (shaded) and residential infrastructure. Drawing by the author from the project materials developed in Sverdlovskgrazhdanproekt in 1968 under supervision of Konstantin Uzkikh and engineer P.E. Zundbland.

WWII cities across the world. In Western Europe, the previous war-driven industrialisation was quickly replaced by the demand for mass production of consumption goods and prompted economic and industrial restructuring along with creation of new and stricter environmental policies in the cities. The new zoning rules aimed to separate the industry and residential areas to much greater distances than before (Hatuka and Ben-Joseph 2017:10–24). Such separation of enterprises from the places of living along with rapid demographic growth and suburbanisation required provision of new engineering and social services to growing urban communities. This demand placed extra pressure on existing old infrastructural networks, many of which required a lot of financial expenditure for their modernisation.

Facing the growing infrastructural crisis, some governments (in Anglo-American world) started the process of privatisation of previously state-owned cities operational infrastructure, what, on the one hand, allowed its upgrading through the private investments and, on the other hand, signified the beginning of fragmentation of cities infrastructures and processes of their management. Additionally, this period is also characterised by the shift in attitudes to large, comprehensive and rational city plans as they became "inflexible, unwieldy and failed to deliver necessary infrastructure networks" (Graham and Simon 2001:104).

A Soviet development of this period had similar needs for reconstruction of a post-war industrial infrastructure, and this process of restructuring was also characterised by a relative release in centralised control; however, it had different political reasoning than processes of decentralisation in Western Europe. The Krushchev's reform of 1957, which proposed the organisation of regional Sovnarkhozes (*Sovety Narodnogo Khozyaistva*, Councils of National Economy), provided an opportunity for temporary local consolidation of power in certain sectors of economy. The intention of the reform was to reduce the centralised power of the ministers by transferring control to the regional Sovnarkhozy. Such an initiative, although recalled later, allowed local authorities and planners to get involved in the economic and planning processes which previously were outside of their jurisdiction. Specifically, in the Ural region such economic and political release allowed local authorities to receive access to previously closed industrial infrastructures within the cities and undertake assessments of ongoing urban-industrial processes.

A number of architectural conferences were organised to address problematic issues related to the planning and function of industrial zones. One of the conferences dedicated to the problems of industrial construction in the Urals' conditions was held in Sverdlovsk in 1960. The panellists discussed a set of problems that were characteristic of industrial development in the city, including the problem of the irrational use of land for industrial construction, the inadequate length of engineering and industrial transportation networks and the use of excessive and overweight structural elements in the construction of industrial facilities (Vilesov and Vilesova 2010:117).

Questions relating to the building and function of industrial infrastructure were partially delegated to the architects and planners, who were supposed to produce detailed plans of the existing and new industrial zones, prepare proposals for the standardised and cost-effective methods of the reconstruction of industrial facilities and develop strategies for the improvement of working conditions in the industrial enterprises.

In the previous period, the city authorities had been restricted in their opportunity to control the spatial and functional organisation of industrial sites and did not normally produce and own planning documentation related to industrial development. All initiatives in industrial construction and control over the operation of industrial infrastructure were concentrated in the hands of the state industrial agencies, which were administered directly by the Ministry of Industrial Development.

The new theoretical and practical efforts in the systematisation of the urban-industrial planning of that period can be traced in the work of Sverdlovsk's team of architects, who proposed a new complex approach in the planning of industrial sites. The director and architect of the Sverdlovsk Masterplan studio in "Sverdlovskgrazhdanproekt" (the Sverdlovsk Institute of Civic Planning), Konstantin Uzkikh, and the engineer P.E. Zundbland developed the first Detailed Plan Project of the industrial zone Sortirovochnaya (Sorting Station) in Sverdlovsk in 1959. The project was commissioned by the Department of Construction and Architecture of the City Executive Committee.[10] This was one of the first detailed planning projects of the industrial zone developed in the Soviet Union with the extensive participation of architects and city planners.

The project proposal by Uzkikh was approved by the authorities and was discussed in the Soviet professional press. The importance of such local planning control was also recognised by the central Gosstroi Committee, which imprinted it into a normative act. A new local state institution, Promstroiproekt (the Institute of Industrial Construction and Design), was created to undertake the development of detailed planning projects of industrial zones in Sverdlovsk.[11]

In 1962 Uzkikh presented the work of his team dedicated to the planning of industrial and servicing zones on the example of Sverdlovsk at a conference in Brazil. The manuscript of his speech demonstrates the approach to the theoretical systematisation of industrial development and describes the process of industrial planning that architects were proposing at the time.

Uzkikh stated that an industrial zone should be defined by its size, by the amount of industrial enterprises it hosts and by their technological character. He suggested that all enterprises should be combined into groups based on the similarities of their production processes and their potential for technological cooperation. It was also proposed that all industrial zones in the city be classified into five major categories depending on the level of their negative ecological impact. These categories should define the minimum spatial distances to be set between the industries and residential areas. For instance, Category 1 required no less than 1,000 metres of a health-protecting zone to be provided between production site and residential area, while Category 5 required such a zone of a width of no less than 50 metres. Uzkikh also emphasised the importance of the architectural organisation of the boundaries of industrial sites, stating that they should be planned through the creation of plazas in front of the main entrances to the industrial enterprises and through the architectural treatment of external side of the industrial zone (Uzkikh 1962).

Further, Uzkikh discussed the system of the subdivision of industrial territories, adopting the principle used in the subdivision of non-industrial areas in the city and proposing the organisation of industrial zones, industrial districts and industrial microdistricts (industrial *mikrorayony*). The industrial zone was the largest industrial element in the system and could occupy between 400 and 1,000 hectares of land and include two or more industrial districts, a railway station, a transportation hub and several roads. The industrial district could have a size of between 100 and 400 hectares and should be distinguished by a single system of transportation, a single system of engineering services and a system of connected enterprises. An industrial microdistrict could occupy less than 100 hectares of land and was defined by the absence of large transportation roads and railways. Its territory should be planned to be predominantly pedestrian and should have a leisure centre with servicing facilities for the workers containing a shop, cafés, canteen, sports area and park (Uzkikh 1962).

The standard package of planning documentation for the detailed planning of an industrial district included the site plan with the detailed analysis of the existing industries, the masterplan indicating adjacent residential districts, the "red lines" plan showing the boundaries of industrial zones and axes of the roads, railways and sections of the streets. There was also a combined plan showing buildings and engineering services and facilities, annotated for the development and use of an industrial district.

The presentation by Konstantin Uzkikh in Brazil demonstrates the specifics of the local urban-industrial planning strategies embraced during that period, which aimed to systematise and include the previously unrestricted and largely independent development of industrial zones under the general control of the city planning process. This initiative also marked the beginning of the consolidation of the planning power under the single umbrella of the local authorities. The main purpose of such restructuring was to ensure a more sustainable development of industrial zones, their less negative environmental impact on the city's environment and the provision of better and healthier working conditions.

Thus, in the 1960s period of relaxation in centralised control, the industrial territories partially became a planning concern of architectural specialists. This initiative opened up new design and planning opportunities for the local architects in the sphere of industrial development and partially compensated for the diminishing scope of architectural works in the sphere of residential building due to the reduced number of individual projects and the increase of standardisation methods in design and construction.

Practically, architects and planners became involved in the assignments dedicated to surveying and mapping the existing industrial zones and creation of proposals for their improvement and modernisation. Thus, among many tasks was the aim to reorganise the transportation system within the industrial sites, advocating for the planned separation of different types of transport and for the classification of the roads into industrial transportation roads, railroads, private transport roads and independent safe pedestrian walkways, which should not intersect with the major transportation routes. Normally, a territory of a large factory would extend to several square kilometres and contain various distributed production facilities. Once entered to the factory through the main gate, the workers would need to walk long distances before reaching their places of work, while their cut routes would lie through unsafe places and structures such as crossing railways, pipelines, high voltage lines and operating warehouses (Yakovlev 2006:332). The planners suggested better-organised and safer pedestrian walkways avoiding hazardous infrastructures and created legible navigation. They also proposed various support facilities for the workers, distributed throughout factories including canteens, administrative and recreational facilities (Behtenev 1990:115–22).

Another task aimed in the analysis and improvement of the problematic production zones. Thus, it was proposed to reorganise and relocate certain workshops, storage facilities, engineering networks, roads and entrances to the factories in order to achieve safer logistics, efficient functions and better ecological conditions internally within the enterprises and minimise collision and negative environmental footprint externally within the city fabric. The reorganisation of production zones was considered through juxtaposition of their work with the function of the city outside. Thus, the understanding of manufacturing processes and their technological extensions informed city planners on the nature of their immediate ecological impact on the city and allowed monitoring of an air, water, noise and light pollution (Behtenev 1990; Yakovlev 2006).

The engagement of architects in industrial development created the necessity for the preparation of a particular type of architectural and planning specialist. The new faculty of industrial architecture opened in Sverdlovsk in 1967 and became one of the first programmes in architectural education to specialise in industrial planning and design. On the one hand, the new educational programme provided knowledge of the basics of technological and industrial processes, the typology of industrial buildings and the specifics of industrial logistics. On the other hand, it allowed architecture students to undertake the new type of design and planning assignments, which would normally lie outside of architectural scope of work, including spatial analysis of the production sites and project proposals incorporating specific industrial forms and functions (Popov 2007).

After graduation, the students of the faculty of industrial architecture had the opportunity to join the local state design and planning institutions, which were extensively involved in the development of new industrial projects and were conducting the reconstruction of the existing industrial buildings and industrial sites. This initiative to bring architectural and city planning methods into industrial construction also encouraged interdisciplinary collaboration and established a new local branch of knowledge in the sphere of industrial architecture. Various theoretical studies dedicated to the problems of industrial design and planning proliferated in the 1970s and 1980s in Sverdlovsk. New publications and educational materials emerged during that time, offering a methodology for the planning of industrial sites, design of industrial facilities and techniques for the improvement of the interior spaces of industrial buildings, as well as new principles of functional design of industrial administrative buildings and utility complexes (*Administrativno-Bytovoy Kompleks*) intended for the workers and administrative staff of industrial enterprises (Korotich 1989:48–55; Behtenev 1990:115–22).

One of the showcases of collaborative approach in industrial city planning was presented in the Complex Development Plan of Industry and Other Economic Spheres in Sverdlovsk, which was created with the participation of more than 20 local design and scientific institutes. The plan addressed the issues of the haphazard growth of the industrial city, formulated the strategies of industrial modernisation and proposed procedures for the future progressive development of different economic spheres of the city. It also addressed the questions of the improvement of labour conditions and methods of increasing industrial productivity. Significantly, the Complex Development Plan proposed the relocation of the most environmentally damaging industries outside of the city and outlined requirements for the reduction of water and air pollution as well as land contamination. The importance of this collaborative work was also in its mainly local initiative, or initiative from "below", which was approved by the higher authorities (the Decree No. 14 of the Gosplan of the USSR, 1971) and created a precedent for the development of similar planning strategies in other industrial cities across the Soviet Union.

The Complex Development Plan set the economic and strategic foundations for the creation of the new Sverdlovsk Masterplan, which formulated the main social, economic and ecological targets and quantitative indicators. The new Sverdlovsk Masterplan was developed in 1971–72 in the studio of the Masterplan "Sverdlovskgrazhdanproekt" under the lead of architects K. Uzkikh and V. Piskunov, engineer A. Tseykinskaya and economist N. Barbarskaya. Various state organisations and institutions also took a consulting part in preparation of the Masterplan, including Gosstroi, the Ural Industrial Construction Scientific Research Institute (Promstroinyiproekt), Moscow's Giprogor and Sverdlovsk's Institute of Sanitation and Epidemiology. The masterplan defined the perspective development of the city's territory, set the priorities for urban planning, designated the strategies in construction and proposed the schemes of revitalisation of the city for the following 25–30 years.

The masterplan indicated that the city is distinguished by the high proportion of its residents occupied in the industrial production sector of the city's economy (54% of the total one million population of the city at the time) as well as by the growing number of its scientific and technological enterprises. The document also describes the city's comparatively compact spatial structure with maximum linear dimension being 25 kilometres. It proposed controlling the growth of the city's population, indicating the maximum number of residents for the planning period of 25–30 years as 1.3 to 1.5 million. Such control was to be achieved by limiting the construction of new industrial enterprises in Sverdlovsk. In future, the construction of new large industries should take place instead in the smaller new city-satellites, which would be built around Sverdlovsk, followed by the development of a transport network to connect them. This strategy was given the name the Progressive System of Group Resettlement and assumed better cooperation between the industries, gradual relocation of 25 environmentally harmful enterprises from the city and a simultaneous increase in labour efficiency on the part of the remaining enterprises through modernisation and the use of automated technologies (Uzkikh 1973(a):8 – 9; Uzkikh 1973(b)).

The masterplan also proposed the relocation of the busiest transportation roads outside of the main city territory. All the roads were classified according to their functional use (industrial or for public transport) and their speed limits. The masterplan also indicated the normative time of travel by public transport from home to work for an average resident as a maximum of 35–40 minutes (Sanok 2013:25).

One of the most important formulations in the written section stated that the Session of the City Council required "all industrial organisations, transport and communication organisations, scientific-experimental institutes, district councils and other enterprises to conduct all types of development and construction in the city in strict accordance with the approved Masterplan and in the case of any deviations to hold official consultations with the city authorities".[12] Such a formulation aimed to ensure that the creation of the new masterplan will end previous haphazard growth of the city and would allow centrally regulate its further development by planning authorities.

The masterplaning process of 1972 revealed a number of problematic areas where existing industrial sites spatially collided with the new extended residential districts and created not just multiple rigid boundaries between different functional zones but also situations of clustering and enclosure of one type of infrastructure by another. Some smaller industrial enterprises happened to be surrounded by the larger residential development, or vice versa, with certain residential districts being cut off by the industrial belts and industrial transportation. The long rigid spatial boundary was evidently present on the north of the city and was formed by the Uralmash and El'mash industrial sites. Another significant industrial boundary was created by the industrial railway line running through the whole city from southeast to southwest, where it expands into a large industrial transportation hub.

In this situation, local architects and city planners became particularly concerned with the ecological condition of residential mikrorayons. Unlike general approach in the Soviet city planning at the time, which followed the method of so-called "open plan" (*svobodnaya planirovka*) of mikrorayon (meaning that residential blocks and communal infrastructure were arranged on a specific plot of land without following a particular regular pattern), the emphasis in the local research and practice was placed on a system of calculated factors of environmental influence. Thus, the studies of dominating winds in the region and spreads of industrial emissions were supposed to determine the position of buildings allowing aeration of inner areas of mikrorayons and subsequently reduction of concentration of harmful substances in the air of internal courtyards (*dvors*).

The concept of a "protective city barrier" was also suggested to separate the internal areas of mikrorayons from the negative impact of noisy roads and industrial railways, proposing the location of higher and longer residential blocks against such sources of noise and pollution to form a physical protective barrier. The orientation of bedrooms and living spaces in the flats

Ecological planning and spatial assessment of production sites 223

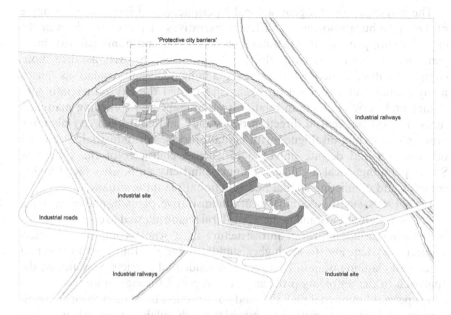

*Figure 12.2* Drawing of a model of Mikrorayon Sinie Kamni in Yekaterinburg surrounded by industrial railways. Drawing by the author from archival materials of Museum of Architecture and Design (UrGAKHU) (Muzei Arkhitektury i Dizaina) in Yekaterinburg.

of these residential blocks was proposed to be inwards, while only secondary facilities, such as lift shafts, staircases and kitchens were located on the sides facing the roads and industries. The peripheral placement of "protective city barriers" enabled creation of comparatively noise-free internal spaces within the residential mikrorayons. The planning of Mikrorayon Sinie Kamni in Sverdlovsk surrounded by industrial sites and railways is an illustrative example of the "protective city barrier" concept at work. The long residential blocks proposed along the railways intend to protect internal residential areas from noise and air pollution (Figure 12.2).

**Conclusion**

This chapter has addressed the notion of the periphery by discussing the role of industrial production sites in the regional city planning process during state socialism. It questioned a stereotype image of a socialist industrial city as a strictly planned and centrally regulated settlement which is attached to or contains at its margins industrial enterprises. Addressing the example of Yekaterinburg, it revealed an opposite situation of haphazardly positioned and sprawling manufactures which have taken a dominant role in shaping the spatial and functional constructs of the city, largely shifting and cutting its infrastructure into many disconnected districts situated on the fringes of giant industrial sites.

The investigation of a geographically peripheral and less studied example of Yekaterinburg also reveals different narratives of power relations in the city planning process. It illustrates how economic and administrative independence of industries from the local authorities as well as their noncompliance development with the local planning goals and regulations caused many spatial and ecological problems and turned the city rather into subsidiary and peripheral site in relation to dominant and sprawling manufactures. The official central direction of "grand" and controlled city planning process was in fact disalligned with actual processes of industrialisation and urbanisation and despite universal instructive character of the system of Soviet planning, local architects, planners and city authorities faced highly complex and conflicting spatial and functional planning situations.

In the 1960s–80s, the continual local administrative, planning and educational efforts in Yekaterinburg and in the whole Ural region allowed to chart and assess the existing urban-industrial infrastructure as a whole complex and interconnected system and to set higher standards for the future ecological and functional planning of industrial sites and residential districts. Specifically, the approach to the unfolding problems was proposed through more transparent and supervised industrial planning and construction in relation to city's development. This was supposed to be achieved through collaborative work of various professionals including architects, technologists, engineers and managers.

## Notes

1 VKP(b) – (Russian abbreviation) All-Union Communist Party (Bolsheviks) existed until 1952 and later transformed into a Communist Party of the Soviet Union (CPSU).
2 'Pis'mo Korbyuz'e k Ginzburgu i Otvet Ginzburga' in *Sovremennaya Arkhitektura* (SA), 1930, No. 1–2 (Yekaterinburg: Tatlin, 2010), p. 60.
3 The concept of Howard's English Garden cities was popularised in 1910s in Russia by architect Vladimir Semenov and remained popular throughout the early Soviet planning practices.
4 *Bol'shoi Sverdlovsk Masterplan. Kratkoe Opisanie Skhemy Pereplanirovki Goroda*, 1930, p. 12
5 Ibid., p. 10.
6 Tailorism – a scientific management theory that became widespread in the beginning of the 20th century in America and proposed the application of studies of human efficiency in the management of industrial production processes. Fordism – a system of mass production, which utilises division of labour, technological automation and standardisation of processes. See Graham S. and Marvin S., 2001, p. 67
7 A term proposed by Gentile M. and Sjöberg Ö.
8 SNiP (Russian abbreviation) (*Stroitel'nye Normy i Pravila*) Building Construction Regulations.
9 PDK (Russian abbreviation) (Predel'no Dopustimye Kontsentratsii Vrednykh Veshchestv v Vozdukhe).
10 Sverdlovskgrazhdanproekt was the successor of the Ural regional institute Uralgiprogor established in 1931 in Sverdlovsk. In 1993 the organisation was renamed "Uralgrazhdanproekt".

11 A note attached to the description of detailed planning of industrial zone, Sortirovochnaya written by Architect Piskunov in 1994. Source: Museum of Architecture and Design UrGAHU, Sverdlovsk.

12 'Spravka o Rassmotrenii General'nogo Plana Goroda Sverdlovska v Mestnikh Instantsiyakh i Utverzhdenie ego Sovetom Ministrov RSFSR, 1971' ['The Report about the Revision of the Materplan of Sverdlovsk in the Local Governmental Agencies and its Approval by the Council of Ministers of RSFSR, 1972], p. 5. Museum of Architecture and Design UrGAHU, Sverdlovsk. Fond No. 3. Khr. 1. 156/33.

## References

Alferov, I., Belyankin, G., Kozlov A., and Korotkovsky, A. 1980. *Sverdlovsk*. Moscow: Stroiizdat.

Behtenev, V. 1990. Arkhitekturno-Planirovochnaya Organizatsiya Sotsyal'no-Kul'turnogo Kompleksa Promyshlennoi Zony. In: Sanok, S. and Stryapunina, O. (Eds.), *Regional'nye Problemy Planirovki, Zastroiki i Blagoustroistva Naselennykh Mest Urala*. pp. 115–122. Sverdlovsk.

Gentile, M., and Sjöberg, Ö. 2006. Intra-urban landscapes of priority: the Soviet legacy. *Europe-Asia Studies* 58(5):701–729. DOI: 10.1080/09668130600731268

Gobova, N. 2020. *The History of the Socialist City of Yekaterinburg (Formerly Sverdlovsk): Planning, Construction, Social Urban Development and Architectural Design, 1920s –1980s*. London: UCL (University College London). Unpublished Ph.D thesis.

Graham, S., and Simon, M. 2001. *Splintering Urbanism: Networked Infrastructures, Technological Mobilities and the Urban Condition*. London: Routledge.

Hatuka, T., and Ben-Joseph, E. 2017. Industrial urbanism: typologies, concepts and prospects. *Built Environment* 43(1):10–24.

Highmore, B. 2010. Streets in the air: Alison and Peter Smithson's doorstep philosophy. In: Crinson, M., and Zimmerman, C. (Eds.), *Neo-avant-garde and Postmodern Postwar Architecture in Britain and Beyond*. New Haven: The Yale Centre for British Art.

Korotich, A. 1989. Formirovanie Arkhitekturnogo Obraza Promyshlennikh Zdanii. In: Kholodova, L., and Kuleshova, L. (Eds.), *Razvitie Promyshlennoi Arkhitektury Urala*. Moscow.

Kosenkova, Yu. 2005. Dve Skhemy Izmeneniya Tsennostnikh Orientirov v Sovetskom Gradostroitel'stve (1937–1938 i 1947–1948 gody). In: Bondarenko I. (Ed.), *Arkhitectura v Istorii Russkoi Kul'tury. Vypusk 6: Perelomy Epokh*. Moscow: KomKniga.

Lakhtin, V. 1977. *Sistema Rasseleniya i Arkhitekturno-Planirovochnaya Struktura Gorodov Urala*. Moskva: Stroiizdat.

Milyutin N. 1930. *Problema Stroitel'stva Sotsialisticheskikh Gorodov. Osnovnye Voprosy Ratsional'noi Planirovki i Stroitel'stva Naselennykh Mest [The Problems of Construction of Socialist Cities]*. Moscow and Leningrad: Gosudarstvennoe izdatel'stvo.

Popov, A. 2007. Promyshlennaya Spetsializatsyya v Ural'skoi Arkhitekturnoi Shkole. In *Arkhitekton*, n. 2, June 2007, http://archvuz.ru/2007_2/6/ (accessed 12 February 2022).

Sanok S. 2013. General'nyi Plan Sverdlovska 1972 goda [Masterplan of Sverdlovsk of 1972. In: Goloborodsky, M., Tokmeninova, L., and Sanok, S., *Istoriya General'nogo Plana Ekaterinburga 1723-2013*. Yekaterinburg: Tatlin.

Tokmeninova, L. 2013. Rabota nad General'nym Planom Sverdlovska 1920–1960 godov. In: Goloborodsky, M., Tokmeninova, L., and Sanok, S., *Istoriya General'nogo Plana Yekaterinburga 1723–2013*. Yekaterinburg: Tatlin.

Uzkikh, K. 1962. O Proektirovanii Promyshlenno-Skladskikh Raionov Gorodov [About Design of Industrial and Warehousing Zones in the cities]. Unpublished paper of a presentation in Brazil in 1962. The Museum of Architecture and Design UrGAHU, Sverdlovsk.

Uzkikh, K. 1973a. General'nyi Plan Goroda Sverdlovska [Mastrplan of Sverdlovsk city]. Sverdlovsk: Sverdlovskgrazhdanproekt. Museum of Architecture and Design UrGAHU, Sverdlovsk.

Uzkikh, K. 1973b. *Arkhitektura SSSR*, General'nyi Plan i Budushchee Sverdlovska [Masterplan and the future of Sverdlovsk], n. 10:7–9.

Vilesov, A., and Vilesova, N. 2010. O Vozniknovenii, Stanovlenii i Razvitii Obshchestvennoi Tvorcheskoi Organizatsii Sverdlovskikh Arkhitektorov Yekaterinburg: Tatlin.

Yakovlev, V. 2006. *Razvitie Arkhitekturno-Planirovochnikh Struktur Malykh Matallurgicheskikh Zavodov Urala*. Yekaterinburg: UrGAHA, Dissertatsiya Kandidata arkhitektury. Unpublished Ph.D thesis.

# 13 Peripheral landscapes
## Ecology, ideology and form in Soviet non-official architecture

*Masha Panteleyeva*

### Nature, city and capital

The historiography of Soviet architecture in its relationship to natural sciences, which largely remained peripheral since Stalin's death, distances itself from the 1960s–70s Western environmentalism and its ecological ideologies as a product of entirely different social conditions. The formal exploration of this relationship within the late Soviet context and its political dimensions presents a complex intertwining of themes of formal and personal freedom in design as well as expression of concern for the future ecological well-being of cities.

In the 1930s, the process of collectivisation initiated by Stalin transformed the Soviet rural landscape into a physical site of class struggle. The Soviet state's new agenda and the emancipation of proletariat in a condition of rapid industrialisation called for eradication of inequality between town and country. Although early Marxist theorists who saw nature as an autonomous pre-condition called for a more guided and productive relationship with the environment, a particular dominating attitude towards the natural world, seen as peripheral to the notion of the state, arose from often erroneous interpretations of such theories. The state's attempts to forcefully "regulate" natural resources also resulted in a massive displacement of the population from their native lands that brought an array of negative consequences for country's social fabric and ecology.

In the cultural sphere, starting in the late 1920s, the use of aerial views in Soviet cinema confirmed this newly acquired control over the country's landscape. The authoritative gaze from above provided a stark contrast to the horizontal perception of the country from the train. In his article "Through the cloudy eyeglasses" for *Novyi LEF* Sergei Tret'iakov (1928) described the new nature of Soviet social space as a hierarchical relationship between consumer and product and as a "dangerous" shift towards the Western economic oppression: he defined Soviet landscape as "nature in the eyes of a consumer" pointing out a transition from the authoritative domination of man over nature to a more passive "consumption" of its formal characteristics through the act of "consuming" nature as an asset to a more comfortable life. A technological achievement was needed to create a sense of a unified space of the Soviet landscape.[1]

After the Russian Revolution in 1917, in light of the nationwide discussion on eliminating the dramatic difference between city and country that was intensified by the rapid industrialisation in new Soviet state as well as Lenin's engagement in creating stronger environmental policy in the future, the interest in garden cities briefly returned to the forefront of Soviet agenda. In *Marx's ecology: materialism and nature* sociologist John Bellamy Foster (2000) discusses Lenin's deep involvement in environmental conservation, stating that he understood that "human labor could not simply substitute for the forces of nature and that a 'rational exploitation' of the environment, or the scientific management of natural resources in accord with the principles of conservation, was essential".[2]

Lenin's concern for environmental protection had reached its zenith in the establishment of multiple *zapovedniki* [nature preserves] across the country that acted as centres of ecological research and helped establish a new public mentality in regards to the subject. Being officially supported by the government, *zapovedniki* also established the new standards for Soviet environmental sciences. Further initiatives followed, including development of multiple urban parks and green recreational areas surrounding large cities. As part of the scholarship uncovering the ecological movement in the East, Douglas R. Weiner (2000), in his *Models of nature: ecology, conservation, and cultural revolution in Soviet Russia*, emphasises the turn to conservation after the revolution as signifying a shift in the political culture. He claims that Soviet political leaders "greeted" ecology as it appealed to "socialism's double mission [of] enlightenment and the rational organisation of social and economic life on the basis of science".[3]

In his recent book, eco-socialist writer Paul Burkett (2014) argues against the common understanding of Marxism as purely productivist in its relationship to nature, claiming that Marx "always recognised nature as an inherent component of human wealth" and that "human production, under both capitalism and other systems, is constrained by natural, physical, biological, and even ecological laws".[4] This particular attitude towards nature is evident in pre-Stalinist Soviet philosophical embodiment of Marxist understanding of the human social relationship to the natural world, including the state's approach to urban planning, that was nothing short of a world-level achievement in the field of environmental sciences.

Echoing Marx, Russian geochemist Vladimir Vernadsky (1928), suggested the close relation between biological and human spheres in his writings on noosphere – the term signifying the new era of technological domination of man on earth. His ideas are culturally understood, however, not as the blind celebration or acceptance of man's domination over everything natural, but rather as the assumption of a great responsibility to preserve and sustain nature – a precursor of ecological science. Vernadsky spoke in favour of the independence of science from external constraints such as political ideology, warning about the dangers of shifting into the new paradigm of the technosphere, implying that "statesmen should be aware of the present elemental process of transition of the biosphere into noosphere", that is, the physical and cultural transformation of environment and the formation of a uniform geo-cultural landscape.[5]

Prior to Stalin's consolidation of power in the early 1930s, one of the largest official and well-publicised attempts of the Soviet government to address the environmental agenda in urban planning was the national competition for the design of the "Green City", announced by *Pravda* newspaper in 1929, the initial idea of which belonged to journalist and writer Mikhail Kol'tsov (1929).[6] This competition received multiple submissions from well-known architects and, although the plan was eventually commissioned to Nikolai Ladovsky's team, well-known proposal by Konstantin Melnikov, reflecting the growing concern for ecology, presented a complex technologically experimental architectural "laboratory" that underlined ecological concerns largely overlooked in the state's focus on industrialisation. It is possible that in this new typology Melnikov saw a future prototype unit of urban planning, where Moscow's working class could experience and readjust to the new type of "socialised" life in unison with nature but also in full control of it. The highly formalised architectural organisation of this urban plan seemed to forcefully orchestrate both the bodies of workers and the natural elements surrounding them, into a perfect, scientifically justified, symbiosis. Both the physical properties of architectural and natural elements and the corporeality of the inhabitants were put through a test – light, air and greenery were artificially manipulated and combined in precise ratios, creating a therapeutic environment for the benefit of the body and mind. Thus, this material synthesis of science, life and nature was formalized through architecture. Melnikov had already addressed the necessity of "merging of bodies and [natural] materials" and the spiritual function of glass, when referring to his design of Lenin's sarcophagus. He described the complex glass structure as a "crystal" – a mystical symbol, designed to preserve the eternal spirit of Lenin's deceased body. Enclosed by "a four-sided elongated pyramid cut by two internally opposed inclined planes of glass that by their intersection formed a strict horizontal diagonal" that once again broke up "the static rectangle of the casket into two lively acute triangles" (Starr 1981).[7]

On the grander scale, this dynamic symbiosis of bodies and materials was manifested in the 1925 Soviet Pavilion at the Paris *Exposition Internationale des Arts Décoratifs et Industriels Modernes*, where the entire building volume was centred around the diagonal central stair: "The visitors who pass by the storefront of the pavilion, do no enter; they will, however, [truly] enter it, if they will be *like* my pavilion: these glass walls and [wooden] stair, so practical for channeling the crowd [...] allow to extend the flow of life itself". Such lateral movement across architectural space, specific to the oblique or diagonal function, both in planar and three-dimensional space of Melnikov's buildings, pertained to both the trajectory of the body and the performance of basic materials. Pulling the public in, and through the exhibit, the stairs' double-diagonal effect served as the pavilion's main dynamising element. This particular attention to a collective movement is later summarised in the master plan of the Green City. At the bottom of the explanatory diagram, submitted to the jury of the competition, Melnikov outlines a series of principal *movements*, orchestrating the public life of the entire city. Complete with a motto "The Power of the Green City is in the System of Movement", these diagrams present four prescribed ways of coordinating the crowds of workers in urban space: "to the centre", "to the central sector", "to the laboratories" and "for long-term stay" (Adamov, 2006).[8]

**Thawing landscapes**

Stalin's death in 1953 did not bring many changes to the official environmental politics and its management of the habitat, both natural and urbanised. Khrushchev's government continued to push industrialisation and urbanisation to the forefront of Soviet politics, initiating projects such as the construction of multiple hydroelectric power stations on the Volga River and the agricultural Virgin Lands Campaign, as well as the massive population relocations causing a damaging separation from the land. Volunteer construction work under the Virgin Lands programme initiated by Khrushchev, for example, signified a new approach to unexplored territories assigning the new generation with a new powerful agency and ability to transform their environment (their motto stated "The students have their own planet—the Virgin Lands"), while at the same time, constituting a productive space of collaboration between the state and the new generation. In the early 1970s, the success of this initiative triggered another phenomenon of mass construction by youth – the Molodezhnyj Zhiloj Kompleks or MZhK (Youth Residential Complexes). MZhKs were the first housing projects for young people that were built by the future tenants themselves and were implemented by the state as a strategy to reduce the housing crisis at the time.

The natural environment continued to be exploited, restructured and altered. However, the phenomenon of the Thaw and the gradual restoration of Soviet civil society triggered some new developments. Soviet ecologists reinvigorated their contacts with Western scientific centres, while the general public once again became concerned with the questions of ecology and the fate of the natural reserves, which at that point symbolically turned into the "islands of freedom" (Zalygin 1999).[9] An entirely new movement in the literary world led by the so-called *derevenshchiki* (Village Prose) writers revealed the degradation of peasant culture as the symptom of urbanisation, idealising the image of Russian traditional village and natural landscape as a truly valuable national attribute (Razuvalova 2015).[10] *Derevenshchiki* believed in the direct connection between the development of civilisation and the process of decay in biosphere, confirmed by their identification of humanity with natural world. Katerina Clark in her seminal book *Soviet Novel: History as Ritual* confirms this ideological juxtaposition stating that "far from celebrating the 'technological revolution', many novels of this time [were] built around some danger of an ecological disaster associated with an overemphasis on technology. Thus we see some slippage between official injunction and the response among writers" (Clark 2000).[11] By the mid-1970s, this "slippage", together with the general public's fear of the nuclear threat during the Cold War, manifested in widespread ecological activism.

Overall, the renewed sense of "freedom" that came during the Thaw marked the formation of the Soviet civil society and the politicisation of the environmental and urban subjects.

To counter this form of stagnation in architectural and urban practices, new informal architectural collectives began to emerge in the late 1970s, attempting to reintroduce social and natural sciences, reevaluating landscape as intrinsically "humanising" as opposed to "productive", and overall foregrounding the role of nature in urban design. Their return to strong organic forms in design also signified a deviation from the state's top-down approach to urban planning towards participatory urbanism. In these designs, the understanding of nature as a material entity, which was previously absent both from constructivist narratives (where nature was seen as purely "social") and from the Stalinist agenda (where it acted as a depositary of natural resources to satisfy state needs) was conceptually rehabilitated. The emergence of socialist ecological studies within Soviet architectural practice and theory in the late 1970s marked the new understanding of "landscape" and "nature" as material and form-defining spatial agents embodying political change, inscribing these alternative practices into a "counter-history" of the state approach to planning history. Architecture's formal turn to sculptural organicism that developed roughly in the late 1960s was closely connected to the rise in ecological concern in urban development (later exemplified by the events such as the Expo'74 along with other environmental proposals of that time such as the multi-disciplinary "Ecopolis" programme, established in the late 1970s by Dmitry Kavtaradze, the director of the ecology laboratory in the biology department of the Moscow University. Through these works it is possible to trace a gradual intensification and plasticisation of form that were evident in the formal deviation from the established canon of architectural drawing and the frequent use of more "pliable" materials (such as plasticine) in constructing architectural models. It could be argued that through this shift architects anticipated the emerging understanding of architecture as environment as well as its formal identification with nature, where the "static" architectural form was no longer relevant in urban planning. Furthermore, focusing on how the ideas of "landscape" and "nature" at large were integrated into the formal language of late Soviet architectural production can help reevaluate them as material agencies of political change.

This change in the formal language of architecture occurred as a reaction to the post-Stalin turn towards the theoretical conceptions of Modernism – a position considered by alternative architectural collectives, such as the Novyi Element Rasseleniia (NER) group, for example, as "excessively" abstract and devoid of natural and social formal references. In developing this idea, it is

232  *Masha Panteleyeva*

useful to discuss NER's shift towards the organic form in the late 1960s in the context of Marxist ecological critique of capitalism,[12] Soviet ecological thought, as well as various environmental movements of the 1960s that had their roots in the 1920s ecological theories (Foster 2000).

NER's very first book *The Ideal Communist City* (Figure 13.1), however, in its description of the new methodology in applying social relationships as the

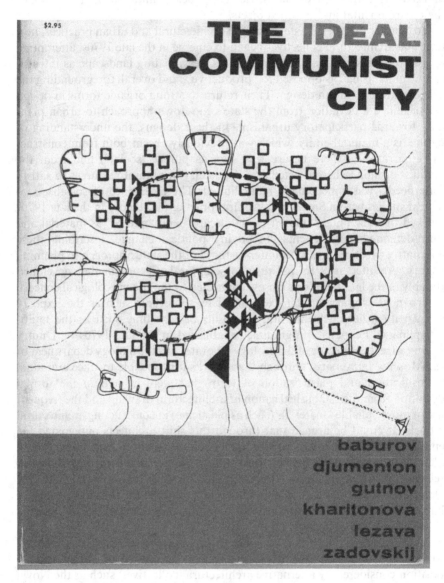

*Figure 13.1* A. Baburov et al., *The Ideal Communist City*, book cover. George Braziller: New York, 1968. Credit: The NER Group Archive.

basis of future cities, shows a somewhat conformist understanding of Marxist theory, referring to its "promethean" approach to nature, seen as "distinct from man's personal confrontation with nature and his own vital processes" (Baburov et al. 1971:16).[13] They also attribute the ability to understand social processes, and moreover, to "shape and control", them, to the recent advancements in new sciences such as "cybernetics, information theory, human engineering, and the aesthetics of technology", entirely omitting natural or social sciences (Baburov et al.:17).

This particular dominating attitude towards the natural world, criticised by contemporary eco-socialist historians like Paul Burkett and John Bellamy Foster, became increasingly more characteristic of the Soviet theoretical interpretation of Marxism during Stalin, who in the late 1940s proposed a "Great Plan for the Transformation of Nature" – a scientific regulation of natural resources as a reaction to the great famine in 1946.[14] This plan, the main incentive of which was to improve Soviet agricultural production, was unprecedented in its scale and, most importantly, in the totality of the nationwide task itself. It included various large-scale interventions into the natural environment: from irrigations systems to a massive programme of afforestation of the entire country to the construction of numerous power plants. Soviet writer Maxim Gorky (1934) excitedly reflected on this utopian forceful conquest of nature in his account of the construction of the Baltic White-Sea Canal:

> Stalin holds a pencil. Before him lies a map of the region. Deserted shores. Remote villages. Virgin soil, covered with boulders. Primeval forests. Too much forest as a matter of fact; it covers the best soil and swamps. The swamps are always crawling about, making life dull and slovenly. Tillage must be increased. The swamps must be drained. The Karelian Republic wants to enter the stage of classless society as a republic of factor and mills. And the Karelian Republic will enter classless society by changing its own nature.[15]

This subjugation of nature in all its specific materiality, readily laid out on the drawing board beneath the human hand, to the ultimate will of the state under the disguise of "science" is well reflected in the infamous statement of the Russian horticulturalist Ivan Michurin (1934): "We cannot expect charity from nature – out task is to take it from her".[16]

This conviction in the power of socialist science and social engineering, based purely, and often too literally, on dialectical materialism, led to the propagation of the concept of "two sciences" – an ideological division into two conflicting systems of knowledge, bourgeois and proletarian. This duality, conceptually splitting the world into two polarities, was exhaustively highlighted by the press and in various propaganda posters. One of the more vivid examples is one by Mikhail Cheremnykh, titled "Two Worlds – Two Plans!" symbolically representing the proletarian world as two Soviet men in front of the Soviet afforestation map. The image is supplemented with an affirmative caption: "we are planting new life", while the image of a "bourgeois" world right below, representing a capitalist and a militarist, featured a rather condemning caption: "they are planting death!"

Despite such "life-affirmative" imagery, Soviet sociologist Oleg Yanitsky (2005), in commenting on the consequences of Stalin's approach to the natural world, suggests that during the 1930s and 1940s the conditions of forced industrialisation and collectivisation of the Soviet society, in addition to the large population transfers initiated by the government, often resulted in a separation of people from their land and thus the loss of cultural roots.[17] Such a new transitory way of life and the increased mobility of Soviet population was gradually becoming the norm and eventually resulted in a different approach to rural living, then seen as transformable and devoid of autonomy, in facing the all-consuming technological progress. This domination of science over nature in Soviet culture has its partial roots in the 19th-century Russian cosmism movement – a synthesis of spiritual beliefs and science – founded by Nikolai Fedorov. His desire to liberate the world from the forces of gravity and instill both spiritual and physical transformation of man resonated in Vladimir Solovyov's concept of "all-unity" and had an immense influence on the Russian avant-garde art and intellectual thought at large. According to Fedorov, through the synthesis of rational science and religion, men could defeat death by awakening the dormant powerful forces within human nature, i.e., overcome natural forces with those of the sciences: "when the earth was considered as the centre, we could be tranquil spectators who take appearance for reality, for the authentic; but as soon as this conviction disappeared, the central position of the thinking human being became the goal, the project" (Fedorov 1906).[18] Stemming from these ideas is Konstantin Melnikov's belief in the "spirituality" of the "passive" architectural materials derived from nature, such as wood, brick and glass, which could be brought to their "higher" function by architectural science. One of the main axioms of totalitarian society is the "imperative of technological progress" where biosphere is being transformed into technosphere (Yanitsky 2005). Douglas Weiner points out the changes in the Soviet interpretation of Marxist philosophy during Stalin, which contributed to the political ban of multiple (especially experimental) biological disciplines:

> While Soviet Marxist philosophers rejected the notion that humans and nature were things apart, real-life Stalin-era attitudes often did set nature apart from people as a hostile force [...] One reason for this antipathy toward and contempt for nature flowed from another impulse within the Russian Marxist (and radical intelligentsia) tradition: its phenomenology of the human being. Viewed as a climax of evolution, human beings were seen as progressively, relentlessly evolving toward total mastery of the course of life on the planet.[19]

*Peripheral landscapes* 235

Indeed, during Stalin's rule, the growing dismissal of theoretical and natural sciences on the official level, supported by the governmental campaigns against Soviet scientific community, such as the infamous purges in scientific communities,[20] re-conceptualised the understanding of the relationship between human and the natural world. *Lysenkoism* was a political campaign led by the biologist Trofim Lysenko with Stalin's support against the scientific studies in experimental biology, genetics in particular, and replaced by the pseudo-scientific conception of inheritance of acquired characteristics developed by Jean-Baptiste Lamarck in the 18th century. Unsurprisingly, these ideological purges had a direct influence on the development of Soviet urban planning policies and the future of research in parallel fields. The ecologist from the late Soviet period Dmitry Kavtaradze – the founder of the movement towards reintroducing ecology to Soviet science in the 1970s – confirms that since the 1930s there was a long-standing "absence of interest from the state in environmental justification of town-planning policy", which had created a historical gap within the discipline (Kavtaradze 2005).[21]

**Forms of nature**

NER's early urban proposals signified a return of such interest to urban planning, at first by simply reconciling multiple urban schemes of the early 20th century that suggested a possible reconciliation of urbanism with the natural domain. Those included Melnikov's Green City and another well-known proposal for the city of Magnitogorsk by Ivan Leonidov, where the architect's deep interest in natural sciences such as crystallography and natural philosophy led him to develop an original understanding of architectural form as derived from morphologies found in nature.[22] NER's own concept of so-called "unified space" similarly anticipated the future transition to a more integrated approach in urban planning. The integrated block or *microrayon* model (residential area unit) implied that all necessary services, such as retail, educational and medical facilities, would be located within walking distance to housing. They envisioned it first of all as a socio-cultural global environment that accounted for a more complex interaction between human beings and their natural environment:

> It is fair to say that functionalism played a revolutionary part in the history of architecture. [...] however, [it] (despite the progressiveness of its conceptions) expresses the crowning achievements of classical architecture. The real revolution comes with planning a unified urban environment. This would be the real and profound revolution in architecture, the beginning of a true 'architecture of space.' What essentially characterises this revolution is the fact that the new architecture will correspond to the social processes that will consciously be reconstructing society.
> 
> (Baburov et al. 1971)[23]

This passage perhaps inadvertently manifests the historical (and spatial) replacement of philosophical conflict of "man versus nature" with what Vladimir Vernadsky, back in the early 20th century, recognised as the transition from biosphere to noosphere. The formal relationship between nature and urban space in NER's work undergoes transformation between their initial linear city model of the late 1950s and the final dissolution of the group in the mid-1970s. These transformations are clearly manifested in the shifting treatment of urban form in both their drawings and publications. An early diagram, illustrating their first collective book, The *Ideal Communist City*, positions nature at the very bottom of the list of essential elements constituting what NER calls the "system of human relationships in communism" (Baburov 1971). In this diagram, nature acts as a material entity in service of man (notably, a separate category from urban parks and agriculture) providing spatial conditions for a set of specific "forms of social relations", including "solitude", "voluntary relationships" and "family and childhood relationships".[24] Such directed assignment of function to the natural material world, primarily as a backdrop to human activities, is an obvious nod to the modernist understanding of nature as an "image", which is to be consumed. In their description of the new type of housing they write about a combination of functionality with psychological comfort, confirming this idea:

> The principle of bilateral orientation is especially important in cases where apartments are located above the ground; in other words, where direct access to nature is replaced by visual contact. Here there is no substitute for a variety of views from windows. Architects at present do not concern themselves with "views." Yet it is precisely the view that has a most powerful influence upon the individual's psychological state when he is living in the limited space of an apartment. The view works as an extension of interior space and becomes an integral part of the world in which the individual is living.[25]

This somewhat Corbusian model would retain its influence throughout the early 1960s, although the group admitted early on that the "villa in the park" model was "an expensive kind of well-being" and a typology that belonged to the realm of the bourgeois society.[26]

What is more significant in NER's more mature understanding of nature's material role in the formation of future socialist "settlements" is the parallel they build between the function of nature in human development and the concept of human *obshchenie* (a hard-to-translate word that conveys both the idea of "socialisation" and that of "communication"). They suggest that the wider the social circle is, the less nature needs to function as a part of society; that is, young children whose social relations are limited to their family, for example, would need the most immediate contact with natural elements. This would in its turn dictate the layout of the housing units (ground floors would be for families, while top-floor apartments with views –

Peripheral landscapes 237

for single adults).[27] In their descriptions of future settlements the NER group also employs formal analogies found in nature as a functional urban prototype – the concept, which would be resolved in their work formally and aesthetically only several years later, during the preparation for the exhibition at the NIITAG (1967) and the subsequent installation during the Milan Triennale (1968). This updated proposal represented an entirely different relationship with the idea of nature:

> The concept [of the future urban environment] is analogous to the growth process in nature by which each element is assured an appropriate site and function. The growth of any complex organism has a definite limit, after which the organism generates a new organism, resembling itself. The chaotic growth of the modern city beyond any limits is comparable only to the growth of a malignant tumour. This kind of growth slows down the normal activity of the organism and ultimately destroys it.[28]

This marked a transitional moment in NER's work and a formal shift away from previously indeterminate formal solutions based on the influences from modernist functionalism and the avant-garde concept of politicised urban communal space, and towards a more organic and ultimately autonomous urban space – a clear attempt to reassign central agency back to nature (Figure 13.2).

This tendency was also present in the Soviet art production of that time: a 2017 exhibition at the Zimmerli Art Museum that explored the subject of nature in Soviet non-conformist art in the 1960s and 1970s suggested that nature was approached "not only as a subject matter or a backdrop to [the art] work, but in some cases as an actor or co-producer", therefore challenging "the link between nature, optimism, and progress, which socialist

*Figure 13.2* The NER group, urban settlement integrated into the surrounding landscape. Plasticine model for Milan Triennale, 1968. Credit: The NER Group Archive.

realist aesthetics had promoted".[29] *Dvizhenie* group that was featured at the exhibition understood (archiectural) form itself as an imprisonment of human space from nature rather than a release into freedom. Their 1968 "Snow Meridian" piece appropriated Malevich's abstract form and rendered it obsolete by positioning the original composition in the actual snow, which was seen as an essentially "national", "natural" and object-like tactile natural material. The materiality of nature in this work, thus, was translated into a palpable art object. These ideas are not unlike those that eventually compelled the NER group to abandon the Corbusian model of the Unité d'Habitation, which, according to the disenchanted in functionalism young architects, looked desolate in the Siberian landscape. The modernity of its form no longer made sense while buried under deep snow.

NER's second and last collective publication, *The Future of the City* (Gutnov and Lezhava 1977), reflected these conceptual changes towards organic form and adopted a new terminology, describing the city's "temporal dimension" as consisting of carcass, fabric and plasma.[30] The authors also included the chapter devoted entirely to urban ecology, titled "Environment Quality", attempting to outline the ecological infrastructure of the future city. As a solution to various ecological problems discussed in the first half of the chapter, the authors suggested the principle of the discontinuity of urban form, which would allow to experience urban space in smaller "doses", with elements of nature "interrupting" the city and reconnecting the urban dweller with the outside landscape via pedestrian links.[31]

> Penetrating into the body of urban space, the outer zone of a large city forms a delimited inner city spaces that provide the formation of various structural units—urban spaces, large complexes, NER-like settlements—and serve as protective membranes for these units. Thus, this developing green zone is transformed into a sort of a large-scale "ecological membrane" of the city, simultaneously dividing and uniting the system consisting of identical unit cells. This zone should be considered as an important and indispensable element of any urban plan, deserving to be implemented no less, and often more than any other part of the city.[32]

In the words of the authors, this ecological structure, that is, a symbiotic network of artificial and natural coexistences, was to once and for all solve the existing conflict between city and nature and to create balanced conditions in which every element of the system existed in a cultivated relationship to the whole. It is remarkable that in this description, city and nature (although both were still seen as highly manipulated environments) finally merged into a single non-dividable form – a unified, mutually supporting symbiosis, responsible not only for ecological balance of urban environment but for its primary morphogenesis itself.

NER's architectural conceptions, as it can be concluded from their awareness of various Western projects inspired by biological forms, most likely were also fuelled by the contemporary works in Soviet experimental biology. During the late 1960s, this field had gradually begun to return to the forefront of Soviet science. The earliest scientific experiments that had the potential influence on NER's closed bio-urban systems were conducted under the umbrella of the late-1950s space programme development. During the mid-1960s, Soviet biophysicist Joseph Gitelson and the team of ecologists worked on a closed ecological systems research aimed to provide human life support in space, which resulted in an experimental prototype BIOS-3 (Gitelson et al. 2003)[33] – one of the world's first closed ecosystem prototypes. Various iterations of the BIOS model were built between 1965 and 1972 at the Krasnoyarsk Institute of Biophysics. It was comprised of an underground structure containing water-recycling and air-purification system and other equipment and allowed to reproduce and demonstrate an insular circulation of elements in a biosphere. Cultivated by the resurgence of Soviet civil society and somewhat in opposition to the environmentally deaf political narrative of the Soviet state, the new outpouring in ecological studies was closely tied to the rehabilitation of sociology as a science. In the relatively open political climate of the late 1960s, the renewed attention to Vladimir Vernadsky's early 20th-century theoretical foundations of interconnected evolution of nature and society, offered a tangible possibility of its practical verification.

As a precursor to these ideas, NER's futuristic plasticine models, prepared for the Milan Triennale in 1968, resembled elaborate terrain models rather than cities. Such formal adherence to biological forms representing an *autonomous social space* within cities proved far more complex than a simple opposition to industrialisation and functionalism in design. Their models constructed primarily with plasticine and celluloid (an unusual choice of materials at that time) pointed at their tendency to *architecturalise* the topography of the natural landscape instead of making urban space more nature-like. In that, NER revealed a new formal paradigm of socialism, giving a recognisable form and an aesthetic charge to the collective space of the city as part of the natural landscape. The centralised organisation of NER's miniature plasticine cities outlined a hierarchical stratification of Soviet society in a form of relaxed symmetries found in nature – an imagery that was fundamentally misaligned with the notion of a homogenous "classless" society provided by the official Soviet discourse. Built with pliable modelling materials that allowed to investigate the "flexibility" of urban form the aesthetics of NER's urban models were driven by the collective act of model making, emphasising the non-linear and "author-less" approach to urban design. In addition, the emphasis on the performative role of modelling materials, found in the hand-made, sculptural quality of the finished models, suggested a new ambiguity of the relationship between urban project and its "miniature" physical representation. While these aspects were often dismissed in Soviet official practice of model making as an afterthought of the design process, in NER's work they acquired a central place as the principal agents of urban research. In fact, the state-official dialectic of "technology vs nature" was *entirely* removed from this process as well as from its final product.

NER's so-called "snail" urban model that represented these tendencies was suspended high above ground in the central space frame structure at the Japan World Expo'70, along prominent Western architects such as Archigram, Moshe Safdie, Yona Friedman, Hans Hollein, Christopher Alexander and Giancarlo de Carlo. The latter was the one who invited Gutnov to participate, signifying NER's more immediate alignment with Western "school of thought" rather than with official Soviet architecture. Responding to the pavilion's theme of *future living*, Hollein, Kurokawa and Safdie focused on the idea of a living "container", while Archigram, with their "Dissolving City," Alexander, with an "Exit Capsule", and de Carlo, with a "City of Participation", went beyond the scope of the domestic environments, focusing on the sociological aspects of future cities (Zhongjie, 2010).[34]

Despite NER's complex social vision of the city, in this dominating context of metabolist-inspired futuristic capsules and molecular structures , NER's snail-shaped model, rigid in its references to classical form and inspired by nature rather than technology, resembled an architectural fossil rather than a "living" environment. Tightly enclosed in a circular box and conventionally resting directly on the floor, it appeared to be ill-fitting within the highly technological environment presented by others. Despite this apparent dissonance, it seemed that NER's proposal in all its complexity of biological references, anticipated disillusionment not only with socialist urban planning but also with the technological progress at large. After all, its limitations were already manifested in Osaka's vision of the future world, symbolically presented as undefined and generic megastructure.

Towards the end of its existence in the mid-1970s, the NER group came to the conclusion that the newly acquired autonomy of late Soviet society, both social and physical, as well as its connection with nature, can in return provide freedom from a defined urban form. This translated into an idea of "personal freedom", meaning that the absence of a distinct materiality of urban space can be equivalent to political autonomy of every inhabitant in this new system. NER's formally ambiguous urban elements borrowed from biology, in reality, were concepts that embraced a fundamentally new Soviet material reality – one that did not pertain to buildings, cites, territories or classless society but to the sphere of ultimate creative autonomy. Through their turn towards the sculptural and symbolic quality of architectural form NER anticipated the acceptance of emerging ecological studies into the architectural field, but they also mediated formal freedom and ideological determination, lending itself as a site for unlimited formal experimentation.

### Notes

1 This major shift in perception of Soviet space as "controllable" territory was reflected in films like "Istrebiteli" [Fighter Planes], 1939. See Sergei Tret'yakov, "Skvoz' neprotertye ochki" [Through the cloudy eyeglasses] Novyi LEF 9 (1928): 20.

2 See John Bellamy Foster, *Marx's Ecology: Materialism and Nature* (New York: Monthly Review Press, 2000), 243.
3 Douglas R. Weiner, *Models of Nature: Ecology, Conservation, and Cultural Revolution in Soviet Russia* (University of Pittsburgh Press, 2000), 231.
4 Paul Burkett, *Marx and Nature: A Red and Green Perspective* (Chicago: Haymarket Books, 2014) xv.
5 Vladimir Vernadsky, "Problems in Biogeochemistry II," in *Transactions of the Connecticut Academy of Sciences* (New Haven: Yale University Press, 1944), 498.
6 Mikhail Kol'tsov, "Dacha – tak dacha!" [Countryside, so countryside!], *Pravda* 24 (1929).
7 Starr, Frederick, *Melnikov: Solo Architect in a Mass Society* (Princeton Architectural Press, 1981), 81.
8 Abramov, Oleg, *K. S. onstantin Melnikov. Arkhitektorskoe slovo v ego arkhitekture* (Arkhitektura-S, 2006), 101
9 Sergei Zalygin, "Otkroveniia ot nashego imeni" [Revelations on our behalf], *Novyj Mir* 10 (1992): 215. Quoted in Douglas R. Weiner, *A little Corner of Freedom: Russian Nature Protection from Stalin to Gorbachev* (Berkley: University of California Press, 1999), 4.
10 See Anna Razuvalova, "Pisateli-'derevenshchiki': literatura i konservativnaya ieologiya 1970kh godov" [Village Prose Writers: Literature and Conservative Ideology in 1970s] (Moscow: Novoe Literaturnoe Obozrenie, 2015).
11 Katerina Clark, "Soviet Novel: History ad Ritual" (Bloomington: Indiana University Press, 2000), 269.
12 For example, see Marx's theory of metabolic rift in John Bellamy Foster's work.
13 Baburov et al., *The Ideal Communist City*, The i Press Series on the Human Environment. G. Braziller, 1971. 16.
14 For more details see Paul Josephson "The Stalin Plan for the Transformation of Nature, and the East European Experience" in ed. Doubravka Olšáková, *In the Name of the Great Work: Stalin's Plan for the Transformation of Nature and its Impact on Eastern Europe* (New York: Berghahn Books, 2016)
15 Maxim Gorky, *Belomor: An Account of the Construction of the New Canal Between the White Sea and the Baltic Sea* (Moscow: OGIS, 1934).
16 Ivan Michurin, *Itogi shestidesyatiletnikh trudov po vyvedenuyu novykh sortov plodovykh rastenij* [The Results of the Sixty-year Research Concerning the Breeding of New Species of Fruit-bearing Plants] (Moscow, 1934).
17 See Oleg Yanitsky, *Ekologicheskaya kul'tura Rossii XX veka* [Ecological Culture in Russia in XX Century] (2005), 141.
18 (Nikolai Fedorov, *Filosofiya obshchego dela* [*The Philosophy of the Common Work]* Vol. I, (Moscow: Vierny, 1906), 293).
19 Weiner, *Models of Nature*, 234.
20 Ibid., 234.
21 Dmitry Kavtaradze, "Urbanizatsiya biosfery" [Urbanization of Biosphere], *Problemy okruzhayushchei sredy I prirodnykh resursov* [Problems of Environment and Natural Reserves], (VINITI 7, 2005), 34.
22 For more see Ekaterina Bukharova's article on Leonidov's landscape designs where she discusses his formal influences found in natural sciences, and in the work of Ernst Haeckel's, in particular. See "Cosmic Architecture by Ivan Leonidov: Narkomtiazhprom's Landscaped Staircase in Kislovodsk Sanatorium," *Academichesky vetsnik UralNIIproekt RAASN* 1 [Academic Journal of Ural Research Institute, Russian Academy of Architecture and Construction] (2011), 55.
23 Baburov et al., *The Ideal Communist City*, 165.
24 Baburov et al., *The Ideal Communist City*, 27.

25 Ibid., 68.
26 Ibid.
27 Ibid., 90.
28 Baburov et al., *The Ideal Communist City*, 112.
29 Online source: *A Vibrant Field: Nature and Landscape in Soviet Nonconformist Art, 1960s–1980s* (04 March 2017–01 October 2017) curated by Anna Rogulina. http://www.zimmerlimuseum.rutgers.edu/dodge-gallery-lower-level/vibrant-field-nature-and-landscape-soviet-nonconformist-art-1970s-1980s#.WPZ-EDdprKA. Accessed on 18 May 2017.
30 Gutnov and Lezhava, *The Future of the City*, 86–107.
31 Ibid., 58.
32 Ibid., 64.
33 See Joseph Gitelson, Genry Lisovsky and Robert D. MacElroy, *Manmade Closed Ecological Systems* (London: Taylor and Francis, 2003).
34 Lin Zhongjie, *Kenzo Tange and the Metabolist Movement: Urban Utopias of Modern Japan* (New York: Routledge, 2010).

## References

Adamov, O. 2006. *K. S. Melnikov. Arkhitektorskoe slovo v ego arkhitekture*.
Baburov, A., and A. Gutnov. 1971. *The ideal communist city*. The i Press Series on the Human Environment: G. Braziller.
Burkett, P. 2014. *Marx and nature: a red and green perspective*. Chicago: Haymarket Books.
Clark, K. 2000. *Soviet novel: history as ritual*. Bloomington: Indiana University Press, p. 269.
Fedorov, N. 1906. *Filosofiya obshchego dela* [*The Philosophy of the Common Work*] Vol. I. Moscow: Vierny, p. 293.
Foster, J.B. 2000. *Marx's ecology: materialism and nature*. New York: Monthly Review Press.
Gitelson, J., Lisovsky, H., and MacElroy, R.D. 2003. *Manmade closed ecological systems*. London: Taylor and Francis.
Gorky, M. 1934. *Belomor: An account of the construction of the new canal between the White Sea and the Baltic Sea*. Moscow: OGIS.
Gutnov, A., and Lezhava, I. 1977. *The Future of the city*. Moscow: Stroiizdat.
Josephson, P. 2016. The Stalin plan for the transformation of nature, and the East European experience. In: Olšáková, D. (Ed.), *In the name of the great work: Stalin's plan for the transformation of nature and its impact on Eastern Europe*. New York: Berghahn Books.
Kavtaradze, D. 2005. Urbanizatsiya biosfery [Urbanization of Biosphere]. *Problemy okruzhayushchei sredy i prirodnykh resursov* [Problems of environment and natural reserves], VINITI, 7:34.
Kol'tsov, M. 1929. "Dacha—tak dacha!" [Countryside, so countryside!], *Pravda* 24.
Michurin, I. 1934. *Itogi shestidesyatiletnikh trudov po vyvedenuyu novykh sortov plodovykh rastenij* [The results of the sixty-year research concerning the breeding of new species of fruit-bearing plants]. Moscow.
Razuvalova, A. 2015. *Pisateli-'derevenshchiki': literatura i konservativnaya ieologiya 1970kh godov* [*Village prose writers: literature and conservative ideology in 1970s*]. Moscow: Novoe Literaturnoe Obozrenie.

Starr, F. 1981. *Melnikov: Solo Architect in a Mass Society*. Princeton Architectural Press.
Tret'yakov, S. 1928. Skvoz' neprotertye ochki [Through the cloudy eyeglasses]. *Novyi LEF* 9: 20.
Weiner, D.R. 2000. *Models of nature: ecology, conservation, and cultural revolution in Soviet Russia*. Pittsburgh: University of Pittsburgh Press, p. 231.
Yanitsky, O. 2005. *Ekologicheskaya kul'tura Rossii XX veka* [Ecological culture in Russia in XX century].
Zalygin, S. 1999. Otkroveniia ot nashego imeni [Revelations on our behalf]. *Novyj Mir* 10 (1992): 215. Quoted in Douglas R. Weiner, *A little Corner of Freedom: Russian Nature Protection from Stalin to Gorbachev*, Berkley: University of California Press, p. 4.
Zhongjie, L. 2010. *Kenzo Tange and the metabolist movement: urban utopias of modern Japan*. New York: Routledge.

# 14 Conceptions of 'nature' and 'the environment' during socialism in Albania

## An ecofeminist perspective

*Dorina Pojani and Elona Pojani*

**Introduction**

Twentieth-century socialist governments in Europe were notorious for their mechanistic treatment of the natural environment. Nature was viewed as a force to be tamed in order to advance agricultural and industrial production. This chapter examines the conceptions of 'environment' and 'nature' during socialism in Albania, a small country of three million people on the Adriatic Sea between the former Yugoslavia and Greece. A peripheral nation within the Eastern Block, Albania is unique among its Eastern European peers for having clung to an Orthodox version of socialism from the end of WWII through the 1980s. When Yugoslavia, the Soviet Union and China in turn introduced market reforms, Albania, in retaliation, cut ties with those "revisionist" countries. Its self-imposed isolation forced the country to strive for economic self-sufficiency at all cost.

During socialism, urban planning was approached in a technocratic manner. All vacant lots and non-residential buildings were expropriated and became public property. Larger houses that had belonged to the upper class were subdivided into two or three sections and rented to working families. A standardised aesthetic of multi-section low-rise apartment buildings was adopted. Housing was scarce, and most of the apartments, which the government let at nominal rents, were small, often packing entire extended families within. Cities had to be compact in order to minimise travel distances, as private car ownership was prohibited, bus systems were poor, and not everyone could afford to purchase bicycles (Pojani 2010). However, industrial pollution was high.

In order to insure workers' allegiance to the new system, the government strove to eliminate differences among social classes and among regions. Opportunities for even the tiniest private businesses were minimal, whereas government commercial space was limited to some ground floors of apartment buildings on main streets. Nonetheless, social disparities were still manifest during the socialist era, although driven mostly by educational and family background rather than residential location or level of disposable income (Pojani 2010). In 1990, the repressive regime was overthrown and replaced by a democratic government.

DOI: 10.4324/9781003327592-19
This chapter has been made available under a CC-BY-NC-SA 4.0 license.

While the socialist regime was toppled more than three decades ago, it is still worth revisiting socialist-era notions of nature. For all its brutality, which is now in disgrace, the socialist conceptions about nature remain Albanians' key reference point in their struggle to understand the present. The method we apply in explaining this history is unusual in planning studies: we analyse the content of symbolic products (literature, film, music, painting) which were served to the public during socialism. We do so because, in that period, in Albania, art and literature practically served as government propaganda (Shatro 2016; Satka Mata 2011). While the creative sector was more "peripheral" than the productive sphere, "the planned symbolic economy represented the expression of the centralised socialist state in its purest form" (Tochka 2016:118). In contrast, elsewhere in the Eastern Bloc, the domain of culture was one of the few sites where resistance occurred.

We adopt ecofeminism as our theoretical lens. A strand of feminism originating in the 1980s, ecofeminism weaves together global patterns of environmental exploitation and accounts of women's oppression. In so doing, it exposes the "master model" that has shaped humans' relationship with nature – in socialist Eastern Europe and farther afield (see Pojani 2021). To ecofeminist commentators, the inferior status of women (under socialism and capitalism alike) matches that of nature and non-human animals. This is attributed to a masculine conception of reality as a hierarchical dichotomy in which one part is superior to the other: men vs women; matter vs mind; culture vs nature; economy vs ecology. It is also due to a mechanistic understanding of history as a linear dialectic in which progress (at the expense of as the expense of the environment, if need be) is both desirable and inevitable (Mies and Shiva 2014). This theoretical lens is suitable for the Albanian context of a long-standing patriarchy that preceded and accompanied socialism and endures into the present. The study is grounded in two socio-historical contexts: Socialist Realism and environmental management in Eastern Europe during the Cold War, which are briefly discussed below.

## Background

### Socialist Realism: purpose and method

Socialist Realism was the official artistic genre in socialist Eastern Europe. Its purpose was not merely to entertain audiences. The primary aim was "to mobilise, to encourage, to enthuse the working population" and "illustrate the miracles of socialist production" (Richter 1997:92). Artists and writers "invented a glorified quasi-reality or wishful reality to transmit a moral" (Richter 1997:92). Works in Socialist Realism style were monumental and holistic, but also schematic and didactic. They usually centred around a glorified hero (the socialist New Man/Woman/Youth), who represented the whole proletariat. Narrative, visual or auditory patterns were prescribed and canonised; characters were clearly evaluated as positive or negative; plots were comprehensible and easy-to-follow; illustrations were clear and legible; music was accessible to the masses; and messages were simplistic and moralistic (Richter 1997; Satka Mata 2011; Nelson 2000; Dado 2010; Beqiri 2020).

Even landscape paintings that aestheticized "the village" and "the peasantry" often contained traces of agricultural work and/or animal husbandry or, more explicitly, propagandistic slogans displayed on hillsides (see Velo 2014). An Albanian commentator notes how bucolic scenes on postage stamps produced in the 1960s and 70s typically contained "tractors ..., locomotives, hydroelectric dams, high-voltage towers, factory chimneys [smokestacks], and happy people in blue work-wear with gears and wrenches in hands" (Leka 2017:456). In Albania, creations were particularly prudish, with few erotic overtones, because here, Socialist Realism was developed in the context of a pervasive cultural conservatism, influenced by Islamic traditions of propriety and female subordination (Woodcock and Ikonomi 2014). Folklore was a frequent source of inspiration (Tochka 2017).

Most countries in the Eastern Bloc gradually liberalised their symbolic economies starting in the 1950s, whereas Albania did not do so until the 1990s. While some Albanian writers and artists were not clear-cut conformists, and the quality of cultural products varied based on individual talents, one cannot speak of "dissidents" in this context. All subversive forms of art and literature were banned here, and censorship was extreme (Shatro 2016; Satka Mata 2011; Tochka 2016). Should any inappropriate works slip through, damning critiques were published afterwards. In addition, gatekeepers sought to block any Western, "decadent" currents – surrealism and cubism, rock and pop, post-modernism and deconstructivism, and so on – from "contaminating" the "pure" Albanian milieu (Satka Mata 2011; Tochka 2016).

Films like Roman Polanski's *Knife in the Water* (1962) or novels like Mikhail Bulgakov's *The Master and Margarita* (1967), which had no positive hero and/or no optimistic message, could not have been produced in Albania; rock bands like Hungary's "Illés" would not have been allowed to exist; plays that, however subtly, criticised the regime, like Václav Havel's *The Memo* (1965), would have never seen the stage. Not only was domestic production under surveillance but foreign works presented to the Albanian public were also carefully curated in terms of ideological content.

Some traditional legends were also recast to fit the Socialist Realism mould. For example, "Rozafa's Legend" tells the story of three brothers at work building a castle (or, in some versions, a bridge). They are making slow progress: what they build during the day mysteriously collapses at night. They are told by a sorcerer that a human sacrifice is required for the building to stand. After a gamble, the brothers bury one of their wives alive at the foundation, and as foretold, the castle finally stands overnight. Ever since the foundation has remained wet as the young mother sheds tears for her orphaned son (Kuteli 1987). This story reveals the misogyny of traditional Balkan societies but also their disruption through city building and modernisation, symbolised by the castle. The legend strikes a cautionary note about the danger of interfering with nature. In contrast, in a Socialist Realist reinterpretation by Ismail Kadare, Albania's foremost writer, "Rozafat becomes ... a trope of the sacrifice and suffering of individuals for the progress of the collectivity" (Morgan 2017:105).

Overall, the Albanian version of Socialist Realism was the obedient handmaiden of politics and the economy (Shatro 2016; Tochka 2017). In its entirety, this genre was necessary to illustrate the utopia of socialist life (Velo 2014; Grgić 2021). It served to mould a "New Albanian Man" – one that in theory embraced the socialist cause unconditionally but in reality was too scared to rebel. In a sense, the entire public was "feminised" and infantilised in that it was presumed to be subservient to the Party – in the way women and children must submit to husbands and fathers. Shaping the collective consciousness through symbols helped the Albanian cult dictator, Enver Hoxha, maintain power for over 40 years (Shatro 2011; Velo 2014; Pojani 2014; Satka Mata 2011).

Creative professionals who deviated from the Socialist Realism frame faced severe punishment, including marginalisation, exile to remote areas, imprisonment and even execution. Notorious "witch hunts" were organised periodically, targeting writers, painters, musicians, filmmakers and even architects or urban designers (Tochka 2016; Pojani 2014). At the same time, individuals who toed the Party line – out of fear, opportunism or true belief in socialism – were rewarded with steady jobs, better housing, equipment, sabbaticals, retreats, prizes as well as repute (Shatro 2016; Tochka 2017). For these reasons, the creative sector was sought after by many – although the productive sphere was prioritised in the economy.

That is not to say that the cultural edification of the population was unimportant to the regime. A major task of intellectuals was to enlighten, uplift, civilise and educate the aesthetic tastes of the masses (Nelson 2000; Tochka 2016), which in Albania, at the end of WWII, were in large part illiterate. If their quality is debatable, symbolic works were certainly produced in high quantity. By the 1970s and 80s, Albania released a dozen films per year, held at least five national song festivals, and published myriad novels, short story volumes, and poetry collections; theatre and dance troupes performed in every city; and the Art Academy in the capital trained professional painters, musicians and actors. Therefore, there is an abundant body of works which can be used for study purposes.

*Environmental management under socialism*

Early Marxists argued that environmental degradation under capitalism was a core reason for societies to embrace socialism. Much hope was entertained that a socialist world would create the conditions for optimal interaction between humans and the rest of nature (Gare 1993). A deep concern for nature and the environment was reflected in symbolic products from the Russian Revolution era. Russian literature was widely translated into

Albanian and served as a model for the introduction of Socialist Realism in Albania. Also, many Albanians spoke some Russian, which was the main foreign language taught in the schools (followed by French).

For example, the sci-fi novel *Red Star*, written at the turn of the 20th century by Alexander Bogdanov, a Soviet author, tells the story of a socialist revolutionary who leaves a backward Earth to end up in Mars. There, he finds an advanced socialist society – classless, genderless and polyamorous – which has overcome all bourgeois constraints and now lives in apparent harmony with the environment. However, soon he discovers that his first impressions were misleading. Mars is in fact suffering from heavy industrial pollution, overpopulation and resource depletion. To cope, Mars's socialist government plans to colonise Earth – in the process exterminating all Earthlings (Bogdanov 1984 [1908]). This story will sound dangerously familiar to contemporary readers. Uniquely, this prescient novel indirectly lays the blame for environmental degradation at the feet of mechanistic and technocratic societies.

Later sci-fi novels were similarly critical of environmental practices but displaced the responsibility from socialist to capitalist interests. For example, in *The Air Seller* the action takes place in Siberia, but the villain is an American industrialist, Bayley. Housed in a vast underground facility in the middle of nowhere, he is slowly sucking the Earth's oxygen, deep-freezing it and storing it for later use. His plan is to sell the oxygen back to the people once he has succeeded in creating a deficit. Bayley is backed by powerful Western imperialists and even claims to having trade relations with Mars. His base is populated entirely by men – workers, researchers and engineers – apart from a single female scientist, who by the end commits suicide. Eventually, the Red Army saves the world by storming the base and destroying Bayley's operation (Belayev 1966 [1956]).

This story does anticipate current practices of privatisation – of beaches, freshwater sources and the like. But privatisation of nature is framed as a foreign notion. In reality, the Soviet Union and its satellites adopted oppressive policies which sought to dominate nature through science and technology. In some cases, attitudes here were more callous than in capitalist systems because social property, such as nature, was seen as "orphaned" and free for the taking (Ziegler 1985).

While patriotic songs and poems eulogising vast forests, green valleys, majestic mountains and pristine rivers were ubiquitous, pre-socialist approaches of worshipping and fearing nature were left aside (Richter 1997). Proletarian science and technology served the economic goals of the Five-Year Plans rather than the principles of ecology and conservation. Nature was anthropomorphised and became a sort of déclassée: just another domestic enemy to defeat (Richter 1997). In an industrialisation frenzy, socialist nations dammed rivers, built canals, paved cities, drained marshes, dug mines and reservoirs, mechanised agriculture, acclimatised exotic flora and fauna and disrupted traditional societies and ecologically fragile

# Conceptions of "nature" and "the environment" during socialism 249

*Figure 14.1* Untitled, Emi Skënderi, 2022. This drawing illustrates the contrast between environmental ideals and realities under socialism.
Source: Authors' private collection.

environments (Gare 1993; Richter 1997). To the extent possible, the earth was redesigned, remoulded and exploited according to the will of the Party (which replaced old Gods in the local psyches).

Possibly, socialist governments realised that this was a risky approach from an economic perspective. Yet they entertained the notion that resources were sufficient into the distant future and pollution was a temporary anomaly; all problems would be resolved once socialism was sufficiently advanced. Meanwhile, people were mostly silent – out of ignorance, indifference, fear or faith in the political myths propagated by local socialist parties. Out of need, the populace even joined the governments in plundering, misusing and degrading nature: illegal poaching, cutting of firewood and waste dumping were not uncommon (Ziegler 1985). Figure 14.1, a drawing by a six-year-old girl living in Albania, illustrates the contrast between environmental ideals and realities under socialism. While the governments painted an idyllic picture of living conditions, the reality was much darker.

**Study method**

This chapter analyses the content of 13 symbolic works from the socialist era in Albania (Table 14.1). All touch upon environmental themes, even if those are secondary. Also, most deal with gender issues in an explicit or subtle way.

Table 14.1 Analysed works (in chronological order)

| Title | Genre | Author | Year | Link |
|---|---|---|---|---|
| "Mrika" (*Mrika*) | Opera | Llazar Siliqi (librettist) | 1958 | bksh.al/details/113867 |
| "The swamp" (*Këneta*) | Novel | Fatmir Gjata | 1959 | bksh.al/details/57735 |
| "Lil' gypsy woman" (*Kurbatka*) | Poem | Dritëro Agolli | 1961 | bksh.al/details/34688 |
| "The great river" | Short story | Mitrush Kuteli | 1964 | bksh.al/details/9295 (2009 ed.) |
| "The globe" (*Globi*) | Popular song | Fatos Arapi (lyrics) | 1969 | youtube.com/watch?v=aqNVepLfiEs |
| "Planting trees" (*Mbjellja e pemëve*) | Painting | Edi Hila | 1972 | galeriakombetare.gov.al/en/collection/img/edi-hila-mbjellja-e-pemeve_1.jpg |
| "When it rains" (*Kur bie shi*) | Children's song | Odhise Grillo (lyrics) | 1975 | www.youtube.com/watch?v=Mk_D-NRVLRw (2015 ed.) |
| "City lady" (*Zonja nga qyteti*) | Theatre play | Ruxhdi Pulaha | 1976 | bksh.al/details/109293 |
| "Ben walks on his own" (*Beni ecën vetë*) | Young adult novel | Kiço Blushi | 1979 | bksh.al/details/40010 |
| "The scarecrow" (*Dordoleci*) | Animated film | Bujar Kapexhiu (script) | 1980 | aqshf.gov.al/?movie=1379&lng=en |
| "Joniada" (*Joniada*) | Ballet | Anastas Kondo (librettist) | 1984 | arkiv.uart.edu.al/handle/123456789/393 |
| "Two checkmates" (*Dy herë mat*) | Film (comedy) | Bujar Kapexhiu (script) | 1986 | aqshf.gov.al/?movie=165 |
| "The train departs at 6:55" (*Treni niset në 7 pa 5*) | Film (drama) | Nexhati Tafa (script) | 1988 | aqshf.gov.al/archive/load/id/195 |

Our analysis does not focus on artistic quality but rather seeks to elucidate the content as it relates to nature and the environment, and where possible, gender. In selecting the works, we have sought to mix different artistic and literary genres and to cover as many aspects of environmental management as possible: urbanisation, agriculture, horticulture, pisciculture, mining, energy production, logging, forestry and land reclamation. The geographic coverage is broad, with the works set all around Albania, from the northeastern mountains to the southwestern coast. All works were commissioned by the government. The timeline spans three decades, from the 1950s through the late 1980s. All the authors are renowned and the works themselves are well known to the Albanian public. They were promoted in a variety of media when they were first produced, and most are still in circulation.

**Analysis**

The following analysis is structured into two main themes: (1) triumph of science and culture over nature and (2) emerging environmental consciousness. Seven works are included in the first theme and six works in the second. This distribution does not reflect the actual prevalence of each theme in symbolic production. Also, in the sample, there is no temporal shift from the first theme to the second; the two ran in parallel. We reiterate that the typical approach in socialist-era art and literature, through the end of the 1980s, was to applaud the colonisation of nature as a Labour Party victory rather than to recognise the ecological disaster that occurred. However, some works managed to evade, if not entirely subvert, Socialist Realism conventions, and a selection of those is included here. Where works appear to be influenced or inspired by foreign works, this has been noted in the analysis.

*Theme 1: triumph of science and culture over nature*

*"The swamp"*

This sprawling canonical novel, later adapted into film, fictionalised a real event taking place in the late 1940s: the draining of the Maliq swamp in southeast Albania. The purpose of this project was two-fold: to convert the marshland to agricultural use and to eradicate malaria-causing mosquitoes in the area. If wetlands are drained or cleared, they become a carbon source, releasing stored carbon into the atmosphere; also, biodiversity is lost (EPA 2022). Some Albanian sources claim that local experts in the pre-socialist era had been opposed to drainage for this reason and had endorsed less drastic mosquito management approaches in these sensitive habitats. But this project was key to the socialist regime. The Albanian dictator understood that its success would strengthen and legitimise his rule in the immediate post-war period. To justify the necessity for such a costly undertaking, the swamp itself needed to be vilified, and symbolic products such as this novel helped a great deal in this respect. The story casts the swamp as pure evil: not only is it full of mosquitoes, but it also breeds other "dangerous" or "disgusting" animals such as snakes, frogs or turtles. These represent an enemy to be defeated. No mention is made of wetlands as biologically productive ecosystems comparable to rain forests and coral reefs (EPA 2022). Even the use of the word "swamp" as opposed to "wetland" or "lagoon" highlights the negative connotations attributed to this piece of land. The battle to drain the swamp is also framed as clash between popular beliefs and progressive science and technology. According to a local legend, a monster (similar to Loch Ness's) lives at the bottom of the swamp and has historically thwarted any efforts at drainage. Socialist engineers and proletarians – mostly men – set out to break the old myth, liberating the people from infested waters and irrational beliefs all at once. In reality, much of the work was performed by unpaid political prisoners, dozens of whom lost their lives in the process. In the novel, gruelling labour by maltreated, underfed and ailing slaves lacking appropriate tools and technology is idealised as the creative work of the conscious masses (see Richter 1997). Despite socialist efforts and fanfare, the project's first stage failed due to technical difficulties. To save face with the local populace but also out of paranoia, the regime scapegoated a group of white-collar employees (all men, except for one woman) for the failure. Framed as saboteurs, all were found guilty by a kangaroo court and four were sentenced to death – including the woman, Zyraka Mano. She was executed while pregnant, alongside her husband. This macabre act is exemplary of the violence endured by women and families who fell out of favour with the regime (*Kujto* 2 May 2018). In the novel, Zyraka, who was Yugoslavian and therefore perceived as more sophisticated and forward than local women, is misrepresented as a vamp. While married, she brazenly attempts to seduce one of her co-workers (for no clear motive) but he, being a highly principled New Man, refuses her advances. The rest of the novel endorses and reinforces the official version of the events. After a series of attempts, the swamp was fully drained in the 1960s and served as the local breadbasket for a few decades. A commemorative documentary film (*Toka të përtëritura* [*Rejuvenated lands*]) was commissioned in 1969. Now the area is in fragmented private ownership. The bitter irony is that it regularly experiences devastating floods, placing at risk the livelihoods of local farmers (Shqip Show 2010). With climate change and increased flooding, it is on the way to returning to its pre-socialist state.

*"Mrika"*

Set in Northern Albania, a historically underdeveloped and isolated region, this opera tells a story of emancipated love – at least on the surface. Mrika is a highland girl in love with a local boy, Doda, who is also smitten with her. To marry, the two protagonists must overcome a major barrier: Mrika's forced engagement to another local boy (the antagonist), which was arranged by her family when she was still "in the cradle", following local custom. The resolution of the conflict involves breaking old taboos around marriage and women's conduct. Mrika's opportunity for emancipation is presented when she is offered work as a builder at a hydroelectric power station which is under construction in a nearby town. This allows her to leave her family and spend time with her beloved, who also works at the construction site. (Mrika's betrothed, in contrast, shuns the work and even plots to sabotage it.) Romance flourishes in the foreground of a socialist megaproject. Mrika represents the typical socialist New Woman. Like the New Man, she is a former farmer converted to an urban or semi-urban proletarian. Mrika exemplifies the positive values of hard work, honesty and authenticity – and in return the socialist system helps liberate her from the patriarchal yoke. Although Mrika is expected to embody some traditional female traits, such as modesty and respect for her elders, she is also masculinised through speech and social position (see Grgić 2021). She is not a simple labourer but rather a forewoman (*brigadiere*). While Doda happily accepts her as a supervisor, her betrothed, who is cast as misogynistic and therefore anti-socialist, calls her a "wild goat" and claims that he will domesticate her into a "docile sheep". However, Mrika is not afraid to talk back and even berate him. Eventually, Mrika's elders let go of past prejudices and agreed to her marriage to Doda. In that sense, unlike the tragic Italian operas of the 18th and 19th centuries, this opera has a "happy ending". But the treatment of nature, as depicted in "Mrika", is hardly positive. The story of the hydroelectric power station construction is significant and based on reality. It refers to a real station called Karl Marx, which was commissioned in 1957 and is still active. Its reservoir is fed and drained by the Mat River. On the cover of "Mrika"'s libretto, no sign of a love story can be found. Instead, we see a young woman clad in folk costume holding a shovel while a river dam is delineated in the background. Throughout the opera, characters sing praise to the construction works, which are being "happily" carried out by "volunteers" from all over Albania. In reality, workers were forced to participate without compensation in what amounted to modern slave labour. Hence Mrika's power as a woman comes at the expense of nature as well as a mass of helpless persons. Hydropower energy, while "cleaner" than electricity generated from coal, has had significant environmental impacts in Albania. By diverting and slowing river flows, dams have disrupted local ecosystems in major way (EcoAlbania 2017). These problems are, of course, not mentioned in the libretto. Instead, the lyrics make a virtue of the damage to nature. Workers are set to "turn Mat's waves into light", "cut through hills" and "overthrow mountains". The metaphor of light appears throughout the text. Light refers to both the electrification of Northern Albania (and therefore its escape from the "Dark Ages") but also to the enlightenment of the Albanian people under the guidance of the Party. The Party, like the power plant, is said to produce an inextinguishable light which shines Albania's path to socialism. The opera concludes with an ode to the hydropower plant and the Party rather than the lovers, leaving no doubt about the priority of these themes over Mrika and Doda's romance.

*"Joniada"*

"Joniada"'s story is quite similar to "Mrika"'s, although the two works were produced more than two decades apart. This demonstrates the endurance of Socialist Realism schemata in Albanian art and literature. Joniada is a typical socialist heroine who also serves as *brigadiere* in an *aksion* (compulsory summer labour camp for youth). The action is set in a coastal town in Southern Albania. The task at hand – quite common at the time – is to terrace the hillsides lining the sea in order to turn them into productive olive and orange groves. In the libretto, this is referred to as "beautifying the coastline". Terrace farming can be beneficial in preventing soil erosion and conserving water in addition to increasing food production. However, poorly designed or managed terraces can lead to serious environmental problems such as runoff and soil loss (Deng et al. 2021). Even the economic benefits of using the coast for agricultural activities instead of tourism and recreation are questionable. But the socialist era motto was: "let's render mountains as fertile as fields" – which illustrates the aggression with which nature was treated. The ballet libretto starts by reporting verbatim one of the Albanian dictator's directives: "not a single patch of land is to remain fallow". Later, the libretto describes how workers equipped "with sledgehammers and crowbars, cut through rock, fill the stone cavities with soft soil, and plant the seedlings". On one hillside, the new trees are arranged so as to spell a Party slogan. While *aksion* is described as "fun", in reality work and hygiene conditions were quite poor. Youth worked long hours under the scorching sun and then slept in tents or rudimentary cabins with no toilets. Food and water were rationed and breakouts of infectious diseases such as dysentery were common. To sustain the myth of overwork as positive and desirable, the sun is referred to in the libretto as "the most ancient worker of the cosmos". Yet many Albanian people remember *aksion* days fondly as generation-building exercises (Tochka 2016). In fact, this is where Joniada met her future husband, Shpend. The barrier that the amorous couple must overcome in this case is the gossip about Joniada's moral character, which deters her beloved at first (eventually he comes around, and the ballet ends on a happy note). While her role as *brigadiere* puts Joniada in contact (and even close physical proximity) with men and gives her power over men, it also exposes her to slander. The love between Joniada and Shpend is expected to be platonic. While young women of that era were encouraged to choose their future husbands – usually among their classmates or co-workers – they were far from being sexually liberated. As noted, the Albanian version of socialism retained many elements of the pre-existing Ottoman culture of protecting female chastity as another way to keep the populace under control. The promotion of love marriages was merely a tool to break family/clan ties so as to recruit young women in the ranks of a scarce labour force. In this case, Joniada's mother is quite supportive of the couple from the outset and hopes that Joniada's life will be happier and easier than her own.

Like Mrika's mother, Joniada's mother is a widow who has brought up her children alone in extreme poverty. Both older women are painted as angelic – but no alternative life (e.g., remarriage) is allowed for them. These mothers' self-sacrifice for the family echoes Albanians' expected self-sacrifice for the socialist nation. "Joniada" includes a flashback showing what happened to the protagonist's absent father. We see him migrate from Albania to France in search of employment. Having worked as a coal miner for 20 years, without ever returning home, he dies in a mining accident. Thus, coal extraction is framed as evil – but only because this story's mine is located in a foreign land. In reality, work conditions were abysmal in Albanian mines. In the story, economic emigrants, such as Joniada's father, bid farewell to their desolate wives by an old olive tree on a hill, which for this reason is known as "the olive of tears". This tree – described as ugly and lumpy – symbolises Albania's underdeveloped past and is contrasted to the "joyous" young olive saplings that are now being planted under the Party's guidance. Actually, olives are renowned throughout the Mediterranean as beautiful, resilient and expressive trees, and their branches have been a symbol of peace and friendship since ancient Greek mythology. This type of myth (re)making illustrates how the meaning of nature was distorted to fit socialist ideology.

*"City lady"*

Unlike the two preceding works, which centred on young women, the protagonist of this humorous play is an older woman named Ollga. Same as the mother figures in "Mrika" and "Joniada", Ollga is widowed and has a grown daughter. But in contrast to those long-suffering and placid women, who represented the idealised socialist farmer, Ollga is high-strung and outspoken. At the start of the play, she and her daughter, Meli, live in an elegant house in Korçë, an affluent and cultured city in eastern Albania. Clearly, they derive from the urban bourgeoisie – a class which the socialist regime sought to eradicate. The action starts when Meli – a nurse – is assigned to work in the countryside. Not wanting to separate, mother and daughter relocated together. While Meli, as a representative of the New Socialist Youth, takes the transfer in stride, Ollga is quite unhappy with her new circumstances. She regards the peasantry as inferior and the country as the epicentre of backwardness. This attitude clearly does not align with socialist values. However, owing to her status as an older, retired woman, Ollga is not considered as a real threat to the Party's cause. Her snappy comments and complaints are laughed off as old lady grumpiness and eccentricity. Essentially, she is a "paper tiger" (to use a Mao Zedong expression), who can be easily tamed through consistent role modelling and support provided by surrounding socialist characters. As expected, Ollga relents and even comes to enjoy rural living. However, this is not because she cares for the proximity to nature or any environmental benefits (fresh air, quietness and so on) afforded by this lifestyle. Primarily, her conversion owes to the realisation that the socialist village offers the same material comforts as the city. Animal transport has been replaced by bicycles; households own a variety of electric appliances such as TV sets, fridges and washing machines; couples can afford lavish weddings; and facilities such as kindergartens, clinics and cultural centres are readily available. Well before the term "planetary urbanisation" was coined in urban geography, this play presented a condition where the urban has invaded the rural, based on the socialist ethos of "turning villages into cities" (Leka 2017). While blurring divisions between civilisation and wilderness was the socialist aspiration, the "elevated" village presented in this play was illusory. The play's popularity grew once it was adapted into film – and, in large part, the film's success depended on its setting in Tushemisht, a particularly picturesque village on Lake Ohrid. (The Albanian dictator even had a summer residence nearby.) The socio-economic reality of most Albanian villages at the time was very different, and major urban-rural inequalities have persisted into the present.

*"Two checkmates"*

This is another popular comedy which has endured in the national imaginary thanks to its accessible slapstick humour, an ensemble cast of well-known comedians, a series of memorable one-liners and a sympathetic child actress in a lead role. Its setting is in Saranda, an attractive tourist city on the Ionian Sea (now scarred by over-construction). The action centres around a simultaneous chess exhibition which is being organised with much fanfare in the city. Chess has an important connection to socialist ideology: in the Soviet Union, it was promoted as a key tool in training military men and raising the cultural level of workers (Hudson 2013); several movies with a chess theme were produced, starting with silent-era's "Chess Fever". In the Albanian film, the grandmaster, Ilo Pinci, is a pompous researcher and chess enthusiast from the capital, whereas the other players are local amateurs. All players are adult men except for Rudina, a pre-teen girl, who is a chess prodigy. After a series of humorous mishaps, the simul takes place and, despite her young age and female gender, Rudina defeats the grandmaster. But given the film's farcical style, it is clear that the girl's win is not to be accepted at face value. Apart from Rudina, the other female characters in the film are in supporting roles (all variations on the "wife" theme: the supportive wife, the domineering wife, the enabling wife and so on). The film's secondary storyline involves Albania's aquaculture industry. Ilo Pinci and Rudina's father are both pisciculture experts, engaged in a scientific-ideological battle over their opposing approaches to mussel farming in nearby Lake Butrint. The question is not whether mussels should be farmed or not, but whether the spat should be domestic or imported. Rudina's father eventually wins the argument – hence the film's title, two checkmates. This portion of the story is based on reality. Mussels have been farmed in Lake Butrint (a stratified saltwater lagoon south of Saranda) since the late 1960s. Where mussel production is responsible, it can have a positive environmental impact. However, farming can also modify the physical and chemical characteristics of the benthic environments and is vulnerable to threats such as algae, weather, diseases, predators, water quality and pollution (Avdelas et al. 2021). A few years before the film's plot was conceived, Butrint mussels had become extinct due to the lake's stagnant water with excessive salinity, a development that had caused a national stir. The issue was temporarily resolved by rechannelling the flow of Bistrica River – which had naturally fed the lake until 1958 but had then been deviated for irrigation and hydropower generation purposes (Vjeri 2021). Of course, these vicissitudes are not mentioned in the film. Both pisciculture experts are in agreement that shellfish production is important, and the mussel is celebrated as "queen of the lake" for its economic value. Symbolically, Rudina uses a mussel as the black queen in her chess game. Now, farmed mussels, as well as other creatures in the Lake Butrint ecosystem, are further threatened by global warming, sewer discharges and poor maintenance (*Telegraf* 18 April 2017; *Panorama* 29 April 201; *ArgjiroLajm* 25 January 2020).

*"The scarecrow"*

These two final works included in this theme target children. In "The scarecrow", a group of pupils are in charge of managing their school's experimental parcel. To discourage birds from feeding on the recently cast seed, they have installed a scarecrow, represented as a young, dishevelled male. But the scarecrow is lazy, incompetent and somewhat cowardly; he even falls asleep on the job. He is the ultimate "bad worker" in need of reform – a common Socialist Realism trope. Not only are crows unafraid of the powerless straw man, they also mock and bully him by stealing his hat and clothes. The character seems inspired by the brainless and naïve scarecrow in L. Frank Baum's classic 1900 novel "The Wonderful Wizard of Oz". But while Baum's scarecrow acquires wisdom gradually during his adventures with Dorothy, this film's scarecrow is "upgraded" on the spot through socialist technology and science. The children envision replacing the scarecrow with a metallic robot, which they attempt to build in the school's workshop. As this proves too complicated, they settle for a simpler mechanical helix attached to the scarecrow's hat, which allows him to fly or at least levitate off the ground. This gives him an edge over the hovering crows. Perhaps unintentionally, this solution mirrors socialist industrialisation efforts in Albania, which sought to control nature but often fell short of expectations due to technological constraints.

*"When it rains"*

This song, still widely taught to kindergarten-age children, is about superstitions, common throughout the Balkan region, which socialism sought to overcome through logic and reason. The story is framed as an intergenerational conflict between Granny, an elderly, uneducated woman standing for the old generation mired in "old wives tales", and her grandson Petrit, a star student, representing the New Youth. The song opens with a cat meowing and grooming itself. Granny believes that this behaviour presages rain. Petrit is better schooled than Granny and therefore more lucid and progressive. His empirical reasoning is that, in the absence of clouds and thunders, rain is unlikely – regardless of cat's grooming behaviour. A cheerful choir in the background reinforces Petrit's lecture, further refuting Granny's proposition. Socialist science, as taught in the grade schools of the era, prevails. The lyrics place humans on a pedestal for possessing intelligence and reasoning skills which other animals lack. This resolution also mirrors "Two checkmates", which portrays a child as smarter and more knowledgeable than an adult. But there is something amiss here. Not only is Petrit's posturing sexist and ageist but his "scientific" conclusions may also be wrong. Research on animal responses and adaptation to weather and climate is ongoing, but many species are known to prepare for storms and other weather events (Buchholz et al. 2019). Meanwhile, the song encourages children to treat non-human animals as inferior. Dismissing folklore is objectionable in another sense. Superstitions, even if proven wrong, are proof of people's efforts to understand the mysteries of nature and existence, and as such, should not be ridiculed.

*Theme 2: emerging environmental consciousness*

*"Lil' gypsy woman"*

This contradictory poem sits at the cusp between the two overall themes discussed in this chapter. The main character is an itinerant Roma woman (i.e., a member of a historically marginalised community), whereas the narrator is the poet himself, Dritëro Agolli, a white man and renowned public figure in Albania. This constitutes a power imbalance from the outset – although, in a later interview, Agolli claimed to have had intimate contact with Romani communities in his youth and to have been quite fascinated by their unique culture and nomadic lifestyle (*Shqiptarja* 1 September 2018). The poem is written in the second person singular. The poet addresses the woman directly, telling her story on her behalf while the woman remains silent. Even the diminutive form of address ("lil' gypsy woman" or *kurbatka* rather than a proper noun) is patronising although meant to be affectionate. However, the author's approach is quite sympathetic to the woman. Strong and vital, she likes falling asleep under the stars and waking up to the sound of rustling leaves. Practically, she is a symbol of wild nature itself. Her being a single mother, with no male partner in sight, further highlights her "untamed" state. Unlike white Albanian women of the era, whose physical mobility was limited and surveilled and whose marriages were typically arranged, this Roma woman appears to be physically and sexually liberated. This brings to mind

Prosper Mérimée's Carmen. While *kurbatka* has received no formal education in socialist schools, she is intellectually emancipated. Although alone, she is a good, deeply caring mother to her son. The bond between the two is described in tender detail. However, the poet does not paint Romani lifestyle as consistently positive. He notes that it involves battling the elements (scorching sun and torrential rain) as well as financial hardship. (The woman earns a meagre income from basket weaving.) The most emotional part is the separation between mother and son. The latter is lured away by culture. A talented self-taught musician, he joins the local *estrada* (a socialist type of variety theatre) and moves into a stable home in the city. Music, as a universal language, is the link that connects the natural and urban realms in this case. Eventually, the son wants to take in his ageing mother. At this point, the poem abandons its romantic style and adopts the didactic tone of Socialist Realism. The poet/interlocutor scolds the woman over her drifting habits; he even claims that the unsettled Romani spirit is obsolete and the future rests on "concrete and steel and plastic" – in other words, urban permanence. This reproachful stanza, which refutes the rest of the poem, was perhaps grafted on to please censors. The story ends in a compromise between nature and civilisation: the woman agrees to spend the winter in her son's home in the city but, come spring, she responds to "the call of the wild".

*"Beni walks on his own"*

This young adult novel, expanded from the script of a multi-award-winning film, is a cautionary tale about the danger of curtailing human contact with nature. It tells the story of Beni, an only child of about ten, who lives a highly sheltered life in a city apartment. His mother, an educated professional, is a "helicopter parent", as well as a germophobe. She controls Beni's every move, fears for his health and safety and rarely allows him to play outside with other children. Consequently, Beni has had very little contact with the outdoors and has grown to be an anxious and co-dependent child. He is scared of plants and especially animals, including pets. The idea of encountering wild animals causes him nightmares. It may be argued that Beni's fears were produced not only by helicopter parenting but also by an absurd biophobia promoted in Socialist Realism art and literature. One commentator observes that socialist-era Albanian poetry had room only for celebrative, symbolic, productive and/or stately plants and animals such as laurels, poppies, oaks, sycamores, cows and eagles. Certain carnivore, venomous, reptilian or nocturnal animals including wolves, foxes, jackals, lynxes, snakes, spiders, tortoises, frogs, owls and bats were attributed negative connotations borrowed from pre-socialist myths. Mentions of inclement or potentially depressing weather (storms, black clouds, fog, thunder, hail and gales) were only allowed in epic poems about the heroic deeds of Albanian historical figures (Leka 2017). Therefore, neither parents nor readings have done much to instill a love of nature in Beni. The story escalates when an uncle visits from the countryside. He immediately takes stock of the situation and offers to take Beni to the country during the summer holidays. His concern is not just Beni's unfamiliarity with nature but also his being a "mama's boy" – which makes him prone to bullying. A gender element is evident here: the well-meaning uncle expects Beni, as male, to metaphorically "leave home" and toughen up in order to become a fully realised person. For a girl, domesticity, timorousness and attachment to the family might have been more acceptable or even desirable. Beni's parents reluctantly agree to the uncle's request. Beni spends the summer in the country in the company of a group of local peers (all boys). In this masculine realm he learns how to camp, hunt, care for farm animals, pick wild fruit, as well as other rural chores and recreational activities. This constitutes a major growth period for Beni: through contact with nature, he learns self-reliance as well as the importance of physical activity and the body-mind connection. Despite some Socialist Realism tropes, this novel is one of few works from that era which presents the natural environment as a powerful healing force rather than as an amenity under human sovereignty. The novel is also an early critique of urbanisation – in socialist Albania and elsewhere – for the artificial and sanitised environments that it creates for children. Since then, the child-friendly city that offers green spaces for unstructured play has become a central issue in urban planning (Unicef Albania 2017).

*"The train departs at 6:55"*

This film, produced as the socialist regime was waning, takes on the logging industry in an unnamed northern district. This was the first time that a state industry was openly attacked in film. During an audit, Etleva, an accountant at a provincial branch of the national bank, has discovered an irregularity. The local logging enterprise (a public entity, like the bank) has been cutting an excess number of trees, which it lacks the capacity to process or transport to a sawmill. Numerous felled trunks are left to rot in the ground, which results in an economic loss for the bank. From the perspective of the logging enterprise director, this model makes sense: he is only rewarded for meeting or exceeding tree cutting targets and keeping lumberjacks in full employment. The script thus exposes the perverse incentives of the socialist economic system, which led state managers to ignore externalities. But even here, the critique is advanced primarily on financial rather than environmental grounds. It is not a coincidence that Etleva is a bookkeeper rather than an environmentalist. In fact, the film lacks the lexicon to describe what is occurring in environmental terms. Rather than frame the issue as an environmental disaster, Etleva asks: "should anyone be responsible for this loss, which we cannot call theft since it's not lining anyone's pockets?" As a young and single female employee, Etleva fights an uphill battle, and even risks her job, for her concerns to be taken seriously. The bank's director (her boss) and the logging enterprise director are "buddies" – as well as bullies. They treat

Etleva with hostility and condescendence. To undermine her reputation and credibility, malicious rumours are spread about her being pregnant outside of wedlock. As seen in "Joniada", this was a common ploy to discredit and silence women. Etleva is left traumatised and loses her family's love and respect; her sweetheart abandons her. The conflict is resolved when the bank director takes a closer look at the figures and comes to realise that Etleva is indeed correct in her conclusions. What convinces him to give her the benefit of the doubt at his friend's expense is a (male) doctor's revelation that (a) Etleva is not pregnant but rather (b) has a uterine tumour. This confirms her chastity and elevates her to martyrdom. Had she been considered in any way "improper", the fate of this Albanian Erin Brockovich may have paralleled the fortunes of those felled trees in the forest.

*"Planting trees"*

Unlike the film above which explores deforestation, this painting (Figure 14.2) shows a group of youth planning trees. Neither the title of the painting nor the abstract colours and shapes of the trees clarify whether this is a horticulture

*Figure 14.2* "Planting trees", by Edi Hila (1972). Source: National Art Gallery of Albania. Artwork in the public domain. One can perceive this *aksion* scene as bursting with euphoria and youth energy, or as an agonising cry for help. Nature is as tormented as the people and seeks to escape regimentation.

or forestation programme. As in "Joniada" and "Mrika", the workers are probably partaking in an *aksion*. But there is no clear hierarchy among the characters (as opposed to "Mrika" and "Joniada" which highlight the presence of a forewoman), and the relationships appear to be egalitarian. Gender differences in appearance are also minimal. A few figures are clearly female (as they are shown wearing skirts), but most are gender-neutral. The dynamic shapes of both the trees and human bodies can be construed in different ways. In a romanticised view of *aksion* propagated by Socialist Realism art, the workers may be seen as experiencing a sort of euphoria at the prospect of contributing to the nation and sustaining the Party's cause. One can perceive the painting as bursting with youth energy, and the *aksion* scene may be interpreted as the socialist version of the Garden of Eden. The artist himself has stated in recent interviews that his intention had been to convey optimism (*Shqiptarja* 24 May 2018). But in a contemporary reading with the benefit of hindsight, the contorted, warm bodies may be seen as agonising in pain or crying for help. Their feet seem to be stuck in mud, preventing free movement. The work of putting nature directly in the service of people is not liberating but oppressing. The trees, like the humans, are in a state of vertiginous delirium. They are blue rather than a naturalistic green. Nature is as tormented as the people and seeks to escape regimentation. (Neat rows of newly planted trees are shown on the hills in the background.) The laws of perspective or shadowing are ignored, which results in a merging of humans with the rest of nature. The choice of predominantly cold colours, which seem to reference Hieronymus Bosch's "Garden of Earthly Delights", further conveys feelings of general unease. Art censors in the 1970s understood that the painting was indeed "unrevolutionary" in nature; the painter was banished from the art world and assigned to manual labour in a chicken factory farm. He re-emerged after the fall of socialism and is still active and admired.

*"The great river"*

This short story's environmental theme is made explicit from the outset. The story opens with a reminder of nature's permanence compared to the evanescence of human life. Instead of the typical anthropocentric perspective, the story is told from the Great River's viewpoint (the word "river" is masculine in the Albanian language). A gentle giant, the River shares his plight with the reader: humans are excessively narrowing his bed by grabbing land for construction, gardening and animal husbandry. He loves people and wants to help rather than hurt them, so he patiently tolerates their greed. But he cautions, over and over, that, in case of heavy rains he might do major damage. No one really listens. A small water surge, which should have served as a forewarning, is soon forgotten. Some people visit religious temples to pray for good weather, while others build taller and stronger defences against the River. Efforts to exploit nature continue unabated. While the author is sympathetic to the human characters in the story, he spreads the blame on everyone. Women are no more environmentally friendly than men are – although more men than women are cited in the story. The poor are as selfish as the rich. The local landlord (*beu*) is painted as particularly arrogant – he even belittles the Great River as "mountain stream" – but the peasants are exploitative as well. They confer upon nature what *beu* confers upon them. After a build-up, a devastating flood occurs which knocks down houses, clears away crops and kills animals and people. Humans learn their lessons the hard way. The story ends with peace restored: the shores are left in their natural state. This resolution may not have pleased the censors as it does not perpetuate the socialist ideology of people conquering nature. Also, in the past, the author had been imprudent in dealing with socialist rulers. A leading figure of pre-WWII Albanian letters, Mitrush Kuteli was imprisoned by the socialist regime in 1947 over his political stances. For the next two years, he was put to work draining the infamous Maliq swamp, discussed earlier. Upon his release, he worked as translator of Russian literature – which served as a model for the introduction of Socialist Realism in Albania. He wrote little new material of his own and died in 1967, bereft of the honour and recognition due to a writer of that stature (Elsie 2017). This short story may have been supressed had it not been for its setting in pre-socialist Albania, an era framed by the regime as the root of all backwardness. Also, Kuteli's magic realism style, derived from the oral literature of his native city in eastern Albania but also inspired by Russian authors such as Nikolai Gogol, may have led censors to dismiss this story as a fairy tale (Kuteli was also known as a folklorist and children's author). In reality, this is possibly Albania's first – and perhaps the only – symbolic product to apply a "deep ecology" approach to storytelling.

## "The globe"

This song won first prize at the eighth Festival of Song – a nationwide competition which took place annually and was considered as the most important cultural event of the year. Uniquely for its era and for festival winners, the song is impressionistic, lacking a plot. Nothing much happens: a teenager receives an Earth's globe for his birthday, and the lyrics report his thoughts and feelings as he turns around the educational toy. (We use the "he" pronoun because the original singer was male; over the years, the tune has been rendered by numerous singers, both male and female; it works equally well as the lyrics are gender-neutral.) The boy is compelled to search for Albania in the globe and rejoices when he spots it. Where other songs would have taken this opportunity to eulogise the nation and aggrandise its achievements, this boy is simply filled with awe at the magnificence of the Earth, on which Albania is but a tiny speckle. He marvels at everything the Earth "carries on its back": infinite skies, waters, mountains, trees, birds, cities, harbours, lights and people. The lyrics evoke hope and dynamism. This is in the style of Socialist Realism, but there is no mention of socialism itself.

### Theme 1
**Triumph of science and culture over nature**

- **The swamp**
  Certain natural environments are evil. Pain and suffering is justified in order to eliminate them. Socialism will free people from natural dangers and old beliefs.

- **Mrika**
  Romantic love is linked to socialist megaprojects. Women can be liberated from the patriarchy if they agree to become complicit in subjugating nature.

- **Joniada**
  All nature, including mountains, should be rendered productive. 'Beautiful' equals man-made. Overwork is 'fun'. Women should be chaste and masochistic.

- **City lady**
  The urban should invade the rural. Socialist ethos of 'turning villages into cities' has been realised. Older women are malleable and not a threat to the regime.

- **Two checkmates**
  All nature, including lagoons, should be rendered productive. Non-human animals are celebrated for their economic value.

- **The scarecrow**
  Socialist science can overcome obstacles posed by nature. Traditional arts and crafts should be 'upgraded' through technology.

- **When it rains**
  Children are encouraged to treat non-human animals as inferior. Agist attitudes are acceptable since old generations are backwards.

### Theme 2
**Emerging environmental consciousness**

- **Lil' gypsy woman**
  A Romani woman symbolises wild nature, freedom, and sexual liberation. Eventually she has to find a compromise between a nomadic lifestyle and civilisation.

- **Beni walks on his own**
  Through contact with nature, children (boys) learn self-reliance, the importance of the body-mind connection. Nature is a powerful healing force.

- **The train departs at 6:55**
  An environmental catastrophe and the perverse incentives of the socialist economic system are exposed. A female whistleblower is bullied and victimised.

- **Planting trees**
  The work of putting nature in the service of humans is not liberating but oppressing. Nature and people are tormented and seek to escape regimentation.

- **The Great River**
  A 'deep ecology' approach to storytelling. Nature's permanence is contrasted to the evanescence of human life. Unusual non-anthropocentric perspective.

- **The globe**
  A spiritual ode to the Earth, rendered in gender-neutral language.

*Figure 14.3* Summary of findings.
Source: Authors' work.

Religion having been banned in Albania at the time, there is no Father God looking over the globe, just a mesmerised human. A few verses have revolutionary undertones, employing phrases such as "'peoples' awakening", "fiery hearts" and "freedom fighting". However, there is no direct link to the socialist cause here. If the lyrics were read out of context, they could apply to contemporary social movements such as feminism. By being intentionally vague, the lyricist (Fatos Arapi, still widely acclaimed) managed to create a more universal work than the limits of Socialist Realism allowed. Today, the song can be interpreted as a spiritual ode to the Earth.

**Conclusion**

Albania is a special case compared to other socialist countries in Europe. It was peripheral and extremely isolated even within the Eastern Block, and surveillance of the non-productive sphere was extreme. Here, symbolic products served the purpose of government propaganda – with only few instances of (mild) resistance. Therefore, our analysis has revealed much about the official conception of "nature" and "the environment" – which we have linked to environmental exploitation and women's oppression. A succinct summary of the findings is presented in Figure 14.3.

Today, Albania and other post-socialist societies face the challenge of replacing the anthropocentrism engrained during socialism with biocentrism. Nature can no longer be valued for purely utilitarian reasons as an economic resource. Its lethal power should not be forgotten either. The Albanian people and government need to acknowledge the intrinsic value of nature and recognise the rights of all living creatures. Humans can no longer elevate themselves above nature – as preached by socialist ideology and, before that, theistic religions and even Humanism. As small participants of life on Earth, humans should seek to learn from the intricate and balanced workings of the natural world. These should serve as a template for creating harmony, efficiency and productivity.

**References**

Avdelas, L., Avdic-Mravlje, E., Borges Marques, A.C., Cano, S., Capelle, J.J., Carvalho, N., Cozzolino, M., Dennis, J., Ellis, T., Fernández Polanco, J.M., Guillen, J., Lasner, T., Le Bihan, V., Llorente, I., Mol, A., Nicheva, S., Nielsen, R., van Oostenbrugge, H., Villasante, S., Visnic, S., Zhelev, K., and Asche, F. 2021. The decline of mussel aquaculture in the European Union: causes, economic impacts and opportunities. *Reviews in Aquaculture* 13: 91–118. 10.1111/raq.12465
Belayev, A. 1966 [1956]. *The Air Seller*. Translated from Russian into Albanian by V. Kokona. Tirana: Naim Frashëri.
Beqiri, S. 2020. *Drama kombëtare në vitet 1950-1975* [National Drama in 1950–1975]. Tirana: Botimpex.
Bogdanov, A. 1984 [1908]. *Red Star: The First Bolshevik Utopia*. Translated from Russian into English by C. Rougle. In: Bloomington: Indiana University Press.

Buchholz, R., Banusiewicz, J., Burgess, S., Crocker-Buta, S., Eveland, L., and Fuller, L. 2019. Behavioural research priorities for the study of animal response to climate change. *Animal Behaviour* 150:127–137. 10.1016/j.anbehav.2019.02.005

Dado, F. 2010. *Letërsi e painterpretuar* [Uninterpreted literature]. Tirana: Bota Shqiptare.

Deng, C., Zhang, G., Liu, Y., Nie, Y., Li, Z., Liu, J., and Zhu, D. 2021. Advantages and disadvantages of terracing: a comprehensive review. *International Soil and Water Conservation Research* 9(3):344–359. 10.1016/j.iswcr.2021.03.002

EcoAlbania. 2017. EBRD confirms negative impacts of Albanian hydropower plants on people and the environment. Press release, available at: https://www.ecoalbania.org/en/2017/10/03/ebrd-confirms-negative-impacts-of-albanian-hydropower-plants-on-people-and-the-environment/. Last accessed on 26 May 2022.

Elsie, R. 2017. Mitrush Kuteli, biography. Webpage, available at: http://www.albanianliterature.net/authors/classical/kuteli/index.html. Last accessed on 4 June 2022.

Environmental Protection Agency. 2022. Wetlands. Webpage, available at: https://www.epa.gov/report-environment/wetlands. Last accessed on 2 June 22.

Gare, A. 1993. Soviet environmentalism: the path not taken. *Capitalism Nature Socialism* 4(4):69–88. 10.1080/10455759309358566

Grgić, A. 2021. Building a new socialist art: a short history of Albanian cinema. *Studies in Eastern European Cinema* 12(3):276–292. 10.1080/2040350X.2020.1826693

Hudson, M.A. 2013. Storming fortresses: a political history of chess in the Soviet Union, 1917-1948. PhD dissertation, University of California at Santa Cruz, USA.

Kuteli, M. 1987. *Tregime të moçme shqiptare*. Tirana: Naim Frashëri.

Leka, A. 2017. Ceremonial flowers, political animals and weather-pholyphobias in Albanian Socialist Realism. *Poem* 5(4):419–473. 10.1080/20519842.2017.1389354

Mata-Satka, F. 2011. Albanian alternative artists vs official art under communism. *History of Communism in Europe* 2:73–94.

Mies, M., and Shiva, V. 2014. *Ecofeminism*, 2nd ed. London: Zed Books.

Morgan, P. 2017. *Ismail Kadare: The Writer and the Dictatorship 1957–1990*. London: Routledge.

Nelson, A. 2000. The struggle for proletarian music: RAPM and the cultural revolution. *Slavic Review* 59(1):101–132. 10.2307/2696906

Pojani, D. 2010. Tirana: city profile. *Cities* 27(6):483–495. 10.1016/j.cities.2010.02.002

Pojani, D. 2014. Urban design, ideology, and power: use of the central square in Tirana during one century of political transformations. *Planning Perspectives* 30(1):67–94. 10.1080/02665433.2014.896747

Pojani, D. 2021. *Trophy Cities: A Feminist Perspective on New Capitals*. London: Edward Elgar.

Richter, B.S. 1997. Nature master by man: ideology and water in the Soviet Union. *Environment and History* 3(1):69–96. 10.3197/096734097779555962

Shatro, B. 2011. Aesthetical and political aspects of the relationship between literature and ideology in Albania in dictatorship and in post-communism. *Mediterranean Journal of Social Sciences* 2(2):29–39.

Shatro, B. 2016. *Between(s) and Beyond(s) in Contemporary Albanian Literature*. London: Cambridge Scholars.

Shqip Show. 2010. Këneta [The swamp]. Hosted by R. Xhunga, aired on 19 June.

Tochka, N. 2016. *Audible States: Socialist Politics and Popular Music in Albania.* Oxford, U.K.: Oxford University Press.
Tochka, N. 2017. Singing 'with culture': popular musicians and affective labour in state-socialist Albania. *Ethnomusicology Forum* 26(3):289–306. 10.1080/17411912.2017.1407950
Unicef Albania. 2017. Child-friendly city initiative: improved governance to build better cities and communities for children. Available at: https://www.unicef.org/albania/child-friendly-city-initiative. Last accessed on 1 June 2022.
Velo, M. 2014. *Grafika e realizmit socialist në Shqipëri* [Socialist Realist graphic art in Albania]. Tirana: Emal.
Vjeri, H. 2021. Tekat e një liqeni: kur midhja do ngordhte brenda 48 orësh dhe ç'është urgjente tani. *Saranda City* 24 March.
Woodcock, S., and Ikonomi, L. 2014. Imoraliteti në familje: nxitja e ankesave të grave për të përforcuar pushtetin e Partisë në revolucionin kulturor shqiptar. [Immorality in the family: eliciting women's complaints to further Party power in Albania's cultural revolution.] *Përpjekja* 32:155–182.
Ziegler, C. 1985. Soviet images of the environment. *British Journal of Political Science* 15:365–380. 10.1017/S0007123400004233

# Index

Note: Italicized and bold page numbers refer to figures and tables. Page numbers followed by "n" refer to notes.

abstract symbolism 27
Act no. 43/1968 51–52
Act no. 63/1971 59
Act on the Foundations of the System of Societal Planning and the Societal Plan of Yugoslavia 106
agro-town 22, 29
*Air Seller, The* 248
Ak Orda ("White Horde") 26, 27
Albania 8
Alexander, C. 240
Alexy, T. 56
all-unity 234
Almaty's Republic Square 26
Alton Estate, London 40
antiurbanism 81–83
Archigram 240
architect: role during socialism 5–6
*Architektura* 43
*Arhitektura* 76
Arkhangelsky, V.A. 214
*Àrkhitektura SSSR* (*Architecture of the USSR*) 135, 138, 154
Arman, H. 180
Arrighi, G. 17
Association of Polish Architects (SARP) 44
Association of Urban Planners of Yugoslavia (AUPY) 108
Astana, construction of nation in 25–27
Atatürk 26
Athens Charter 39–42, 58; "functional city model" 108
AUPY *see* Association of Urban Planners of Yugoslavia (AUPY)

Balčiūnas, V. 140
Balėnienė, G. 143, *143*, 147
balkanisation of Akmolinsk's population 23
Baltic White-Sea Canal 233
Basic Communication System of Prague 58
basic organisations of associated labour (BOALs, *osnovne organizacije udruženog rada*) 102, 105
Belgrade Urban Planning Institute: *Urbanizam Beograda* 109
Belluš, E. 54
Benetis, E. 146
"Beni walks on his own" 259–260
Beòuška, M. 56
Berkovits, G. 82
Berlin Wall 93
Bieńkowski, K. 39, 40, 44
BIOS model 239
BOALs *see* basic organisations of associated labour (BOALs, *osnovne organizacije udruženog rada*)
Board of Lithuanian Union of Architects 139
Bochorishvili, T. 7, 193, 196, 200, 203n2
Bogdanov, A.: *Red Star* 248
Bratislava 5, 50–64; City National Committee 55; City Regulation and Development Plan 53; "City-Wide Centre" 57; "Great Change" plan 55; great plans, end of 62–64; as new capital city at Eastern Bloc 51–53; new

Index 269

generation of architects and urbanists plan for 53–56; as "The Hearth of the City" 56–59, *60*
"Bratislava Out Loud" *[Bratislava nahlas]* 62
Bratislava Technical University 55
Brėdikis, V. 142
Brenner, J. 88
Breslau 32
Bruns, D. 180
Bučas, K. 140
Budapest 5; District Councils 85, 86; General Urban Plan of 1960 86; Metropolitan Council 85, 86, 90, 91; paradigm shifts in heritage protection and urban renewal 89–91; postmodern renewal attempt's fiasco in late socialism 91–93, *92, 94*; tabula rasa type renewal, frustration of 85–89, *87, 88*; urban periphery, emergence of 81–83; Urban Planning Department (Metropolitan Council) 86
Budapest Architecture School 91
Budapest Urban Planning Company (BUVÁTI) 86, 88, 90–91, 95n8; "Rehabilitation Concept of the Inner Districts of the Capital, The" 91
Bulgakov, M.: *Master and Margarita, The* 246
Burkett, P. 228
BUVÁTI *see* Budapest Urban Planning Company (BUVÁTI)

camelback urbanisation 83
capitalism 102; global 79–81; Marxist ecological critique of 8
Čekanauskas, V. E. 138, 142
Central Asia urbanism, *longue durée* perspective of 17–18
Central Association for Cooperative Residential Construction 45
Central Scientific Research and Design Institute for Town Planning of the Soviet Union 22
Cerkovnaya Street (Church Street) 23
Chachaj, A. 44
Cheremnykh, M.: "Two Worlds — Two Plans!" 233–234
Cherkasy 7

Cherkasy, everyday public spaces in 167–171
Chernishev, S. E. 179
*Chevengur* village 16
Chkhenkeli, I. 192, 195
Chlomauskienė, N. 143
Chovanec, J. 62, *63*
CIAM *see* Congrès Internationaux d'Architecture Moderne (CIAM)
Cibas, A. 138
Ciborowski, A. 69
"City lady" 255–256
City National Committee of Bratislava (Mestský národný výbor – MsNV) 53
City of Moscow, reconstruction of 36
Clark, K.: *Soviet Novel: History as Ritual* 230–231
closed bio-urban systems 239
Cold War 5, 6, 66–68, 101, 103, 119, 120, 121, 231
COMECON *see* Council of Mutual Economic Assistance (COMECON)
communism 16
Communist Party of the Ukrainian Soviet Socialist Republic 22
*Congrès Internationaux d'Architecture Moderne* (CIAM) 54, 57, 106, 108, 176
Corbusier: super-urban doctrine 36; on urban arrangements 44
core-periphery model 2, 4
Council of Mutual Economic Assistance (COMECON) 95n1, 95n7
Council of the MsNV 55
Crawford, C.E. 167
"cross-acceptance" principle 105
Czechoslovak Communist Party (KSČ) 52
Czechoslovak Socialist Republic 51
Czepczyński, M. 153, 160
Czerechowski, W. 39

Danneberga, D. 25
de Carlo, G. 240
dependency theory 79, 80
*derevenshchiki* (Village Prose) 230
*Die Volksstimme* 52
Domański, B. 156
Dubitsky, A. F. 19
Durkheim, E. 79

*dvizhenie* 238
Dzhan 16

East Central Europe 80, 81, 91, 93
Eastern Europe 68
ecofeminism 8, 244–265, 245
ecological consciousness 7–8
"Ecopolis" programme 231
*Eesti Loodus* 176, *177*
11th AUPY Conference (1963) 109
English Garden Cities movement 210
envelope construction 61
environment 244–265
environmental consciousness 7–8, 258–265, *264*
environmental exploitation 8, 245, 265
environmental management, socialism and 247–249
Enyedi, G. 82
equality: social 2; spatial 2
Estonia: Estonian Museum of Architecture 180; "EstonProject" (National Design Institute) 127, 179; Greater Tallinn (Estonian Art Academy) *see* Greater Tallinn (Estonian Art Academy); militarisation, during the Soviet occupation (1940–1991) 120–122; strategic missile forces expansionm, during Cold War 122–124, *123*; Valga/Valka *see* Valga/Valka; War of Independence (1918–1920) 120
European Enlightenment 68
*Exposition Internationale des Arts Décoratifs et Industriels Modernes* (1925) 229

Fedorov, N. 234
Fingerplan 28
Fisher, J.C. 1
Flint, C. 3
Forbat, F. 20
Ford, H. 210
Fordism 212, 224n6
Foster, J. B.: *Marx's ecology: materialism and nature* 228
Fotlyn, L. 54
Foucault, M. 15
Fragment of the Plan of Mikrorayon Shartash 215, *216*
Franců, D. 56
Frank, A. G. 80

Friedman, Y. 240

Gašparec, M. 56
General Plan for the Reconstruction of the City of Moscow 179
General Scheme of Population Distribution on the Territory of the USSR 155
General Tadeusz Kościuszko's Square 34
GenPlan 28
Gentile, M. 213
Gerchuk, Y. 135
Gierek, E. 43
Ginsburg, M. 210
*glasnost* 102
Gldani 7, 191–203; apartment building extensions 199; commercial activities 201–202; mass-housing districts 191–197, *192*, *196*; physical features 197–198; social features 198–199; today 202
global capitalism 79–81
"globe, The" 263–265
Goce Delčev Student Dormitories 72–74
Gorki's Street, Moscow 36
Gosstrakh (State Insurance Agency) 212
Gosstroi Committee 218
Gradski Trgovski Centar (GTC) 76
"The Grand Ensemble" of Sarcelles, Paris 40
Greater Sverdlovsk Masterplan 210–212, 214
Greater Tallinn (Estonian Art Academy) 7, 135; Asta Palm *182*, 184–185; central planning 178–180; decentralization 178–180; everyday life 184–186; "Gosplan" 180; peri-urban leisure 180–184, *182*; summer house settlements 184–186
*Great Game* 19
"Great Plan for the Transformation of Nature" 233
"great river, The" 262–263
Green City 229, 230, 235
Gretschel, A.: "Revitalisation of the Old Town Historical Complex" 45
Grunwaldzka Axis 37
GTC *see* Gradski Trgovski Centar (GTC)

Habsburg Monarchy 50
Harvey, D. 15–16

Hauskrecht, J. 55
Havel, V.: *Memo, The* 246
Hegedüs, J. 84, 85, 94
Highmore, B. 210
Hila, E.: "Planting trees." 261–262, *261*
Hladký, M. 53–56, 58
Hollein, H. 240
Hoxha, E. 247
Hruška, E. 54
Hungary: Budapest *see* Budapest 83; Communist Party's Budapest Committee 86; "Corvin Quarter" 96n14; market-type relations 83–85; "New Economic Mechanism" 84, 85, 95n3; premature welfare state, reforming 83–85; "Wandel durch Handel" (change through trade) policy 95n7
Husák, G. 52
hypersignificant built-up systems 28

IBA *see* International Building Exhibition (IBA) Berlin (1979–1987)
IFHTP *see* International Federation of Housing and Town Planning (IFHTP)
IMF *see* International Monetary Fund (IMF)
industrial democracy 102
Institute of Economic Sciences of the Academy of Sciences 180
institutional nationalism 142
International Building Exhibition (IBA) Berlin (1979–1987) 91
International Federation of Housing and Town Planning (IFHTP) 106
International Monetary Fund (IMF) 95n7
International Union of Architects (UIA) 106
Izbicki, t. 40

Japan World Expo'70 240
Jasinskas, A. 139
Jędrychowski, S. 46
"Joniada" 254–255

Kalm, M. 138
Kamocki, W. 40
Kasperavičienė, B. *141*
Kaunas (Polytechnic Institute) 135

Kavtaradze, D. 231, 235
Kazakh Soviet Socialist Republic 4, 15–29
Kedro, D. 56
Keppe, O. 180
Khairullina, E. 22
Khrushchev, N. S. 4, 15, 16, 20, 25, 137, 187, 194, 230; on agro-town 22, 29; reform of 1957 217; on rural proletariat 22; Virgin Lands Campaign 15, *17*, 21, 22
Khrushchevka 194
*khrushchyovka* 22, 28
Kisho Kurokawa & Associates 27
Kol'tsov, M. 229
Konček, F. 55
Konrád, G. 82, 94, 95n2
Konstantinovski, G. 73
Konstantinovski's Student Dormitories 75
Kocęciuszko's Residential District 37
Kornai, J. 83, 85; "Economics of Shortage" 84
Koula, J. 54
Kraniauskas, R. 146
Krier, R. 96n11
Krūminis, B. 140, 144, 145
KSČ *see* Czechoslovak Communist Party (KSČ)
Kulić, V. 69, 73

Ladányi, J. 92, 96n13
Ladovsky, N. 229
Lamarck, J.-B. 235
La Miraille housing estate, Toulouse 62
late socialism 8, 91–93
Latin America: underdevelopment 79, 80
Latvian State Urban Design Institute 25
Lausmaa, E. 176, 178
Lazdynai 142–145
Lėckienė, A. *143*
Lefebvre, H. 15
Lengiprogor 214
Lengorstrojproekt 22
Lenin, V. 21, 25, 161, 167, 228, 229
Lenin Square, Tselinograd 26
"Lil' gypsy woman" 258–259
Lindgren, A. 25
Ling, A. 62
Lithuanian mass housing estates: as architectural problem, standardisation of 136–140; regionalist approach to 145–146;

role of architects in fighting the monotony of 134–147; Vilnius (Art Institute), experimental design in 138–140
Lithuanian modernist school 136
Lithuanian Soviet Socialist Republic 6; All-Union Construction Committee 138; Construction Norms and Rules (SNiP, *Stroitel'nye Normy i Pravila*) 136, 137, 184; *Gosstroi* 138
Łowinski, J. 40
Lung, Y. 156
Lurçat, A. 20
Lysenko, T. 235
Lysenkoism 235

Makariūnas, J. 139
Malinowski, Z. 44
market socialism 102
MARS Group 62
Marx, K. 25, 79, 228
Marxism 228
mass housing 7
Master Plan of Belgrade 1972 112–114, *113*
May, E. 20
Mechetnaya Street (Mosque Street) 23
Melnikov, K. 229, 234, 235
Merpert, D. 24
Mester, Á. 88, 89
metabolic city 27
metabolism 27
Meuser, P. 134
Michurin, I. 233
*microraion/microrayon/mikrorayon* (microdistrict) 22–23, 61, 137, 138, 147, 185, 187, 193, 218, 235; critique of 143–145; individuality, shaping 140–145; "open plan" (*svobodnaya planirovka*) 222
Mikučianis, V. 140
Miliutin, N.A.: "*Sotsgorod: The Problem of Building Socialist Cities*" 1–2
Milyutin, N. 210
Ministry of Industrial Development 217
Mironov, K. 24
Mironova, N. 24
Mitrović, M. 77n4
modernisation 79–81
modern world systems 2–3
Molicki, W. 40, 42, 44
Molodezhnyj Zhiloj Kompleks (MZhK, Youth Residential Complexes) 230
Montahaev, K. 26
Moravánszky, Á.: "East West Central, Re-Building Europe 1950–1990" 91
Mrduljaš, M. 69
"Mrika" 252–253
MsNV *see* City National Committee of Bratislava (Mestský národný výbor – MsNV)
Murawski, M. 134
Murray, P. 82
MZhK *see* Molodezhnyj Zhiloj Kompleks (MZhK, Youth Residential Complexes)

Nakhutina, T. 24
Napoleonic Wars 31
Nasvytis, A. 139
Nasvytis, V. 139
National Committee of the capital of Slovakia 51
nature 227–230, 244–265; forms of 235–240; triumph of science and culture over 250–258, **251**
Natusiewicz, R. 39
Nazarbayev 26, 28
NER *see* Novyi Element Rasseleniia (NER)
New Belgrade, centre of the first local commune in *110*
New Economic Mechanism 84, 85, 95n3
New Monumentalism 59
*Nova Makedonija* 71
Novi Beograd, Belgrade 57
Novyi Element Rasseleniia (NER) 231–232, 235–240, *237*; *Future of the City, The* 238; *Ideal Communist City, The* 232–233, *232*, 236
Nowicki, J. 43
Nowy Targ estate 39
Nowy Targ Square 38

*obshchenie* 236
October Revolution 16, 21; 46th anniversary of 23; 58th anniversary of 25
Office of the Chief City Architect 53, 57, 61, 64
Olj, A. 25
oppression: women's 8, 245, 265

Oransky, P. 214
Országos Takarékpénztár (OTP) 95n6
"Osiedle Przyjaźni" 44
OTP see Országos
 Takarékpénztár (OTP)

Palace of Tselinograd Virgin Lands
 Developers 25
Palace of Youth of Tselinograd 24, *24*, 25
Paris's La Défense 57
Pärn, J. 120
Pašiak, J. 56
PDK 215
Pei, I. M. 73
Perczel, A. 92, 96n13
*perestroika* 102, 199
peripheralisation 2
peripherality: dimensions of 156
peripheral landscapes 227–242; nature, city and capital 227–230; thawing landscapes 230–235
periphery 2–5, 75; definition of 3; etymology of 16; during state socialism 3; super-periphery 101; urban peripheries for leisure, planning 176–187; working class on 93–94
Peter the Great 26
Petržalka, southern city district of 59–62, *63*
Piskunov, A. 225n11
PKWN see Polish Committee of National Liberation (PKWN) Square
PKWN Square District 39
planner: role during socialism 5–6
"Planting trees" 261–262, *261*
Platonov, A. 16
Plattenbau 39
Polanski, R.: *Knife in the Water* 246
Polish Committee of National Liberation (PKWN) Square 38
Polish People's Republic (PRL) 33
Polish Republic People's 46
Polish Socialist Party 36
Polish United Workers' Party (PZPR) 33, 35–36
Polish Workers' Party 36
Polyansky, A. 24
Popescu, C. 68
Popovski, Ž. 76
Popowice district, design of 43
Potsdam Conference (1945) 32
Powstańców Warszawy Square 34

Prague Spring 51
*Pravda* 229
Prebisch, R. 80
premature welfare state 83–85
private property 83–85
PRL see Polish People's Republic (PRL)
"Project for Greater Tallinn, The" 7, 178
*Projekt 1977* 56
Promstroiproekt (the Institute of Industrial Construction and Design) 218
protective city barrier 222–223
Przyłęcka, D. 44
Ptaszycka, A. 35, 47n1
Ptaszycki, T. 35
PZPR see Polish United Workers' Party (PZPR)

railway, as territorialist infrastructure 20–21
Real Estate Administration Company 83
Red Army 46
reproduction of places 6–7
Risch, W. 134–135
Rostow, W. W.: "Stages of Economic Growth" 79
"Rozafa's Legend" 246–247
Rudolph, P. 73, 74
rural proletariat 22
Russian Civil War 21
Russian Ministry of Railways 20
Russian Revolution (1917) 228, 247

Saarinen, E. 25
Safdie, M. 240
SARP see Association of Polish Architects (SARP)
Saudi Binladin Group 27
"scarecrow, The" 257
Schmidt, H. 20
*Sea of Youth, The* 16
self-managed interest-driven communities (SICs, *samoupravne interesne zajednice*) 102, 105
Semenov, V. N. 179
semiperiphery 3
Senior, D. 70
Serbian Act on Urban and Regional Spatial Planning 105, 106
Serbian Urban Planners Association: *Arhitektura-Urbanizam* 109
Sert, J. L. 57; "Nine Points on Monumentality" 59

Shkvarikov, V. A. 22
Shubin II, F. K. 19
SICs *see* self-managed interest-driven communities (SICs, *samoupravne interesne zajednice*)
Šipalis, J. 146
Sjöberg, Ö. 213
Skënderi, E. *249*
Skoček, I. 54
*Skopje 2014* 77
Skopje City Wall 5, 66–77, *68*; building 71–76
Sližys, V. 140
Slovak National Council (*Slovenská národná rada*) 51
Slovak Technical University 54
Smithson, A. 91
Smithson, P. 91
"Snow Meridian" (1968) 238
Soans, A. 180
social equality 2
social equity 6–7
socialism: and environmental management 247–249; late 8, 91–93; market 102; role of architect and the planner during 5–6; state 1–8, 52, 81, 84, 120, 223; urban planning during 1–9
socialist district 36
socialist gentrification 94
Socialist Realism 35–37, 47; purpose and method 245–247
socio-spatial relations, non-politics of 6–7
Solovyov, V. 234
Sotsgorod 1
*sotsgorods* 211
South Siberian Railway 21
Soviet Architects' Union Executive Committee 139
Soviet Communist Party's Central Committee 137
Soviet Greater Sverdlovsk 211
Soviet housing issue, stages of 194–195
Soviet Ministry of Health 214–215
Soviet socialist planning 15–29
Soviet Socialist Revolution (1917) 209
Soviet world order 79–81, 94
Sovnarkhozes (*Sovety Narodnogo Khozyaistva*, Councils of National Economy) 217
spatial equality 2
Spurný, M. 53

Spychalski, M. 37, 47n2
Ss. Cyril and Methodius University, Skopje 74, *74*
Stalin, J. 16, 229, 230, 234, 235
Stam, M. 20
state socialism 1–8, 52, 81, 84, 120, 223
*Statyba ir Architektūra* (*Construction and Architecture*) 135
Steller, J. 57–58, 61
Strashko, N.T. 214
super-periphery 101
super-urban doctrine 36
"Sverdlovskgrazhdanproekt" (the Sverdlovsk Institute of Civic Planning) 217, 221, 224n10
Svetko, Š. 56
Svetlík, J. 54
"swamp, The" 250–252
symbiosis 27
Szelényi, I. 82, 83, 94, 95n2

Talaš, S. 62, *63*
Tamoševičius, E. 139, 140
Tange, K. 67, 69, 71
Taylor, P.J. 3
Taylorism 212, 224n6
Tbilisi: First Master Plan of Tbilisi (1934) 191–192; growth of 191–194; largest mass-housing district, birth of 195–197; multiple transitions of 200–201; Second Master Plan of Tbilisi (1953) 192; Third Master Plan of Tbilisi 192
TbilQalaqProject 193
thawing landscapes 230–235
Third Congress of Soviet Architects (1961) 137
Tippel, V. 180
Titl, L. 54
Tosics, I. 84, 85, 94
"train departs at 6:55, The" 260–261
Tret'iakov, S. 227
Tselinograd 4, 21–25
Tsubokura, T. 25, 27
Tumski, O. 31
"Two checkmates" 256–257

UIA *see* International Union of Architects (UIA)
UIA *see* Union inter-nationale des Architectes (UIA)
Ukraine 9, 153–172; central city park 161–164, *163*; large housing

estate's courtyard 164–167; large ordinary cities 154–156; square of power 158–161, *159*; urban planning, data and methods of 156–158; Vinnytsia and Cherkasy, everyday public spaces in 167–171
under-urbanisation 95n2
unified space 235
*Union inter-nationale des Architectes* (UIA) 55; Congress IX of 54
Unité d'Habitation 238
UNIVEX 62
Ural Industrial Construction Scientific Research Institute (Promstroinyiproekt) 221
urban peripheries for leisure, planning 176–187
urban planning: during socialism 1–9
Urban Planning Institute of Belgrade 112
Urban Research and Planning Institute (VÁTI) 92, 96n13
utopianism 16
Uzkikh, K. 217–219, 221

Valga/Valka 119–131; border town development during 1970s (joined general plan) 127–129, *128*; transformation of, in the midst of Cold War 124–127
Valiuškis, G. 139
Valter, E. *177*
Van Eesteren's plan (1934) 22
Vaškevičius, J. 140
VÁTI *see* Urban Research and Planning Institute (VÁTI)
Venturi, R. 91
Verdery, K. 84
Vernadsky, V. 228–229, 236, 239
Vietnam War 69
Vilnius Panel Construction Factory 142
Vilnius Urban Construction Design Institute 135; experimental design in 138–140; Master Plan Brief 1964 140
Vinnytsia 7; everyday public spaces in 167–171
Virgin Lands Campaign 15, *17*, 21, 22
Virgin Lands of Central Asia, urbanising 15–29; Astana, construction of nation in 25–27; *longue durée* perspective of 18–19; mutated perspective of 27–28; railway, as territorialist infrastructure 20–21; tsarist outpost on the Ishim 18–19; Tselinograd, plan for 21–25

Wallerstein, I. 2–3, 80
Warsaw Pact (1968) 50, 61
Warsaw University of Technology 35
Wayne State University 112
Weber, M. 79
Weiner, D. R. 234; *Models of nature: ecology, conservation, and cultural revolution in Soviet Russia* 228
Westerman, F.: *Engineers of the Soul* 16
Western City Gate, Belgrade (Genex Tower) 77n4
Western environmentalism 227
"When it rains" 257–258
World Bank 95n7
Wrocław (formerly Breslau) 4–5; architecture, prefabrication and modernisation of 37–39; first General Development Plan of the City of Wrocław (1945–1949) 34–35, *35*; Old Town, reconstruction of (1949–1956) 35–37; people and politics 32–33; post-WWII damage of 34–35; pre-WWII spatial structure of 31–32; "Regained" 46–47; return to downtown (the 1980s) 45–46; socialist district 36; Stage Plan for the Development of the City of Wrocław 38
Wrocław Design Office 34
Wrocław Large Panel 42–45, 47
Wrocław-Południe district, design of 41–42, *42*, 45

Yanitsky, O. 234
Yekaterinburg (formerly Sverdlovsk) 8, 209–225; Complex Development Plan of Industry and Other Economic Spheres in Sverdlovsk 220–221; general Soviet planning paradigms 209–210; industrial and residential areas, new ecological assessment of (1960s–80s) 214–223, *216*, *223*; planning and construction (1930s–50s) 210–214; Progressive System of Group Resettlement 221
Yugoslavia: 1953 Constitution 104; 1963

Constitution 105; 1974
Constitution 105, 106, 110, 114;
economy 69–71; international
planning ideas 106–108, **107**;
knowledge transfers 69–71;
Liberation War 77n2; Master
Plan of Belgrade 1972 112–114,
*113*; Non-Aligned Movement 69;
physical planning 103–106;
planning discourse 106–108, **107**;
politics 69–71; Skopje City Wall
*see* Skopje City Wall; socio-
economic planning 103–106, **104**;
urban planning 103–106
Yugoslav self-management socialism:
critical reflections 109–112;
methodological approach
108–109; role of urban planners
under 101–115

"Zachód I" housing district 42
"Zachód II" housing district 42
Zachwatowicz, J. 37
Zadorin, D. 134
Záhorie Settlement Belt 55
*zapovedniki* (nature preserves) 228
"*Zhilishche* 2000" 199
Zimmerli Art Museum 237
Zubrus, V. 140
Zundbland, P.E. 217

Printed in the United States
by Baker & Taylor Publisher Services